Y0-BSC-983

OT 55
Operator Theory: Advances and Applications
Vol. 55

Birkhäuser Verlag
Basel · Boston · Berlin

R. R. Akhmerov
M. I. Kamenskii
A. S. Potapov
A. E. Rodkina
B. N. Sadovskii

Measures of Noncompactness and Condensing Operators

Translated from the Russian
by A. Iacob

1992

Birkhäuser Verlag
Basel · Boston · Berlin

Originally published in 1986 under the title "Mery Nekompaktnosti i Uplotnyayushchie Operatory" by Nauka. For this translation the Russian text was revised by the authors.

Authors' addresses:

R.R. Akhmerov
Inst. Comput. Technologies
Lavrentjeva 6
630090 Novosibirsk
USSR

M.I. Kamenskii
B.N. Sadovskii
Voronezh State University
Department of Mathematics
Universitetskaja pl. 1
394693 Voronezh
USSR

A.S. Potapov
Voronezh State Teach. Training Institute
Faculty of Physics and Mathematics
ul. Lenina 86
396611 Voronezh
USSR

A.E. Rodkina
Voronezh Institute of
Civil Engineering
ul. 20 let Oktjabrija 64
394006 Voronezh
USSR

Deutsche Bibliothek Cataloging-in-Publication Data

Measures of noncompactness and condensing operators / R. R. Akhmerov . . . Transl. from the Russian by A. Iacob. – Basel ; Boston ; Berlin : Birkhäuser, 1992
Einheitssacht.: Mery nekompaktnosti i uplotnjajuščie operatory <engl.>
ISBN 3-7643-2716-2
NE: Achmerov, Rustjam R.; EST

Printed in Germany directly from the translator's camera-ready manuscript on acid-free paper
ISBN 3-7643-2716-2
ISBN 0-8176-2716-2

TABLE OF CONTENTS

INTRODUCTION

A condensing (or densifying) operator is a mapping under which the image of any set is in a certain sense more compact than the set itself. The degree of noncompactness of a set is measured by means of functions called measures of noncompactness.

The contractive maps and the compact maps [i.e., in this Introduction, the maps that send any bounded set into a relatively compact one; in the main text the term "compact" will be reserved for the operators that, in addition to having this property, are continuous, i.e., in the authors' terminology, for the completely continuous operators] are condensing. For contractive maps one can take as measure of noncompactness the diameter of a set, while for compact maps can take the indicator function of a family of non-relatively compact sets. The operators of the form $F(x) = G(x, x)$, where G is contractive in the first argument and compact in the second, are also condensing with respect to some natural measures of noncompactness. The linear condensing operators are characterized by the fact that almost all of their spectrum is included in a disc of radius smaller than one.

The examples given above show that condensing operators are a sufficiently typical phenomenon in various applications of functional analysis, for example, in the theory of differential and integral equations.

As is turns out, the condensing operators have properties similar to the compact ones. In particular, the theory of rotation of completely continuous vector fields, the Schauder-Tikhonov fixed point principle, and the Fredholm-Riesz-Schauder theory of linear equations with compact operators admit natural generalizations to condensing operators. Therefore, establishing that a given problem for a differential or integral equation reduces to an equation with a condensing operator yields a considerable amount of information on the properties of its solutions.

The first to consider a quantitative characteristic $\alpha(A)$ measuring the degree of noncompactness of a subset A in a metric space was K. Kuratowski in 1930, in connection with problems of general topology. In the mid Fifties in the works of G. Darbo, L. S. Gol′denshteĭn, I. Gohberg, A. S. Markus, W. V. Petryshyn, A. Furi, A. Vignoli, J. Daneš, Yu. G. Borisovich, Yu. I. Sapronov, M. A. Krasnosel′skiĭ, P. P. Zabreĭko and others various measures of noncompactness were applied in the fixed-point theory, the theory of linear operators, and the theory of differential and integral equations.

This book gives a systematic exposition of the notions and facts connected with measures of noncompactness and condensing operators. The main results are the characterization of linear condensing operators in spectral terms and theorems on perturbations of the spectrum (Chapter 2), and the theory of the index of fixed points of nonlinear condensing operators, together with the ensuing fixed-point theorems (Chapter 3). Chapter 1 is devoted to the main definitions, examples, and simplest properties of measures of noncompactness and condensing operators. In Chapter 4 we consider examples of applications of the techniques developed here to problems for differential equations in Banach spaces, stochastic differential equations with delay, functional-differential equations of neutral type, and integral equations.

In the treatment of the theory itself as well as of its applications we aimed at considering the simplest situation, leaving the comments concerning possible generalizations for the concluding sections or subsections. For additional information the reader is referred to the surveys [10, 28, 160].

The authors use this opportunity to express their gratitude to Mark Aleksandrovich Krasnosel'skiĭ, under whose influence many of the problems discussed here were posed and solved.

CHAPTER 1

MEASURES OF NONCOMPACTNESS

In this chapter we consider the basic notions connected with measures of noncompactness (MNCs for brevity) and condensing (or densifying) operators. We define and study in detail the three main and most frequently used MNCs: the Hausdorff MNC χ, the Kuratowski MNC α, and the MNC β. We derive a number of formulas that enable us to compute directly the value of the Hausdorff MNC of a set in some concrete spaces. We give the general definition of the notion of an MNC, study the so-called sequential MNCs, and establish their connection with MNCs. We define and study the condensing operators, and we give examples of maps that are condensing with respect to various MNCs. And finally, we bring into consideration the ultimately compact operators and K-operators as natural generalizations of the condensing maps.

1.1. THE KURATOWSKI AND HAUSDORFF MEASURES OF NONCOMPACTNESS

In this section we define the Kuratowski and Hausdorff MNCs and study their basic properties. The setting is that of a Banach space E; we let Ω denote subsets of E, and we use $B(x,r)$ and $\overline{B}(x,r)$ to denote the open and respectively the closed ball in E of radius r and center x; $B = B(0,1)$.

1.1.1. Definition. *The Kuratowski measure of noncompactness $\alpha(\Omega)$ of the set Ω is the infimum of the numbers $d > 0$ such that Ω admits a finite covering by sets of diameter smaller than d.*

As usual, by the diameter $\operatorname{diam} A$ of a set A one means the number $\sup\{\|x-y\|: x, y \in A\}$, which for A unbounded [empty] is taken to be infinity [resp. zero].

1.1.2. Definition. *The Hausdorff MNC $\chi(\Omega)$ of the set Ω is the infimum of the numbers $\varepsilon > 0$ such that Ω has a finite ε-net in E.*

Recall that a set $S \subset E$ is called an *ε-net of* Ω if $\Omega \subset S + \varepsilon \overline{B} \equiv \{s + \varepsilon b : s \in S, b \in \overline{B}\}$.

1.1.3. Remarks. (a) In the definition of the Kuratowski MNC one can replace "diameter smaller than d" by "diameter no larger than d"; similarly, in the definition of the Hausdorff MNC it is immaterial how the ε-net is defined —by closed or by open balls of radius ε.

(b) In the definition of the Hausdorff MNC, instead of a finite ε-net one can speak of a totally bounded one, i.e., an ε-net S that has a finite δ-net for any $\delta > 0$.

(c) The definitions of the MNCs α and χ are meaningful not only for Banach, but also for arbitrary metric spaces.

1.1.4. Elementary properties of the Kuratowski and Hausdorff MNCs. We list below some of the properties of the MNCs α and χ that follow immediately from the definitions. The terminology introduced in order to formulate these properties will also be used for other MNCs. For this reason we also include in our list some properties that are straightforward consequences of others (for example, nonsingularity follows from regularity, monotonicity from semi-additivity, continuity from Lipschitzianity).

Thus, the MNCs α and χ (denoted below by ψ) enjoy the following properties:

a) *regularity*: $\psi(\Omega) = 0$ if and only Ω is totally bounded;

b) *nonsingularity*: ψ is equal to zero on every one-element set;

c) *monotonicity*: $\Omega_1 \subset \Omega_2$ implies $\psi(\Omega_1) \leq \psi(\Omega_2)$;

d) *semi-additivity*: $\psi(\Omega_1 \cup \Omega_2) = \max\{\psi(\Omega_1), \psi(\Omega_2)\}$;

e) *Lipschitzianity*: $|\psi(\Omega_1) - \psi(\Omega_2)| \leq L_\psi \rho(\Omega_1, \Omega_2)$, where $L_\chi = 1, L_\alpha = 2$ and ρ denotes the Hausdorff metric (more precisely, semimetric): $\rho(\Omega_1, \Omega_2) = \inf\{\varepsilon > 0 : \Omega_1 + \varepsilon \overline{B} \supset \Omega_2, \Omega_2 + \varepsilon \overline{B} \supset \Omega_1\}$;

f) *continuity*: for any $\Omega \subset E$ and any $\varepsilon > 0$ there is a $\delta > 0$ such that $|\psi(\Omega) - \psi(\Omega_1)| < \varepsilon$ for all Ω_1 satisfying $\rho(\Omega, \Omega_1) < \delta$;

g) *semi-homogeneity*: $\psi(t\Omega) = |t|\psi(\Omega)$ for any number t;

h) *algebraic semi-additivity*: $\psi(\Omega_1 + \Omega_2) \leq \psi(\Omega_1) + \psi(\Omega_2)$;

i) *invariance under translations*: $\psi(\Omega + x_0) = \psi(\Omega)$ for any $x_0 \in E$.

The following two properties are isolated as separate subsections in view of their importance.

1.1.5. Theorem. *The Kuratowski and Hausdorff* MNC*s are invariant under passage to the closure and to the convex hull:* $\psi(\Omega) = \psi(\overline{\Omega}) = \psi(\operatorname{co} \Omega)$.

Proof. The invariance under passage to the closure is obvious. The invariance of χ under passage to the convex hull is also quite readily established: if S is a finite ε-net of the set Ω, then $\operatorname{co} S$ is a totally bounded ε-net of the set $\operatorname{co}\Omega$.

Let us prove that $\alpha(\operatorname{co}\Omega) = \alpha(\Omega)$. Suppose $\Omega = \bigcup_{k=1}^{m} \Omega_k$ and $\operatorname{diam}\Omega_k < d$ for all k. It is readily checked that $\operatorname{co}\Omega$ is the union of all possible sums of the form $\sum_{k=1}^{m} \lambda_k \operatorname{co}\Omega_k$, where the vector $\lambda = (\lambda_1, \dots, \lambda_m)$ runs through the standard simplex σ (i.e., $\lambda_k \geq 0$ and $\sum_{k=1}^{m} \lambda_k = 1$). Let $\varepsilon > 0$. The union of all such sums can be approximated, with arbitrary accuracy $\delta(\varepsilon)$ in the sense of the Hausdorff metric ($\delta(\varepsilon) \to 0$ as $\varepsilon \to 0$), by finite unions of sums of the same form, in which λ runs through a finite ε-net σ_ε of the simplex σ. Now from the properties of the Kuratowski MNC we obtain

$$\alpha(\operatorname{co}\Omega) = \alpha\Big(\bigcup_{\lambda\in\sigma}\sum_{k=1}^{m}\lambda_k \operatorname{co}\Omega_k\Big) \leq \alpha\Big(\bigcup_{\lambda\in\sigma_\varepsilon}\sum_{k=1}^{m}\lambda_k \operatorname{co}\Omega_k\Big) + 2\delta(\varepsilon)$$

$$= \max_{\lambda\in\sigma_\varepsilon} \alpha\Big(\sum_{k=1}^{m}\lambda_k \operatorname{co}\Omega_k\Big) + 2\delta(\varepsilon) \leq \max_{\lambda\in\sigma_\varepsilon}\sum_{k=1}^{m}\lambda_k \alpha(\operatorname{co}\Omega_k) + 2\delta(\varepsilon).$$

As $\operatorname{diam}(\operatorname{co}\Omega_k) = \operatorname{diam}\Omega_k < d$, we conclude that

$$\alpha(\operatorname{co}\Omega) \leq d + 2\delta(\varepsilon),$$

and hence, since ε is arbitrary, that

$$\alpha(\operatorname{co}\Omega) \leq \alpha(\Omega).$$

The opposite inequality is obvious. **QED**

1.1.6. Theorem. *Let B be the unit ball in E. Then $\alpha(B) = \chi(B) = 0$ if E is finite-dimensional, and $\alpha(B) = 2$, $\chi(B) = 1$ in the opposite case.*

Proof. The first assertion follows from the regularity of the MNCs α and χ.

Turning to the second assertion, we first prove it for χ. Clearly, the center of the ball B forms a 1-net for B, and so $\chi(B) \leq 1$. Suppose $\chi(B) = q < 1$. Pick $\varepsilon > 0$ such that $q + \varepsilon < 1$, and let $\{x_1, \dots, x_m\}$ be a $(q+\varepsilon)$-net for B:

$$B \subset \bigcup_{k=1}^{m} [x + (q+\varepsilon)B].$$

From the properties of the MNC χ it follows that

$$q = \chi(B) \leq (q+\varepsilon)\chi(B) = q(q+\varepsilon).$$

But this implies $q = 0$ (because $q + \varepsilon < 1$), which is possible only if B is totally bounded. This contradicts the infinite-dimensionality of the space E.

To prove the second assertion for α we make use of the Lyusternik-Shnirel′man-Borsuk theorem on antipodes (see [84]): if S is a sphere in an n-dimensional normed space and A_k ($k = 1, \ldots, n$) is a cover of S by closed subsets of that space, then at least one of the sets A_k contains a pair of diametrically opposite points, i.e., $\operatorname{diam} A \geq \operatorname{diam} S$. Thus, suppose that E is infinite-dimensional. Clearly, $\alpha(B) \leq 2$. Suppose $\alpha(B) < 2$. Then $B \subset \bigcup_{k=1}^{n} \Omega_k$, where $\operatorname{diam} \Omega_k < 2$ for all $k = 1, \ldots, n$ (with no loss of generality one can assume that all the sets Ω_k are closed). Now taking the section of B by an arbitrary n-dimensional subspace E_n and setting $A_k = \Omega_k \cap E_n$, we arrive at a contradiction with the theorem on antipodes. **QED**

1.1.7. Theorem. *The Kuratowski and Hausdorff* MNC*s are related by the inequalities*

$$\chi(\Omega) \leq \alpha(\Omega) \leq 2\chi(\Omega).$$

In the class of all infinite-dimensional spaces these inequalities are sharp.

Proof. The inequalities themselves are consequences of the following obvious remarks: 1) if $\{x_1, \ldots, x_m\}$ is an ε-net of Ω, then $\{\Omega \cap (x_k + \varepsilon B)\}_{k=1}^{m}$ is a cover of Ω by sets of diameter 2ε; 2) if $\{\Omega_k\}_{k=1}^{m}$ is a cover of Ω with $\operatorname{diam} \Omega_k \leq d$ and if $x_k \in \Omega_k$, then $\{x_1, \ldots, x_k\}$ is a d-net of Ω.

The sharpness of the second inequality follows from Theorem 1.1.6. The following example shows that the first inequality is also sharp. Take for E the space c_0 of sequences of numbers that converge to zero, with the norm $||x|| = \sup |x_i|$, and let $\Omega = \{e_k\}_{k=1}^{\infty}$ be the standard basis in c_0. Since the diameter of any set containing more than one element is equal to 1, $\alpha(\Omega) = 1$. On the other hand, $\chi(\Omega) = 1$ because the distance from any infinite subset of Ω to any element of c_0 is not smaller than 1. **QED**

We should mention here that for some spaces E the inequality $\chi(\Omega) \leq \alpha(\Omega)$ can be improved. For instance, one can show that for the space l_p one has

$$\sqrt[p]{2}\chi(\Omega) \leq \alpha(\Omega).$$

Let us prove one more important property of the MNCs α and χ.

1.1.8. Theorem. *The intersection of a centered system of closed subsets of a Banach space is nonempty if this system contains sets of arbitrarily small Kuratowski (or, which in view of Theorem 1.1.7 is equivalent, Hausdorff)* MNC.

Proof. Let $\mathfrak{M}$ be the given centered system. Notice that if $\mathfrak{M}$ contains a set Ω_0 which has MNC equal to zero, and hence, thanks to the regularity of α, is compact, then the assertion of the theorem is a trivial consequence of the definition of compactness: it suffices to pass to the system $\mathfrak{M}' = \{\Omega \cap \Omega_0 : \Omega \in \mathfrak{M}\}$. In the general case, we pick a sequence $\Omega_n \in \mathfrak{M}$ such that $\alpha(\Omega_n) \to 0$ as $n \to \infty$ and we show that the set $\Omega_0 = \bigcap_{k=1}^{\infty} \Omega_n$ is compact, and that after adding Ω_0 to $\mathfrak{M}$ the system remains centered. As above, this will imply the assertion of the theorem.

The compactness of Ω_0 follows from its closedness and the obvious fact that $\alpha(\Omega_0) = 0$. Now let us show that for any finite subsystem $\mathfrak{M}_0 \subset \mathfrak{M}$ the set $A = (\bigcap_{\Omega \in \mathfrak{M}_0} \Omega) \cap \Omega_0$ is nonempty. We choose a sequence $\{x_n\}$ such that $x_n \in (\bigcap_{\Omega \in \mathfrak{M}_0} \Omega) \bigcap (\bigcap_{k=1}^{n} \Omega_k)$. Since $\alpha(\{x_n\}_{n=1}^{\infty}) = \alpha(\{x_n\}_{n=N}^{\infty}) \le \alpha(\Omega_N) \to 0$ as $N \to \infty$, this sequence is relatively compact. Consequently, the set of its limit points is nonempty. It remains to remark that every limit point belongs to A. **QED**

We next prove a number of formulas that enable us to compute the Hausdorff MNC in the spaces l_p, c_0, C, L_p and L_∞.

1.1.9. The Hausdorff MNC in the spaces l_p and c_0. *In the spaces l_p and c_0 of sequences summable in the p-th power and respectively sequences converging to zero the* MNC χ *can be computed by means of the formula*

$$\chi(\Omega) = \lim_{n \to \infty} \sup_{x \in \Omega} \|(I - P_n)x\|, \tag{1}$$

where P_n is the projection onto the linear span of the first n vectors in the standard basis.

Proof. If Q is a $[\chi(\Omega) + \varepsilon]$-net of Ω, then $\Omega \subset Q + [\chi(\Omega) + \varepsilon]\overline{B}$. Hence, one can represent each $x \in \Omega$ in the form $x = q + [\chi(\Omega) + \varepsilon]b$, where $q \in Q$ and $b \in \overline{B}$. Consequently,

$$\sup_{x \in \Omega} \|(I - P_n)x\| \le \sup_{q \in Q} \|(I - P_n)q\| + [\chi(\Omega) + \varepsilon].$$

Since Q is finite, the first term in the right-hand side tends to zero when $n \to \infty$, and so

$$\lim_{n \to \infty} \sup_{x \in \Omega} \|(I - P_n)x\| \le \chi(\Omega) + \varepsilon,$$

which in view of the arbitrariness of ε yields one of the inequalities needed to establish (1).

To prove the opposite inequality, we notice that

$$\Omega \subset P_n\Omega + (I - P_n)\Omega.$$

Using the properties of χ and the total boundedness of $P_n\Omega$, we obtain

$$\chi(\Omega) \leq \chi(P_n\Omega) + \chi[(I - P_n)\Omega] = \chi[(I - P_n)\Omega] \leq \sup_{x\in\Omega} ||(I - P_n)x||.$$

Since n is arbitrary, this gives

$$\chi(\Omega) \leq \lim_{n\to\infty} \sup_{x\in\Omega} ||(I - P_n)x||. \quad \textbf{QED}$$

1.1.10. The Hausdorff MNC in the space $C[a,b]$. *In the space $C[a,b]$ of continuous real-valued functions on the segment $[a,b]$ the value of the set-function χ on a bounded set Ω can be computed by means of the formula*

$$\chi(\Omega) = \frac{1}{2} \lim_{\delta\to 0} \sup_{x\in\Omega} \max_{0\leq\tau\leq\delta} ||x - x_\tau||, \tag{2}$$

where x_τ denotes the τ-translate of the function x:

$$x_\tau(t) = \begin{cases} x(t+\tau), & \text{if } a \leq t \leq b-\tau, \\ x(b), & \text{if } b-\tau \leq t \leq b. \end{cases}$$

Proof. Pick an arbitrary $\varepsilon > 0$ and construct a finite $[\chi(\Omega)+\varepsilon]$-net Q of the set Ω. Let $x \in \Omega$. Denote by y an element of Q such that $||x - y|| \leq \chi(\Omega) + \varepsilon$. Finally, let $\delta > 0$ and $\tau \in [0, \delta]$. Then

$$||x - x_\tau|| \leq ||x - y|| + ||y - y_\tau|| + ||y_\tau - x_\tau|| \leq 2||x - y|| + ||y - y_\tau||$$

$$\leq 2\chi(\Omega) + 2\varepsilon + \max_{y\in Q} \max_{0\leq\tau\leq\delta} ||y - y_\tau||.$$

Consequently,

$$\sup_{x\in\Omega} \max_{0\leq\tau\leq\delta} ||x - x_\tau|| \leq 2\chi(\Omega) + 2\varepsilon + \max_{y\in Q} \max_{0\leq\tau\leq\delta} ||y - y_\tau||.$$

Letting $\delta \to 0$ and taking into account that the finite family Q is equicontinuous, one obtains

$$\lim_{\delta\to 0} \sup_{x\in\Omega} \max_{0\leq\tau\leq\delta} ||x - x_\tau|| \leq 2\chi(\Omega) + 2\varepsilon,$$

which in view of the arbitrariness of ε yields the inequality

$$\frac{1}{2} \lim_{\delta\to 0} \sup_{x\in\Omega} \max_{0\leq\tau\leq\delta} ||x - x_\tau|| \leq \chi(\Omega). \tag{3}$$

In proving the opposite inequality we shall assume that the functions $x \in \Omega$ are extended from the segment $[a, b]$ to the whole real line by the rule: $x(t) = x(a)$ for $t \leq a$, $x(t) = x(b)$ for $t \geq b$. We define the operators R_h and P_h ($h > 0$) through the formulas

$$(R_h x)(t) = \frac{1}{2}(\max\{x(s): s \in [t-h, t+h]\} + \min\{x(s): s \in [t-h, t+h]\})$$

and respectively

$$(P_h x)(t) = \frac{1}{2h} \int_{t-h}^{t+h} x(s) ds.$$

It is not hard to see that the set $P_h R_h(\Omega)$ is relatively compact in $C[a, b]$. We claim that it constitutes a $(q_{2h}/2)$-net of the set Ω, where $q_{2h} = \sup_{x \in \Omega} \max_{0 \leq \tau \leq \delta} ||x - x_\tau||$. In fact,

$$||P_h R_h x - x|| = \max_{a \leq t \leq b} \left| \frac{1}{2h} \int_{t-h}^{t+h} (R_h x)(s) ds - \frac{1}{2h} \int_{t-h}^{t+h} x(t) ds \right|$$

$$\leq \frac{1}{2h} \max_{a \leq t \leq b} \int_{t-h}^{t+h} |(R_h x)(s) - x(t)| ds. \tag{4}$$

If $|t - s| \leq h$, then obviously

$$\min\{x(\tau): \tau \in [s-h, s+h]\} \leq x(t) \leq \max\{x(\tau): \tau \in [s-h, s+h]\}.$$

Consequently, $|(R_h x)(s) - x(t)| \leq \frac{1}{2h} \max_{0 \leq \tau \leq 2h} ||x - x_\tau||$, whence $|(R_h x)(s) - x(t)| \leq q_{2h}/2$. From this and (4) it follows that $\chi(\Omega) \leq q_{2n}/2$. Letting $h \to 0$ we obtain

$$\chi(\Omega) \leq \frac{1}{2} \limsup_{\delta \to 0} \sup_{x \in \Omega} \max_{0 \leq \tau \leq \delta} ||x - x_\tau||, \tag{5}$$

which completes the proof of equality (2). **QED**

1.1.11. Generalization to the space $C(K, \mathbf{R}^m)$. *Formula (2) admits the following generalization to the case of the space $C(K, \mathbf{R}^m)$ of continuous functions on a compact space K with values in $\mathbf{R}^m$* (see [23]), *equipped with the norm $||x||_C = \max_{t \in K} ||x(t)||$:*

$$\chi(\Omega) = \sup_{\mathcal{B}} \inf_{V \in \mathcal{B}} \sup_{x \in \Omega} \operatorname{rad} x(V), \tag{6}$$

where $\mathcal{B}$ is a basis of neighborhoods of some point of K and $\operatorname{rad} x(V)$ denotes the infimum of the radii of all balls in $\mathbf{R}^m$ that contain $x(V)$.

1.1.12. Generalization to the space $L_\infty([a, b], \mathbf{R}^m)$. *Let $L_\infty([a, b], \mathbf{R}^m)$ be the space of equivalence classes x of measurable, essentially bounded functions $\xi: [a, b] \to \mathbf{R}^m$,* *endowed with the norm* $||x|| = \operatorname{vrai}\sup_{t\in[a,b]}||\xi(t)|| = \inf_{\xi\in x}\sup_{t\in[a,b]}||\xi(t)||$. *Then formula* (6) *remains valid in* $L_\infty([a,b],\mathbf{R}^m)$ *for an appropriate interpretation of the notations involved, namely,* $\mathcal{B}$ *stands for an arbitrary maximal filter of measurable sets in* $[a,b]$ *and* $x(V)$ *stands for* $\bigcap \xi(\tilde{V})$, *where* ξ *runs through* x *and* $\tilde{V}$ *runs through the set of all subsets of* V *of full measure.*

1.1.13. The Hausdorff MNC in $L_p[a,b]$. *In the space* $L_p[a,b]$ *of equivalence classes* x *of measurable functions* $\xi: [a,b] \to \mathbf{R}$ *with integrable p-th power, endowed with the norm* $||x|| = (\int_a^b |x(t)|^p dt)^{1/p}$, *the Hausdorff* MNC *can be computed by means of the formula*

$$\chi(\Omega) = \frac{1}{2}\lim_{\delta\to 0}\sup_{x\in\Omega}\max_{0\leq\tau\leq\delta}||x - x_\tau||, \tag{7}$$

where x_τ *denotes the* τ*-translate of the function* x (see **1.1.10**) *or, alternatively, the Steklov function*

$$x_\tau(t) = \frac{1}{2\tau}\int_{t-\tau}^{t+\tau} x(s)ds$$

(here x *is extended outside* $[a,b]$ *by zero).*

We conclude this section by describing how the notions of Kuratowski and Hausdorff MNC can be extended to uniform (in particular, locally convex) spaces.

1.1.14. The Kuratowski and Hausdorff MNCs in uniform spaces. Let E be a uniform space, P a family of pseudometrics that are uniformly continuous on $E \times E$, $\mathfrak{M}$ the set of all subsets of E that are bounded with respect to any pseudometric $p \in P$, and A the set of all functions $a: P \to [0,\infty)$, endowed with the uniform structure generated by pointwise convergence and with the natural partial order: $a_1 \leq a_2$ means $a_1(p) \leq a_2(p)$ for all $p \in P$.

1.1.15. Definition. *The Kuratowski* [resp. *Hausdorff*] *measure of noncompactness* on the space E generated by the family of pseudometrics P is the function $\alpha: \mathfrak{M} \to A$ [resp. $\chi: \mathfrak{M} \to A$] defined as $[\alpha(\Omega)](p) = \inf\{d > 0: \Omega$ admits a finite partition into subsets whose diameters with respect to the pseudometric p are no larger than $d\}$ [resp. $[\chi(\Omega)](p) = \inf\{\varepsilon > 0: \Omega$ has a finite ε-net with respect to $p\}$].

The properties described in **1.1.14** can be reformulated in an obvious manner for the Kuratowski and Hausdorff MNCs in uniform spaces.

1.1.16. The inner Hausdorff MNC. It is readily seen that the Kuratowski MNC of a set Ω is an "intrinsic characteristic" of the metric space (Ω,ρ), where ρ is the metric induced by the norm on E. In contrast, the Hausdorff MNC depends on the "ambient" space E, specifically, on how much freedom one has in the choice of the elements of an ε-net. The definition of the Hausdorff MNC can be modified so that it becomes an intrinsic characteristic of sets. Specifically, we define the *inner Hausdorff measure of noncompactness* $\chi_i(\Omega)$ of the set Ω to be the number

$$\chi_i(\Omega) = \inf\{\varepsilon > 0 : \Omega \text{ has a finite } \varepsilon\text{-net in } \Omega\}.$$

Clearly, the set-function χ_i is nonsingular, Lipschitzian, semi-homogeneous, algebraically semi-additive, and invariant under translations. It is also readily seen that χ_i is regular and invariant under passage to the closure. At the same time, χ_i *is not invariant under passage to the convex hull.* In fact, let $E = m$, the space of bounded numerical sequences, and let $\Omega = \{x_n, -x_n\}_{n=1}^{\infty}$, where $x_n = (\underbrace{1, \dots, 1}_{n-1}, -1, 1, 1, \dots)$. Clearly, $\chi_i(\Omega) \geq 2$, since the distance between any two points of Ω is equal to 2. On the other hand, $\operatorname{co}\Omega$ contains the zero element of E, and the distance from zero to any point of $\operatorname{co}\Omega$ does not exceed 1. Hence, $\chi_i(\Omega) \leq 1$. On the same example one can check that χ_i is neither monotone, nor semi-additive.

1.2. THE GENERAL NOTION OF MEASURE OF NONCOMPACTNESS

In this section we give an axiomatic definition of the notion of a measure of noncompactness and consider a number of examples. As we already mentioned, an important role is played by the invariance of the Kuratowski and Hausdorff MNCs under passage to the convex closure.

1.2.1. Definition. *A function ψ, defined on the set of all subsets of a Banach space E with values in some partially ordered set $(Q, \leq)$, is called a measure of noncompactness if $\psi(\overline{\operatorname{co}}\,\Omega) = \psi(\Omega)$ for all $\Omega \subset E$.*

As we established above, the Kuratowski and Hausdorff MNCs satisfy the condition of this general definition. On the contrary, the inner Hausdorff MNC (see **1.1.16**) is not an MNC in the sense of the general definition.

The term "measure of noncompactness" is not completely appropriate for such a general definition, because only the functions that were said above to be regular serve to actually measure the degree of noncompactness. However, in various problems it is convenient to use different MNCs, including nonregular ones (and which nevertheless have some connection with noncompactness), and it turns out that the only property they share is precisely the invariance under passage to the convex closure. All the notions enumerated in **1.1.4** carry over to general MNCs. In defining Lipschitzianity [continuity] it is natural to require that Q be endowed with a metric [resp. topology], and in defining semi-homogeneity and algebraic semi-additivity it is necessary that an operation of multiplication by scalars and respectively one of addition be given on Q.

1.2.2. Elementary examples of MNCs. Clearly, the functions

$$\psi_1(\Omega) = \begin{cases} 0, & \text{if } \Omega \text{ is totally bounded,} \\ 1, & \text{otherwise,} \end{cases}$$

and

$$\psi_2(\Omega) = \text{diam } \Omega$$

are MNCs (in the sense of Definition 1.2.1). Notice that if E is infinite-dimensional then ψ_1 is not continuous (with respect to the natural topology on the set where it takes its values, the real line). It also clear that ψ_2 is not a regular MNC.

1.2.3. Products of MNCs. Let $\psi_1, \dots, \psi_n$ be MNCs in E with values in $Q_1, \dots, Q_n$, respectively, and let $F: Q_1 \times \dots \times Q_n \to Q$ be a map. Then, as is readily verified, the function $\psi(\Omega) = F(\psi_1(\Omega), \dots, \psi_n(\Omega))$ is an MNC; its properties are determined by the properties of the measures $\psi_1, \dots, \psi_n$ and of the map F.

In the following subsections we give a number of examples of MNCs in concrete spaces. Most of them will be needed later.

1.2.4. MNCs in $C([a, b], E)$. Let E be a Banach space with norm $||\cdot||$ and ψ be a monotone MNC on E. Let $C([a, b], E)$ denote the space of all continuous E-valued functions on $[a, b]$ with the norm $||x|| = \max_{t \in [a,b]} ||x(t)||$. Define a scalar function ψ_C on the bounded subsets of $C([a, b], E)$ by the formula

$$\psi_C(\Omega) = \psi(\Omega[a, b]),$$

where $\Omega[a, b] = \{x(t): x \in \Omega, t \in [a, b]\}$. ψ_C is an MNC in the sense of the general definition. In fact,

$$(\overline{\text{co}}\, \Omega)[a, b] \subset \overline{\text{co}}(\Omega[a, b])$$

for any bounded subset $\Omega \subset C([a, b], E)$, whence $\psi_C(\overline{\text{co}}\, \Omega) \le \psi_C(\Omega)$. The opposite inequality is obvious.

We should remark here that if the MNC ψ is regular and the set Ω is equicontinuous, then by the Ascoli-Arzelà theorem $\psi_C(\Omega) = 0$ is equivalent to Ω being relatively compact.

In Chapter 4 we shall need two other MNCs in $C([a,b],E)$, ψ_C^1 and ψ_C^2 (see **4.1.6** and **4.2.6**), defined as follows. Let $\mathfrak{M}[a,b]$ and $\mathfrak{N}[a,b]$ denote the partially ordered linear space of all scalar functions defined on $[a,b]$ and respectively the subspace of $\mathfrak{M}[a,b]$ consisting of all nondecreasing functions. The MNCs ψ_C^1 and ψ_C^2 on $C([a,b],E)$ with values in $\mathfrak{M}[a,b]$ and $\mathfrak{N}[a,b]$, respectively, are defined by means of the formulas

$$[\psi_C^1(\Omega)](t) = \psi[\Omega(t)]$$

and

$$[\psi_C^1(\Omega)](t) = \chi_t(\Omega_t),$$

where $\Omega(t) = \{x(t): x \in \Omega\} \subset E$, $\Omega_t = \{x_t = x|[a,t]: x \in \Omega\} \subset C([a,b],E)$, and χ_t is the Hausdorff MNC in the space $C([a,t],E)$. The MNC ψ_C^2 is monotone, nonsingular, invariant under translations, semi-additive, algebraically semi-additive, and continuous. If ψ enjoys the properties enumerated, then so does the MNC ψ_C^1. If ψ is continuous and the set Ω is equicontinuous, then $\psi_C^1(\Omega) \in C[a,b]$.

1.2.5. An MNC in $C^1([a,b],E)$. Let $C^1([a,b],E)$ denote the Banach space of the continuously differentiable functions $x: [a,b] \to E$, equipped with the norm $||x||_{C^1} = ||x||_C + ||x'||_C$. The $\mathfrak{M}[a,b]$-valued function ψ_{C^1}, defined on the bounded subsets of $C^1([a,b],E)$ by the formula

$$[\psi_{C^1}(\Omega)](t) = \psi[\Omega'(t)],$$

where $\Omega'(t) = \{x'(t): x \in \Omega\}$, is an MNC. If the set $\Omega' = \{x': x \in \Omega\}$ is equicontinuous and the MNC ψ is continuous, then $\psi_{C^1}(\Omega) \in C[a,b]$.

1.2.6. An MNC in $C^n([a,b],E)$. Let $C^n([a,b],E)$ denote the Banach space of the n-times continuously differentiable functions $x: [a,b] \to E$, endowed with the norm $||x||_{C^n} = \sum_{i=0}^n ||x^{(i)}||_C$. Then each MNC ψ on $C([a,b],E)$ generates an MNC ψ_{C^n} on $C^n([a,b],E)$ by the rule

$$\psi_{C^n}(\Omega) = \psi(\Omega^{(n)}),$$

where $\Omega^{(n)} = \{x^{(n)}: x \in \Omega\}$.

1.2.7. An MNC in c_0. Let c_0 be the Banach space of the numerical sequences that converge to zero, with the norm $||x|| = \max |x_i|$. Let $n(x)$ denote the number of coordinates of the vector x which are larger than or equal to 1. For an arbitrary bounded set $\Omega \subset c_0$ we put

$$n(\Omega) = \min_{x \in \Omega} n(x), \quad \psi(\Omega) = \frac{1}{n(\Omega)+1}.$$

Clearly, $n(\overline{\text{co}}\,\Omega) \leq n(\Omega)$. On the other hand, if the first k coordinates of any vector $x \in \Omega$ are larger than or equal to 1, then the same holds true for the first k coordinates of any vector $y \in \overline{\text{co}}\,\Omega$, and so $n(\overline{\text{co}}\,\Omega) \geq n(\Omega)$. This means that the function n, and together with it the function ψ, are MNCs. Next, it is readily seen that

$$n(\Omega_1 \cup \Omega_2) = \min\{n(\Omega_1), n(\Omega_2)\},$$

and consequently

$$\psi(\Omega_1 \cup \Omega_2) = \max\{\psi(\Omega_1), \psi(\Omega_2)\},$$

so that the MNC ψ is semi-additive. Notice, however, that ψ is not nonsingular, is not invariant under translations, and is not regular.

We conclude this section by describing yet another MNC which, roughly speaking, is different from zero on bounded noncompact sets only in nonreflexive spaces.

1.2.8. The measure of weak noncompactness. Let E be a Banach space and B the unit ball in E. The function $\omega: 2^E \to [0, \infty)$, defined as $\omega(\Omega) = \inf\{\varepsilon > 0: \Omega$ has a weakly compact ε-net in $E\}$, is called the *measure of weak noncompactness.*

The measure of weak noncompactness is an MNC in the sense of the general definition provided E is endowed with the weak topology. This assertion follows from the obvious invariance of ω under the passage to the convex hull and the invariance under the passage to the weak closure, established below:

$$\omega(\Omega) = \omega(\text{wcl}\,\Omega), \tag{1}$$

where wcl stands for weak closure.

Thus, let us prove (1). Suppose $\omega(\Omega) < \varepsilon$. By the definition of ω, there exists a weakly compact set $C \subset E$ such that $\Omega \subset C + \varepsilon\overline{B}$. By the Kreĭn-Shmul′yan theorem (see [34]), the set $\overline{\text{co}}\,C$ is weakly compact, and hence weakly closed. Consequently, $\overline{\text{co}}\,C + \varepsilon\overline{B}$ is also weakly compact. Now the inclusion $\Omega \subset \overline{\text{co}}\,C + \varepsilon\overline{B}$ implies

$$\text{wcl}\,\Omega \subset \overline{\text{co}}\,C + \varepsilon\overline{B},$$

which in turn yields $\omega(\text{wcl}\,\Omega) \leq \varepsilon$. In view of the arbitrariness of $\varepsilon > \omega(\Omega)$, this gives

$$\omega(\text{wcl}\,\Omega) \leq \omega(\Omega).$$

The opposite inequality follows from the obvious monotonicity of ω. This establishes equality (1).

As one can readily verify, the MNC ω is nonsingular, semi-additive, algebraically semi-additive, invariant under translations, and satisfies the inequality $\omega(\Omega) \leq \chi(\Omega)$. The regularity property for ω has the following meaning: $\omega(\Omega) = 0$ if and only if $\mathrm{wcl}\,(\Omega)$ is weakly compact (see [31]).

1.2.9. Theorem. *The measure of weak noncompactness of the unit ball in E is equal to zero if E is reflexive, and to one in the opposite case.*

Proof. Recall that the unit ball B is weakly compact if and only if E is reflexive. The first assertion of the theorem is obvious. Now suppose E is not reflexive. The inclusion $\overline{B} \subset \{0\} + 1 \cdot \overline{B}$ gives $\omega(\overline{B}) \leq 1$. Suppose $\omega(\overline{B}) < 1$. Then there exist an $\varepsilon \in (0,1)$ and a weakly compact set C such that $\overline{B} \subset C + \varepsilon \overline{B}$. But then $\overline{B} \subset \overline{\mathrm{co}}\, C + \varepsilon \overline{B}$, and consequently

$$(1-\varepsilon)\overline{B} + \varepsilon \overline{B} \subset \overline{\mathrm{co}}\, C + \varepsilon \overline{B}.$$

Now we use the following assertion [139]: if $\Omega_1 + \varepsilon \overline{B} \subset \Omega_2 + \varepsilon \overline{B}$, and if Ω_2 is convex and closed, then $\Omega_1 \subset \Omega_2$. In our case this yields

$$(1-\varepsilon)\overline{B} \subset \overline{\mathrm{co}}\, C.$$

Therefore, $(1-\varepsilon)\overline{B}$ is a weakly closed subset of the set $\overline{\mathrm{co}}\, C$, which is itself weakly compact by the Kreĭn-Shmul′yan theorem. Consequently, $(1-\varepsilon)\overline{B}$ is weakly compact, and then so is $\overline{B}$, which contradicts the nonreflexivity of E. **QED**

1.3. THE MEASURE OF NONCOMPACTNESS β

In this section we describe and study yet another MNC which is useful in applications.

1.3.1. Definition. *The measure of noncompactness $\beta(\Omega)$ of the subset Ω of the Banach space E* is the infimum of the numbers $r > 0$ for which Ω does not have an infinite r-lattice or, equivalently, the supremum of those $r > 0$ for which Ω has an infinite r-lattice.

We remind the reader that a set Ω_1 is called an *r-lattice*, or a *lattice with parameter r*, if $\|x - y\| \geq r$ for all $x, y \in \Omega_1$. A set $\Omega_1 \subset \Omega$ with this property is called an *r-lattice of Ω*. Every r-lattice $\Omega_1 \subset \Omega$ that is maximal (i.e., such that it cannot be enlarged to an r-lattice inside Ω) is obviously an r-net of Ω.

1.3.2. Remarks. It is not hard to see that the MNC β is regular, monotone, semi-additive, semi-homogeneous, and invariant under translations and under passage to the closure of the set. Further, it is a straightforward matter to check that the MNCs α, χ and β are related by the inequalities

$$\chi(\Omega) \leq \beta(\Omega) \leq \alpha(\Omega).$$

Also, a simple argument establishes the Lipschitzianity of β:

$$|\beta(\Omega_1) - \beta(\Omega_2)| \leq 2\rho(\Omega_1, \Omega_2).$$

Less obvious properties of the MNC β are established in the next subsections.

1.3.3. Theorem. *The* MNC β *is algebraically semi-additive.*

Proof. Let Ω_1 and Ω_2 be arbitrary bounded subsets of E. Fix some $\varepsilon > 0$. By the definition of β, the set $\Omega_1 + \Omega_2$ has a countable $[\beta(\Omega_1 + \Omega_2) + \varepsilon]$-lattice $\{z_i\}$. Write $z_i = x_i + y_i$, where $x_i \in \Omega_1$ and $y_i \in \Omega_2$. Then for all $i \neq j$,

$$\beta(\Omega_1 + \Omega_2) - \varepsilon \leq ||z_i - z_j|| \leq ||x_i - x_j|| + ||y_i - y_j||. \tag{1}$$

Let us show that from the sequence $\{x_i\}$ one can extract a subsequence $\{u_i\}$ such that

$$||u_i - u_j|| \leq \beta(\Omega_1) + \varepsilon. \tag{2}$$

Then in view of (1) the corresponding subsequence $\{v_i\}$ of $\{y_i\}$ will be a $[\beta(\Omega_1 + \Omega_2) - \beta(\Omega_1) - 2\varepsilon]$-lattice of Ω_2, and consequently

$$\beta(\Omega_1 + \Omega_2) - \beta(\Omega_1) - 2\varepsilon \leq \beta(\Omega_2).$$

In view of the arbitrariness of ε, this implies the needed inequality $\beta(\Omega_1 + \Omega_2) \leq \beta(\Omega_1) + \beta(\Omega_2)$.

To extract the subsequence $\{u_i\}$ consider some maximal $[\beta(\Omega_1) + \varepsilon]$-lattice of the set $\{x_1, x_2, \dots\}$. It is finite and, as remarked above, it forms a $[\beta(\Omega_1) + \varepsilon]$-net of $\{x_1, x_2, \dots\}$. Hence, there is a term of the sequence $\{x_i\}$ such that its closed $[\beta(\Omega_1) + \varepsilon]$-neighborhood contains an infinite subsequence of $\{x_i\}$. Denote the first term with this property x_1^1 and then take the terms of the subsequence described above that lie after x_1^1 in the original sequence and relabel them as x_i^1, $i = 2, 3, \dots$. Thus, we extracted from $\{x_i\}$ an infinite subsequence, all of whose terms lie at distance $\leq \beta(\Omega_1) + \varepsilon$ from its first term. In exactly the same manner, from the sequence $\{x_i^1 : i \geq 2\}$ we can extract a subsequence $\{x_i^2 : i \geq 1\}$ with the same property, and then continue to produce subsequences $\{x_i^3\}, \{x_i^4\}$, and so on.

Now define the sought-for sequence to be $u_i = x_i^i$, $i = 1, 2, \ldots$. The recipe used guarantees that inequality (2) holds for all i and j. **QED**

1.3.4. Theorem. *The function β is invariant under passage to the convex hull, and hence it is an* **MNC** *in the sense of the general definition.*

Proof. In view of the monotonicity of β it suffices to prove the inequality $\beta(\operatorname{co}\Omega) \leq \beta(\Omega)$. For an unbounded set it is obvious, so we shall assume that Ω is bounded. Suppose that the needed inequality does not hold for some Ω, and pick numbers b and c such that $\beta(\Omega) < b < c < \beta(\operatorname{co}\Omega)$.

To reach a contradiction, we consider certain sequences of sets, functions, and numbers, described as follows. We choose in $\operatorname{co}\Omega$ an infinite r_1-lattice $\tilde{Y}_1$ $(r_1 > c)$ and fix in it an arbitrary element $\tilde{y}_1$. Now $\tilde{y}_1$ is a finite convex combination of elements of Ω. As is readily verified, among the latter one can find $\tilde{x}$ such that $||\tilde{x} - y|| \geq r_1$ for y belonging to an infinite set $\tilde{\tilde{Y}}_1 \subset \tilde{Y}_1$. Setting $y_1 = \tilde{x}, Y_1 = \{y_1\} \cup \tilde{\tilde{Y}}_1$, we get the first two objects in our construction.

Further, consider the sets $\Omega_1^0 = \Omega \cap \overline{B}(y_1, b)$, $\Omega_1^1 = \Omega \setminus \Omega_1^0$, and notice that any $y \in \operatorname{co}\Omega$ can be represented as $y = (1 - \mu_1)u_1^0 + \mu_1 u_1^1$, with $u_1^i \in \operatorname{co}\Omega_1^i$, $i = 0, 1$ and $\mu_1 \in [0, 1]$. For each y fix such a representation, defining in this manner two functions: $\mu_1 = \mu_1(y)$ and $u_1^i = u_1^i(y)$. Denote the set $\{u_1^i : y \in Y_1\}$ by U_1^i and define a binary indicator a_1 as follows: $a_1 = 1$ if $\beta(U_1^1) > c$ and $a_1 = 0$ otherwise.

In the second case, when $a_1 = 0$, one necessarily has $\beta(U_1^0) > c$. This follows from the inclusion $Y_1 \subset \operatorname{co}_2(U_1^0 \cup U_1^1)$ (here $\operatorname{co}_N\Omega$ stands for the set of all convex combinations of at most N elements of Ω), the inequality

$$\beta(\operatorname{co}_N\Omega) \leq \beta(\Omega), \tag{3}$$

which is established in the next subsection, the monotonicity and semi-additivity of the function β, and the fact that $\beta(Y_1) > c$. Therefore, in either of the two cases, $\beta(U_1^{a_1}) > c$.

Proceeding in analogous manner we construct objects $Y_n, y_n, \Omega_n^i, \mu_n, u_n^i, U_n^i$ $(i = 0, 1)$, a_n for each positive integer n, taking care at each new step to single-out a set $\tilde{Y}_n$ in $U_n^{a_n}$. In more detail, Y_{n+1} is an r_{n+1}-lattice $(r_{n+1} > c)$ in $\operatorname{co}\Omega_n^{a_n}$, $y_{n+1} \in Y_{n+1} \cap \Omega_n^{a_n}$, $Y_{n+1} \setminus \{y_{n+1}\} \subset U_n^{a_n}$, $\Omega_{n+1}^0 = \Omega_n^{a_n} \cap \overline{B}(y_{n+1}, b)$, $\Omega_{n+1}^1 = \Omega_n^{a_n} \setminus \Omega_{n+1}^0$, the functions $\mu_{n+1} = \mu_{n+1}(u_n^{a_n})$, $u_{n+1}^i = u_{n+1}^i(u_n^{a_n})$, are defined for $u_n^{a_n} \in \operatorname{co}\Omega_n^{a_n}$ by some fixed decompositions $u_n^{a_n} = (1 - \mu_{n+1})u_{n+1}^0 = \mu_{n+1}u_{n+1}^1$, $U_{n+1}^i = \{u_{n+1}^i : u_n^{a_n} \in Y_{n+1}\}$, $a_{n+1} = 1$ if $\beta(U_{n+1}^1) > c$, and $a_{n+1} = 0$ in the opposite case. Throughout the construction $\beta(U_{n+1}^{a_{n+1}}) > c$.

Notice that if $a_n = 1$ and $m \geq n + 1$, then $||y_m - y_n|| > b$. In fact, by construction,

$y_m \in \Omega_m^{a_m} \subset \Omega_n^{a_n} = \Omega_n^1$, i.e., $y \notin \overline{B}(y_n, b)$. This immediately implies that the set $\{y_n : a_n = 1\}$ is a b-lattice in Ω. Since $b > \beta(\Omega)$, the set $\{n : a_n = 1\}$ is finite.

Thus, there is a k such that $a_n = 0$ for $n \geq k$, and hence $Y_n \setminus \{y_n\} \subset U_{n-1}^0$ for $n > k$. Let $m > k$. Since u_{n+1}^i is a function of $u_n^{a_n}$ and $a_n = 0$ for $n \geq k$, u_m^0 is a function of u_k^0. Consider the set $Y_{k+1}^m = \{u_k^0 \in Y_{k+1} : u_m^0 \in Y_m\}$, which, like Y_{k+1}, is an infinite r_{k+1}-lattice; also, $\beta(Y_{k+1}^m) \geq r_{k+1} > c$.

We claim that every element $u_k^0 \in Y_{k+1}^m$ is representable in the form

$$u_k^0 = \sum_{j=1}^{m-k} \delta_j u_{k+j}^1 + \delta u_m^0, \tag{4}$$

where $\delta_j \geq 0$, $\delta = \prod_{j=1}^{m-k}(1 - \mu_{k+j})$, $\sum_{j=1}^{m-k} \delta_j + \delta = 1$. Indeed, for $m = k+1$ this is precisely the representation $u_k^0 = (1 - \mu_{k+1})u_{k+1}^0 + \mu_{k+1}u_{k+1}^1$, and the step from m to $m+1$ is made by substituting in (4) the analogous representation for u_m^0:

$$\begin{aligned} u_k^0 &= \sum_{j=1}^{m-k} \delta_j u_{k+1}^1 + \delta[(1 - \mu_{m+1})u_{m+1}^0 + \mu_{m+1}u_{m+1}^1] \\ &= \sum_{j=1}^{m-k} \delta_j u_{k+j}^1 + \delta\mu_{m+1}u_{m+1}^1 + \delta(1 - \mu_{m+1})u_{m+1}^0. \end{aligned}$$

This is precisely a representation of the needed form. It is convenient to recast (4) in the form

$$u_k^0 = \sum_{j=1}^{m-k-1} \delta_j u_{k+j}^1 + (\delta_{m-k} + \delta)u_m^1 + \delta(u_m^0 - u_m^1).$$

From this equality it follows that

$$Y_{k+1}^m \subset \mathrm{co}_{m-k}\Big(\bigcup_{j=1}^{m-k} U_{k+j}^1\Big) + \delta\overline{B}(0, d),$$

where d is the diameter of $\mathrm{co}\,\Omega$. We next show that the μ_i admit the bound $\mu_i \geq p > 0$, and consequently δ can be made arbitrarily small by taking m sufficiently large. Then with the aid of (3) and (4) we conclude that for some $j \geq 1$,

$$\beta(U_{k+j}^1) > c,$$

which contradicts the equality $a_{j+k} = 0$.

The bound $\mu_i \geq p = (c - b)/(d - b)$ follows from the relation

$$u_{i-1}^{a_{i-1}} - y_i = (1 - \mu_i)(u_i^0 - y_i) + \mu_i(u_i^1 - y_i)$$

and the inequalities $\|u_{i-1}^{a_{i-1}} - y_i\| \geq r_i > c$ $(u_{i-1}^{a_{i-1}} \in Y_i)$, $\|u_i^0 - y_i\| \leq b$ and $\|u_i^1 - y_i\| \leq d$. Notice that the denominator $d - b$ is strictly positive because $d \geq \beta(\operatorname{co}\Omega) > b$.

To complete the proof it remains to establish inequality (3).

1.3.5. Lemma. *For any nonnegative integer N,*

$$\beta(\Omega) \geq \beta(\operatorname{co}_N \Omega),$$

where $\operatorname{co}_N \Omega$ denotes the set of all convex combinations of at most N elements of Ω.

Proof. We use the representation

$$\operatorname{co}_N \Omega = \bigcup_{\lambda \in \sigma} (\lambda_1 \Omega + \lambda_2 \Omega + \ldots + \lambda_N \Omega),$$

where $\lambda = (\lambda_1, \lambda_2, \ldots, \lambda_N)$ runs through the standard simplex σ. As in the proof of Theorem 1.1.5, the above union can be approximated, with arbitrary accuracy $\delta(\varepsilon)$ in the Hausdorff metric, by a finite union of the same form, where now λ runs through a finite ε-net σ_ε of σ. Using the Lipschitzianity, semi-additivity, algebraic semi-additivity, and semi-homogeneity of the function β, we get

$$\beta(\operatorname{co}_N \Omega) \leq \beta\Big[\bigcup_{\lambda \in \sigma_\varepsilon} (\lambda_1 \Omega + \lambda_2 \Omega + \ldots + \lambda_N \Omega)\Big] + 2\delta(\varepsilon)$$

$$= \max_{\lambda \in \sigma_\varepsilon} \beta(\lambda_1 \Omega + \lambda_2 \Omega + \ldots + \lambda_N \Omega) + 2\delta(\varepsilon)$$

$$\leq \max_{\lambda \in \sigma_\varepsilon} \Big(\sum_{i=1}^{N} \lambda_i \beta(\Omega)\Big) + 2\delta(\varepsilon) = \beta(\Omega) + 2\delta(\varepsilon).$$

Since $\varepsilon > 0$ is arbitrary and $\delta(\varepsilon) \to 0$ as $\varepsilon \to 0$, this yields the needed inequality (3). **QED**

1.4. SEQUENTIAL MEASURES OF NONCOMPACTNESS

To this point we took as the domain of definition of an MNC a colection of sets which, together with any of its members Ω, contains the closure of its convex hull. Now let us consider functions of countable sets. Of course, a collection of countable sets does not satisfy the aforementioned requirement; nevertheless, the functions studied here are in many respects analogous to the MNCs. Below E continues to denote a Banach space.

1.4.1. Definition. *Let SE be the collection of all bounded and at most countable subsets of the space E. A function $\psi: SE \to [0, \infty)$ is called a sequential measure of noncompactness if it satisfies the following condition: $\Omega_1, \Omega_2 \in SE$ and $\Omega_1 \subset \overline{\mathrm{co}}\,\Omega_2$ implies $\psi(\Omega_1) \le \psi(\Omega_2)$.*

The notions of monotonicity, semi-additivity, nonsingularity, regularity, continuity, Lipschitzianity, algebraic semi-additivity, semi-homogeneity, and invariance under translations for sequential MNCs are defined exactly in the same way as for ordinary MNCs.

It is an immediate consequence of the definition that every sequential MNC enjoys the monotonicity property.

The next theorem shows that every sequential MNC ψ generates in E an ordinary MNC $\tilde{\psi}$, which "inherits" the properties of ψ.

1.4.2. Theorem. *Let ψ be a sequential* MNC *in E. Then the rule*

$$\tilde{\psi}(\Omega) = \sup\{\psi(C)\colon\ C \in SE,\ C \subset \Omega\}, \tag{1}$$

yields a monotone MNC *in E, defined (and finite) on all bounded sets. Moreover, if ψ has any one of the properties enumerated in the preceding subsection, then $\tilde{\psi}$ also has that property.*

Proof. We first show that the function $\tilde{\psi}$ takes finite values on the bounded subsets $\Omega \subset E$. Let $C_n \subset \Omega$ $(n = 1, 2, \dots,)$ be countable sets such that

$$\lim_{n \to \infty} \psi(C_n) = \tilde{\psi}(\Omega).$$

Let $C = \bigcup_{n=1}^{\infty} C_n$; then clearly

$$\tilde{\psi}(\Omega) = \lim_{n \to \infty} \psi(C_n) \le \psi(C) \le \tilde{\psi}(\Omega).$$

Consequently, $\tilde{\psi}(\Omega) = \psi(C) < \infty$.

En route we showed that in (1) one can always replace sup by max. The monotonicity of $\tilde{\psi}$ is plain.

Now let us show that $\tilde{\psi}$ is an MNC. Suppose Ω is a bounded subset of E. The inequality $\tilde{\psi}(\Omega) \le \tilde{\psi}(\overline{\mathrm{co}}\,\Omega)$ follows from the monotonicity of $\tilde{\psi}$. To prove the opposite inequality consider an arbitrary countable set $C \subset \overline{\mathrm{co}}\,\Omega$ and arrange its elements in a sequence $\{y_n\}$. The inclusion $y_n \in \overline{\mathrm{co}}\,\Omega$ is equivalent to y_n having a representation

$$y_n = \lim_{m \to \infty} \sum_{k=1}^{r(m)} \alpha_{nmk} x_{nmk},$$

with $x_{nmk} \in \Omega$, $\alpha_{nmk} \geq 0$, $\sum_{k=1}^{r(m)} \alpha_{nmk} = 1$. Denote the set of all elements x_{nmk} by C_1. Clearly, C_1 is countable and $C \subset \overline{\text{co}}\, C_1$, so that $\psi(C) \leq \psi(C_1)$. Thus, for any countable set $C \subset \overline{\text{co}}\, \Omega$ there exists a countable set $C_1 \subset \Omega$ whose sequential MNC is at least equal to that of C. This yields the needed inequality $\tilde{\psi}(\overline{\text{co}}\, \Omega) \leq \tilde{\psi}(\Omega)$, and thus the first part of the theorem is proved.

The proof of the second part is tedious, but trivial, and we omit it. **QED**

1.4.3. The MNC ψ^0. Any MNC ψ, defined on the bounded subsets of E, generates in a natural manner a sequential MNC ψ^0 (the restriction of the original measure to the collection of all at-most-countable subsets of E). One is naturally led to asking: under which conditions does the MNC $\tilde{\psi}^0$, constructed by means of formula (1) from the sequential MNC ψ^0, coincide with the original MNC ψ? A complete answer to this question is not know. A partial answer for the case of the Hausdorff MNC χ is given in the following two subsections.

1.4.4. Theorem. *The equality*

$$\tilde{\chi}^0(B) = \chi(B)$$

holds for any ball B in the Banach space E.

Proof. From the properties of the Hausdorff MNC and Theorem 1.4.2 it follows that the MNC $\tilde{\chi}^0$ is monotone, invariant under translations, semi-homogeneous, and semi-additive. Hence, by Theorem 1.1.6, it suffices to verify that $\tilde{\chi}^0(B)$, where B is the unit ball, is equal to 0 or 1 according to whether the dimension of E is finite or infinite. If $\dim E < \infty$, then obviously $\tilde{\chi}^0(B) = 0$. Now suppose $\dim E = \infty$. Since

$$\tilde{\chi}^0(B) = \max\{\chi(C) \colon C \in SE,\ C \subset B\} \leq \chi(B) \leq 1,$$

it suffices to show that $\tilde{\chi}^0(B) \geq 1$. Suppose this is not the case: $\tilde{\chi}^0(B) = q < 1$. Fix an $\varepsilon \in (0, 1-q)$. Let $C \in SE, C \subset B$, and let A be a finite $(q+\varepsilon)$-net of C in E, i.e., $C \subset \bigcup_{x \in A}[x + (q+\varepsilon)B]$. Then $\tilde{\chi}^0(C) \leq (q+\varepsilon)\tilde{\chi}^0(B)$ (here we used the aforementioned properties of the MNC $\tilde{\chi}^0$). Consequently,

$$\tilde{\chi}^0(B) = \max\{\chi(C) \colon C \in SE,\ C \subset B\}$$

$$= \max\{\tilde{\chi}^0(C) \colon C \in SE,\ C \subset B\} \leq (q+\varepsilon)\tilde{\chi}^0(B),$$

whence $\tilde{\chi}^0(B) = 0$, because $q+\varepsilon < 1$. We therefore conclude that any countable set $C \subset B$ is totally bounded, which of course is not the case if the space is infinite-dimensional. **QED**

1.4.5. Theorem. *The function $\tilde{\chi}^0$ does not necessarily coincide with χ. However, the inequalities*

$$\frac{1}{2}\chi(\Omega) \leq \tilde{\chi}^0(\Omega) \leq \chi(\Omega) \tag{2}$$

hold for any bounded set $\Omega \subset E$.

Proof. First let us provide an example where the MNCs $\tilde{\chi}^0$ and χ are distinct. Let A be the set of ordinals of countable power. Let E denote the set of all bounded functions $x: A \to \mathbf{R}$ that satisfy the following condition: for any $x \in E$ there exists an $a_x \in A$ such that $x(a) = 0$ for all $a \geq a_x$. Clearly, E is a linear space (with respect to the natural linear operations). It is readily verified that endowed with the norm

$$||x|| = \sup\{|a(x)|: \ a \in A\}$$

E is a Banach space. Now let

$$\Omega = \{x \in E: \ 0 \leq x(a) \leq 1 \text{ for all } a \in A\}.$$

Let us show that

$$\tilde{\chi}^0(\Omega) = 1/2, \tag{3}$$

whereas

$$\chi(\Omega) = 1. \tag{4}$$

This will establish the first assertion of the theorem as well the as the fact that in the inequalities (2) (if they hold) the constant 1/2 is sharp.

Let $C = \{x_x\}_{n=1}^{\infty}$ be an arbitrary countable subset of Ω . Clearly, one can find an element $a^* \in A$ such that $a^* \geq a_x$ for all n (indeed, A is not countable). Then one can readily check that the element $x^* \in E$ defined by the formula

$$x^*(a) = \begin{cases} 1/2, & \text{if } a \leq a^*, \\ 0, & \text{if } a > a^*, \end{cases}$$

provides an 1/2-net for the set Ω. Consequently, $\tilde{\chi}^0(\Omega) \leq 1/2$. The opposite inequality follows from (4) and inequalities (2), which will be established below.

Let us prove (4). First of all, $\chi(\Omega) \leq 1$, since Ω is a subset of the unit ball. Next, let $\{y_i\}_{i=1}^k$ be an arbitrary finite collection of elements of E. Pick $a_0 \in A$ such that $a_{y_i} < a_0$ for all $i = 1, \ldots, k$, and define $x_0 \in \Omega$ by the rule

$$x_0(a) = \begin{cases} 0, & \text{if } a \neq a_0, \\ 1, & \text{if } a = a_0. \end{cases}$$

Obviously, $||y_i - x_0|| = 1$ for all $i = 1, \dots, k$. Hence, no finite collection of elements can form an ε-net of the set Ω for $\varepsilon < 1$, i.e., $\chi(\Omega) \geq 1$.

Finally, let us prove (2). The inequality $\tilde{\chi}^0(\Omega) \leq \chi(\Omega)$ is plain. It remains to verify that

$$\frac{1}{2}\chi(\Omega) \leq \tilde{\chi}^0(\Omega). \tag{5}$$

The case $\chi(\Omega) = 0$ is trivial. Suppose $\chi(\Omega) > 0$. Then for any given $\varepsilon > 0$, the set Ω has no finite $[\chi(\Omega) - \varepsilon]$-net in E. Therefore, one can produce a countable set $C \subset E$ such that

$$||x - y|| \geq \chi(\Omega) - \varepsilon$$

for all $x, y \in C$, $x \neq y$. But in this case, as it is easily seen,

$$\chi(C) \geq \frac{1}{2}[\chi(\Omega) - \varepsilon],$$

and so

$$\tilde{\chi}^0(\Omega) \geq \frac{1}{2}[\chi(\Omega) - \varepsilon].$$

Since ε is arbitrary, this yields inequality (5) and completes the proof of the theorem. **QED**

1.5. CONDENSING OPERATORS

In this section we introduce the condensing operators and study some of their properties.

1.5.1. Definitions. Let E_1 and E_2 be Banach spaces and let ϕ and ψ be MNCs in E_1 and E_2, respectively, with values in some partially ordered set $(Q, \leq)$. A continuos operator $f: D(f) \subset E_1 \to E_2$ is said to be *(ϕ, ψ)-condensing* (or *(ϕ, ψ)-densifying*) if $\Omega \subset D(f)$, $\psi[f(\Omega)] \geq \phi(\Omega)$ implies Ω is relatively compact. The operator f is said to be *(ϕ, ψ)-condensing* (or *(ϕ, ψ)-densifying*) *in the proper sense* if $\psi[f(\Omega)] < \phi(\Omega)$ for any set $\Omega \subset D(f)$ with compact closure (in a partially ordered set $(Q, \leq)$ the strict inequality $a < b$ means that $a \leq b$ and $a \neq b$). If the set Q is linearly ordered, then clearly the two notions of condensing operator coincide.

Suppose that on Q there is defined an operation of multiplication by nonnegative scalars. A continuous operator f is said to be *(q, ϕ, ψ)-bounded* if

$$\psi[f(\Omega)] \leq q\phi(\Omega)$$

for any set $\Omega \subset D(f)$. Whenever $E_1 = E_2$ and $\phi = \psi$ we shall simply say "*ψ-condensing*" and "*(q,ψ)-bounded*". In the case $q < 1$, (q,ψ)-bounded operators are sometimes referred to as *ψ-condensing with constant q*.

1.5.2. Elementary examples. (a) Any compact operator defined on a bounded subset of a Banach space is obviously ψ_1-condensing, where ψ_1 is the MNC introduced in **1.2.2** [Translator's note: throughout this book the Russian "completely continuous" will be translated as "compact"; thus, here a "compact" operator will be one that is continuous *and* compact in the sense that it maps bounded subsets of its domain into compact sets; operators that have only the second property will be explicitly said to do so]; similarly, any contractive operator on a bounded subset is ψ_2-condensing (with ψ_2 as defined in **1.2.2**). However, the MNCs ψ_1 and ψ_2 often turn to be not so convenient to work with, as they do not possess sufficiently nice properties; for instance, as we remarked earlier, ψ_2 is not regular.

(b) A more meaningful example is provided by the compact and the contractive operators, which, as one can readily see, are condensing with respect to the Kuratowski MNC α.

(c) Obviously, any compact operator on a bounded set is also condensing with respect to the Hausdorff MNC χ.

1.5.3. Remark. Contractive operators are not necessarily χ-condensing. In fact, let $\{p_n\}, \{q_n\}, \{r_n\}, \{s_n\}$ be sequences of numbers in the interval $(0,1)$ satisfying

$$s_{n+1} < p_n < q_n < r_n < s_n \to 0 \quad (n \to \infty),$$

and let $\{a_n\}$ and $\{b_n\}$ be sequences of piecewise-linear functions, defined on $[0,1]$, whose values at $0, 1$, and the break points p_n, q_n, r_n, s_n are shown in the following table:

$$\begin{array}{rcccccc} t = & 0 & p_n & q_n & r_n & s_n & 1 \\ a(t) = & 1 & 1 & 1 & -1 & -1 & -1 \\ b(t) = & 0 & 0 & 1 & -1 & 0 & 0. \end{array}$$

Let $\Omega = \{a_n\} \subset C[0,1]$ and define the operator $f: \Omega \to C[0,1]$ by the formula $f(a_n) = b_n$. Clearly, $||a_n - a_m|| = 2$ and $||b_n - b_m|| = 1$ $(m \neq n)$, so that f is contractive (with contractivity constant $1/2$). At the same time it is easy to show (using, say, formula (2) in **1.1.10**) that $\chi(\Omega) = \chi[f(\Omega)](= \chi(\{b_n\}) = 1$.

1.5.4. Elementary properties of condensing operators. In the assertions given below it is assumed that Q is a closed cone in a Banach space and "$\leq$" is the partial order relation defined by Q (see [85]).

(a) *If the* MNC ψ_1 *is regular, then any* (q, ψ_1, ψ_2)*-bounded operator with* $q < 1$ *is* (ψ_1, ψ_2)*-condensing in the proper sense.*

(b) *The composition* $f_1 \circ f_2$ *of a* (q_1, ψ_1, ψ_2)*-bounded operator* f_1 *and a* (q_2, ψ_2, ψ_3)*-bounded operator* f_2 *is a* $(q_1 q_2, \psi_1, \psi_3)$*-bounded operator.*

(c) *If the* MNC ψ_2 *is algebraically semi-additive and monotone, then the sum* $f_1 + f_2$ *of a* (q_1, ψ_1, ψ_2)*-bounded operator* f_1 *and a* (q_2, ψ_1, ψ_2)*-bounded operator* f_2 *is a* $(q_1 + q_2, \psi_1, \psi_2)$*-bounded operator.*

(d) *If* f_1 *is a* (ψ_1, ψ_2)*-condensing operator and* f_2 *is a* (ψ_2, ψ_3)*-condensing operator that maps totally bounded sets into totally bounded ones,* ψ_1 *and* ψ_3 *are regular* MNCs*, and* $Q = [0, \infty)$*, then the composition* $f_2 \circ f_1$ *is a* (ψ_1, ψ_3)*-condensing operator.*

(e) *If* $Q = [0, \infty)$ *and* ψ_2 *is semi-additive, then the set of all* (ψ_1, ψ_2)*-condensing operators is convex.*

Proof. (a) In fact, the inequalities $\psi_1(\Omega) \leq \psi_2[f(\Omega)] \leq q\psi_1(\Omega)$, $q < 1$, imply $\psi_1(\Omega) = 0$, which in view of the regularity of ψ_1 guarantees the relative compactness of Ω.

(b) Obviously,

$$\psi_3\{f_2[f_1(\Omega)]\} \leq q_2\psi_2[f_1(\Omega)] \leq q_1 q_2 \psi_1(\Omega).$$

(c) In view of the monotonicity and algebraic semi-additivity of ψ_2,

$$\psi_2[(f_1 + f_2)(\Omega)] \leq \psi_2[f_1(\Omega) + f_2(\Omega)] \leq \psi_2[f_1(\Omega)] + \psi_2[f_2(\Omega)] \leq (q_1 + q_2)\psi_1(\Omega).$$

(d) Suppose $\overline{\Omega}$ is noncompact. Then $\psi_2[f_1(\Omega)] < \psi_1(\Omega)$. If $\overline{f_1(\Omega)}$ is also noncompact, then $\psi_3\{f_2[f_1(\Omega)]\} < \psi_2[f_1(\Omega)]$, which in conjunction with the preceding inequality gives $\psi_3\{f_2[f_1(\Omega)]\} < \psi_1(\Omega)$. If, however, $f_1(\Omega)$ is totally bounded, then, by hypothesis, $f_2[f_1(\Omega)]$ is also totally bounded. Consequently, $\psi_3\{f_2[f_1(\Omega)]\} = 0$, since ψ_3 is regular. On the other hand, $\psi_1(\Omega) > 0$, because Ω is not totally bounded and ψ_1 is regular. Thus, in this case, too,

$$\psi_3\{f_2[f_1(\Omega)]\} < \psi_1(\Omega).$$

(e) Let f_1 and f_2 be (ψ_1, ψ_2)-condensing operators and $\lambda \in [0, 1]$. Consider the operator $f_\lambda = \lambda f_1 + (1 - \lambda) f_2$ and suppose that for some Ω,

$$\psi_2[f_\lambda(\Omega)] \geq \psi_1(\Omega). \tag{1}$$

Clearly, $f_\lambda(\Omega) \subset \mathrm{co}[f_1(\Omega) \cup f_2(\Omega)]$. Using the semi-additivity of ψ_2, we get

$$\psi_2[f_\lambda(\Omega)] \leq \max\{\psi_2[f_1(\Omega)], \psi_2[f_2(\Omega)]\}. \tag{2}$$

Since the set where ψ_2 takes its values is linearly ordered, the right-hand side in (2) is equal to either $\psi_2[f_1(\Omega)]$, or $\psi_2[f_2(\Omega)]$. Suppose, for definiteness, that the first case occurs. Then (1) and (2) imply the inequality

$$\psi_2[f_1(\Omega)] \geq \psi_1(\Omega),$$

which in view of the fact that f_1 is (ψ_1, ψ_2)-condensing means that Ω is relatively compact. **QED**

1.5.5. Condensing families. The definitions of the notions of (ϕ, ψ)-condensing and (q, ϕ, ψ)-bounded operators admit natural extensions to *families* of operators $f = \{f_\lambda : \lambda \in \Lambda\}$; in this case $f(\Omega)$ is understood as $\bigcup_{\lambda \in \Lambda} f_\lambda(\Omega)$. Often a family of operators $f = \{f_\lambda : \lambda \in \Lambda\}$ is regarded as an operator of two variables, $f: \Lambda \times E_1 \to E_2$ ($f(\lambda, x) = f_\lambda(x)$). Then instead of speaking of a "(ϕ, ψ)-condensing" or a "(q, ϕ, ψ)-bounded" family one speaks of a *"jointly (ϕ, ψ)-condensing"* or a *"jointly (q, ϕ, ψ)-bounded"* operator.

1.5.6. An example of a condensing family of operators (condensing homotopy). *Suppose the operators $f_0, f_1: M \subset E_1 \to E_2$ are (ϕ, ψ)-condensing, the set where the MNCs ϕ and ψ take their values is linearly ordered (as a consequence of which f_0 and f_1 are (ϕ, ψ)-condensing in the proper sense), and ψ is semi-additive. Then the family of operators $f = \{f_\lambda : \lambda \in [0, 1]\}$, where $f_\lambda(x) = (1 - \lambda) f_0(x) + \lambda f_1(x)$, is (ϕ, ψ)-condensing.*

The **proof** is essentialy the same as for assertion (e) in **1.5.4**. It suffices to remark that $f(\Omega) \subset \mathrm{co}[f_0(\Omega) \cup f_1(\Omega)]$.

The next theorem gives what apparently is the most common test for an operator to be condensing; we state and prove it for families of operators.

1.5.7. Theorem. *Suppose the operators in the family $f = \{f_\lambda : \lambda \in \Lambda\}$ are continuous and admit a diagonal representation $f_\lambda(x) = \Phi(\lambda, x, x)$ through an operator $\Phi: \Lambda \times M \times E_1 \to E_2$ (here E_1 and E_2 are Banach spaces, $M \subset E_1$, and Λ is an arbitrary set). Suppose further that for any $y \in E_1$ the set $\Phi(\Lambda \times M \times \{y\})$ is totally bounded, and that for any $\lambda \in \Lambda$ and any $x \in M$ the operator $\Phi(\lambda, x, \cdot)$ satisfies the Lipschitz condition with a constant $q < 1$ that does not depend on λ and x. Then the family f is (q, χ)-bounded.*

Proof. We have to show that the inequality

$$\chi[f(\Omega)] \leq q\chi(\Omega) \tag{3}$$

holds for any set $\Omega \subset M$, where in the left- [resp. right-] hand side χ denotes the Hausdorff MNC in the space E_2 [resp. E_1]. If Ω is not bounded, then (3) is obvious, since $\chi(\Omega) = +\infty$. Now suppose Ω is bounded and let S be a finite $[\chi(\Omega) + \varepsilon]$-net of Ω in E_1 (where $\varepsilon > 0$ is arbitrary). Consider the set $S_1 = \Phi(\Lambda \times \Omega \times S)$. It is totally bounded, being the union of the finite collection of sets $\Phi(\Lambda \times \Omega \times \{y\})$ $(y \in S)$. We claim that S_1 is a $q[\chi(\Omega) + \varepsilon]$-net of the set $f(\Omega)$ in E_2. In fact, let $z \in f(\Omega)$, i.e., $z = \Phi(\lambda, x, x)$ for some $\lambda \in \Lambda$ and $x \in \Omega$, and let $y \in \Omega$ be such that $||x - y|| \leq \chi(\Omega) + \varepsilon$. Then $z_1 = \Phi(\lambda, x, y)$ belongs to S_1 and

$$||z_1 - z|| = ||\Phi(\lambda, x, y) - \Phi(\lambda, x, x)|| \leq q||x - y|| \leq q[\chi(\Omega) + \varepsilon].$$

Thus, $\chi[f(\Omega)] \leq q[\chi(\Omega) + \varepsilon]$, which in view of the arbitrariness of ε yields (3). **QED**

1.5.8. Corollaries. (a) *Under the hypotheses of Theorem 1.5.7, if the set M is bounded and $q < 1$, then the family f is χ-condensing.*

(b) *The sum $f + g$ of a compact operator $f: E_1 \to E_2$ and a contractive operator $g: E_1 \to E_2$ is a χ-condensing operator on any bounded set $M \subset E_1$.*

The following result on the derivative of a (q, χ)-operator is often useful.

1.5.9. Theorem. *The Fréchet derivative $f'(x_0)$ of any (q, χ)-bounded operator f is itself (q, χ)-bounded.*

Proof. Let $f: D(f) \subset E_1 \to E_2$, x_0 be an interior point of $D(f)$, and $A = f'(x_0)$. Then, for sufficiently small $h \in E_1$,

$$Ah = f(x_0 + h) - f(x_0) + \omega(h),$$

where $\omega(h)/||h|| \to 0$ when $h \to 0$. Consequently, for any bounded set $\Omega \subset E_1$ and $\varepsilon > 0$ sufficiently small,

$$A(\Omega) = \frac{1}{\varepsilon A}(\varepsilon\Omega) \subset \frac{1}{\varepsilon}[f(x_0 + \varepsilon\Omega) - f(x_0) + \omega(\varepsilon\Omega)].$$

Using properties of the Hausdorff MNC and the (q, χ)-boundedness of f we obtain

$$\chi[A(\Omega)] \leq q\chi(\Omega) + \chi[\omega(\varepsilon\Omega)/\varepsilon].$$

Letting $\varepsilon \to 0$, we obviously get

$$\chi[A(\Omega)] \leq q\chi(\Omega). \quad \textbf{QED}$$

1.5.10. Remark. It is readily verified that the preceding theorem remains valid if χ is replaced by an arbitrary monotone, semi-homogeneous, algebraically semi-additive, translation-invariant MNC.

To demonstrate how the notion of a condensing operator is used we prove a fixed-point theorem for such operators. Here we give it in a particular formulation, while in the next subsection and in Chapter 3 we describe various generalizations.

1.5.11. Theorem. *Suppose the χ-condensing operator f maps a nonempty convex closed subset M of the Banach space E into itself. Then f has at least one fixed point in M.*

Proof. We construct a transfinite sequence of closed convex sets Ω_α as follows: $\Omega_0 = M$, and for $\alpha > 0$,

$$\Omega_\alpha = \begin{cases} \overline{\text{co}}\, f(\Omega_{\alpha-1}), & \text{if } \alpha - 1 \text{ exists,} \\ \bigcap_{\beta<\alpha} \Omega_\beta, & \text{otherwise.} \end{cases}$$

Clearly, this nonincreasing (with respect to inclusion) sequence stabilizes starting with some index $\alpha = \delta$: $\Omega_\alpha = \Omega_\delta$ for $\alpha \geq \delta$. Since $\Omega_{\delta+1} = \Omega_\delta = \overline{\text{co}}\, f(\Omega_\delta)$ and f is condensing, Ω_δ is compact. If we show that $\Omega_\delta \neq \emptyset$, then the assertion of the theorem reduces to the Schauder principle.

Let $x_0 \in M$ and $x_n = f^n(x_0)$ $(n = 1, 2, \dots)$. The sequence $\{x_n\}$ is relatively compact, since $\chi[f\{x_n\}] = \chi(\{x_n : n \geq 2\}) = \chi(\{x_n\})$ and f is condensing. Hence, its limit point set K is not empty. It remains to observe that $f(K) = K$, and consequently $\Omega_\alpha \supset K$ for all α. **QED**

1.5.12. Generalizations. In the preceding proof we used only the following two properties of the MNC ψ:

(a) $\chi(\overline{\text{co}}\, \Omega) = \chi(\Omega)$. According to our general definition, this property is enjoyed by any MNC.

(b) $\chi(\{x_n : n \geq 1\}) = \chi(\{x_n : n \geq 2\}$. This equality holds for any semi-additive (i.e., such that $\chi(A \cup B) = \max\{\chi(A), \chi(B)\}$) nonsingular (i.e., such that $\chi(\{x\}) = 0$) MNC, and also for any additively-nonsingular (i.e., such that $\chi(A \cup \{x\}) = \chi(A)$) MNC.

Thus, in Theorem 1.5.11 χ may stand for an arbitrary (not necessarily real-valued) MNC that satisfies condition (b) for any sequence $\{x_n\}$. This condition, too, can be discarded if it is known beforehand that there is a nonempty set $K \subset M$ such that $\overline{\text{co}}\, f(K) \supset K$.

1.6. ULTIMATELY COMPACT OPERATORS

The classes of condensing operators defined relative to distinct MNCs are of course distinct in general, but they nevertheless share a number of general properties. In this section we examine one of them, namely, the ultimate compactness property, which can be formulated without resorting to MNCs. In the next section we shall study a chain of strenghtened versions of this property, each characterizing a certain class of condensing-type operators.

1.6.1. Definition of the sequence T_α. Let M be a subset of a Banach space E. Given an operator $f: M \to E$, we construct a *transfinite sequence of sets* $\{T_\alpha\}$ by the following rule:

$$\begin{aligned} T_0 &= \overline{\text{co}}\, f(M); && \text{(1a)} \\ T_\alpha &= \overline{\text{co}}\, f(M \cap T_{\alpha-1}), \text{ if } \alpha - 1 \text{ exists} && \text{(1b)} \\ T_\alpha &= \bigcap_{\beta<\alpha} T_\beta, \text{ if } \alpha - 1 \text{ does not exist.} && \text{(1c)} \end{aligned} \tag{1}$$

Recall that $\overline{\text{co}}\,\Omega$ denotes the closed convex hull of the set Ω. The main properties of the sequence $\{T_\alpha\}$ are established in the following lemma.

1.6.2. Lemma. (a) *Each T_α is closed and convex.*
(b) $f(M \cap T_\alpha) \subset T_{\alpha+1}$.
(c) *If $\eta < \alpha$, then $T_\alpha \subset T_\eta$.*
(d) *Each T_α is invariant under f in the sense that $f(M \cap T_\alpha) \subset T_\alpha$.*
(e) *There is an ordinal number δ such that $T_\alpha = T_\delta$ for all $\alpha \geq \delta$.*

Proof. Assertion (a) is an obvious consequence of formulas (1), while assertion (b) is a consequence of (1b). Next, it is readily seen that (d) is a straightforward consequence of (b) and (c). However, for us it will be more convenient to establish (c) and (d) simultaneously by induction on α. For $\alpha = 0$ assertion (c) is trivial, while (d) follows from the inclusions:

$$T_0 = \overline{\text{co}}\, f(M) \supset f(M) \supset f(M \cap T_0).$$

Now suppose both assertions hold true for all $\alpha < \alpha_0$, and let us show that they remain true for $\alpha = \alpha_0$. We have to examine two possible cases.

1. The ordinal number $\alpha_0 - 1$ exists. If $\eta < \alpha_0$, then $\eta \leq \alpha_0 - 1$, and so, by the inductive hypothesis, $T_{\alpha_0-1} \subset T_\eta$. Using formula (1b), assertion (d) for $\alpha = \alpha_0 - 1$, and

the fact that T_{α_0-1} is closed and convex, we obtain

$$T_{\alpha_0} = \overline{\mathrm{co}}\, f(M \cap T_{\alpha_0-1}) \subset T_{\alpha_0-1} \subset T_\eta,$$

so that assertion (c) holds for $\alpha = \alpha_0$. Next, from the inclusion $T_{\alpha_0} \subset T_{\alpha_0-1}$ it follows that $M \cap T_{\alpha_0} \subset M \cap T_{\alpha_0-1}$, whence $f(M \cap T_{\alpha_0}) \subset \overline{\mathrm{co}}\, f(M \cap T_{\alpha_0-1}) = T_{\alpha_0}$, i.e., (d) also holds.

2. The ordinal number $\alpha_0 - 1$ does not exist. Then, according to (1c), $T_{\alpha_0} = \bigcap_{\beta<\alpha_0} T_\beta$, so that the validity of assertion (c) for $\alpha = \alpha_0$ is plain. From the same formula it follows that $f(M \cap T_{\alpha_0}) \subset \overline{\mathrm{co}}\, f(M \cap T_\beta)$ for any $\beta < \alpha_0$. Using assertion (d) for $\alpha = \beta$, we obtain $f(M \cap T_{\alpha_0}) \subset T_\beta$. Consequently,

$$f(M \cap T_{\alpha_0}) \subset \bigcap_{\beta<\alpha_0} T_\beta = T_{\alpha_0}.$$

Thus, assertions (c) and (d) are proved. To prove assertion (e) it suffices to remark that if the power of the ordinal number δ is larger than the power of the set of all subsets of the space E, then in the transfinite sequence $\{T_\alpha : 0 \leq \alpha \leq \delta\}$ there must be some repetitions; but for a nonincreasing sequence this is equivalent to stabilization. **QED**

1.6.3. Definition. The limit set T_δ of the transfinite sequence (1) is called the *ultimate range* (or *limit range*) *of the operator* f *on the set* M and is denoted by $f^\infty(M)$. The operator $f: M \to E$ is said to be *ultimately compact* (or *limit compact*) if the set $f[M \cap f^\infty(M)]$ is relatively compact. In particular, an operator f is ultimately compact on M whenever $f^\infty(M) = \emptyset$.

The next lemma lists elementary properties of the ultimate range and of ultimately compact operators.

1.6.4. Lemma. (a) $f^\infty(M) = \overline{\mathrm{co}}\, f[M \cap f^\infty(M)]$.

(b) *If* $M_1 \subset M$, *then* $f^\infty(M_1) \subset f^\infty(M)$.

(c) *If* f *is ultimately compact on* M *and* $M_1 \subset M$, *then* f *is ultimately compact on* M_1.

(d) *The operator* $f: M \to E$ *is ultimately compact if and only if its ultimate range* $f^\infty(M)$ *is compact.*

(e) *If the range* $f(M)$ *of the operator* f *is relatively compact, then* f *is ultimately compact.*

(f) *The operator* $f: M \to E$ *is ultimately compact if and only if the equality* $\overline{\mathrm{co}}\, f(M \cap \Omega) = \Omega$ *implies that* Ω *is compact.*

Proof. Since $f^\infty(M) = T_\delta$ and $T_{\delta+1} = T_\delta$, assertion (a) follows from (1b). If $M_1 \subset M$, then using (1) it is easily seen that $T'_\alpha \subset T_\alpha$ for all α, where $\{T'_\alpha\}$ is the sequence (1) for f on M_1. This yields (b), and then also (c), since $f[M_1 \cap f^\infty(M_1)] \subset f[M \cap f^\infty(M)]$. To establish (d) it suffices to remark (see (a)) that $f[M \cap f^\infty(M)] \subset f^\infty(M)$, and consequently $f[M \cap f^\infty(M)]$ is relatively compact if $f^\infty(M)$ is compact. Conversely, if $f[M \cap f^\infty(M)]$ is relatively compact, then $f^\infty(M) = \overline{\text{co}}\, f[M \cap f^\infty(M)]$ is compact. Next, assertion (e) follows from the obvious inclusion $f[M \cap f^\infty(M)] \subset f(M)$. Finally, in assertion (f) the "if" part follows immediately from (a) and (d), and the "only if" part—from (c) and (d). **QED**

Assertion (d) provides a trivial example of ultimately compact operator. The basic examples are the condensing operators.

1.6.5. Theorem. *Let M be a closed set, ψ a monotone* MNC *in E, and $f: M \to E$ a ψ-condensing operator. Then f is ultimately compact on M.*

Proof. By assertion (a) of Lemma 1.6.4,

$$\overline{\text{co}}\, f[M \cap f^\infty(M)] = f^\infty(M). \tag{2}$$

Hence,

$$\overline{\text{co}}\, f[M \cap f^\infty(M)] \supset M \cap f^\infty(M).$$

Taking into account the monotonicity of ψ, we conclude that

$$\psi(f[M \cap f^\infty(M)]) \geq \psi(M \cap f^\infty(M)). \tag{3}$$

But then from the definition of a condensing operator it follows that the set $M \cap f^\infty(M)$ is relatively compact, and since M and $f^\infty(M)$ are closed, it is compact. Together with it the set $f[M \cap f^\infty(M)]$ is also compact (since f is continuous), i.e., f is ultimately compact. **QED**

1.6.6. Theorem. *Suppose the operator f is condensing with respect to an arbitrary* MNC *ψ and maps the closed convex set M into itself. Then f is ultimately compact on M.*

Proof. It suffices to remark that, under the assumptions of the theorem, $f^\infty(M) \subset M$. Therefore, (2) immediately implies (3) (with the equality sign) with no need of assuming that ψ is monotone. **QED**

1.6.7. Remark. The definition of an ultimately compact operator extends in the usual manner to families $f = \{f_\lambda : \lambda \in \Lambda\}$; in this case the transfinite sequence $\{T_\alpha\}$ is constructed following the same rules (1) as above, where now by $f(\Omega)$ one means $\bigcup_{\lambda\in\Lambda} f_\lambda(\Omega)$. If the family $f = \{f_\lambda : \lambda \in \Lambda\}$ is given by a function of two variables $f: \Lambda \times M \to E$, then one often says that f is a *jointly ultimately compact operator*.

We wish to emphasize right away the following circumstance: if $f_0, f_1 : M \to E$ are ultimately compact operators, then the family $f = \{f_\lambda : \lambda \in [0,1]\}$, $f_\lambda(x) = (1-\lambda)f_0 + \lambda f_1(x)$, which effects a homotopy from f_0 to f_1, is not necessarily ultimately compact.

1.6.8. Example (a linear homotopy that is not ultimately compact). Let E be the Banach space c_0 of the sequences that converge to zero, with the norm $||x|| = \max\{|x_i| : i = 1, 2, \dots\}$, and let B be the closed unit ball in E. Define the operators f_0 and f_1 by the formulas

$$f_0(x) = (0, 0, \dots)$$

and

$$f_1(x) = f_1(x_1, x_2, \dots, x_n, \dots) = (1, x_1, x_2, \dots, x_n, \dots).$$

It is readily verified that f_0 and f_1 are ultimately compact on B. In fact, $f_0^\infty(B) = T_0 = \{0\}$ is a compact set. For the operator f_1 the sets T_α with $\alpha < \omega$ (where ω is the first transfinite number) can be described as

$$T_\alpha = \{(\underbrace{1, \dots, 1}_{\alpha+1}, x_1, x_2, \dots) \in c_0 : |x_i| \le 1 \ (i = 1, 2, \dots)\}.$$

Consequently, $f_1^\infty(B) = T_\omega = \bigcap_{\alpha<\omega} T_\alpha = \emptyset$ (indeed, the sequence $(1, \dots, 1, \dots)$ does not belong to c_0), which means that f_1 is ultimately compact.

Next, for the family $f = \{(1-\lambda)f_0 + \lambda f_1\}$ the sets T_α with $\alpha < \omega$ are described as

$$T_\alpha = \{(y_0, \dots, y_\alpha, x_1, x_2, \dots) \in c_0 : 0 \le y_j \le 1 \ (j = 0, 1, \dots, \alpha), \ |x_i| \le 1 \ (i = 1, 2, \dots)\}.$$

Hence, the set T_ω for the family f, where, obviously, the stabilization of the sequence $\{T_\alpha\}$ obviously begins, has the form

$$T_\omega = \{(x_1, x_2, \dots) \in c_0 : 0 \le x_i \le 1 \ (i = 1, 2, \dots)\}.$$

Therefore, $f^\infty(B) = T_\omega$ is not compact, and so the family f is not ultimately compact on B.

In connection with the definition of the ultimate range we notice the following two interesting facts. First, to reach the ultimate range we may have indeed to construct a genuine transfinite sequence—an example is given below in **1.6.9**. On the other hand, if the operator in question is condensing with respect to a sufficiently nice MNC and if one is interested in the occurrence of the first compact set in the sequence $\{T_\alpha\}$ rather than in its stabilization, then one can always restrict oneselves to the first transfinite number—the proof of this fact is given in **1.6.10–1.6.13**.

1.6.9. Example. Let A denote the set of all ordinals smaller than or equal to a fixed ordinal number $\delta > \omega$ of second kind, i.e., such that $\delta - 1$ does not exist. Let $m(\mathrm{A})$ be the Banach space of all bounded functions $x\colon \mathrm{A} \to \mathbf{R}$, endowed with the norm $||x|| = \sup\{|x(\alpha)|: \alpha \in \mathrm{A}\}$, and let B be the closed unit ball in $m(\mathrm{A})$ centered at zero. If $x \in m(\mathrm{A})$ and $\alpha \in \mathrm{A}$, let $S_\alpha x$ denote the element of $m(\mathrm{A})$ defined by the rule:

$$(S_\alpha x)(\beta) = \begin{cases} x(\beta) & \text{if } 0 \le \beta < \alpha, \\ 0, & \text{if } \alpha \le \beta \le \delta \end{cases}$$

(in particular, $S_0 x = 0$). Define the operator $f\colon B \to m(\mathrm{A})$ by the rule

$$(fx)(\alpha) = ||S_\alpha x|| x(\alpha) \quad (\alpha \in \mathrm{A}).$$

Notice that $f(B) \subset B$. Next, if $x, y \in B$ and $\alpha \in \mathrm{A}$, then

$$|(fx)(\alpha) - (fy)(\alpha)| = |\,||S_\alpha x|| x(\alpha) - ||S_\alpha y|| y(\alpha)\,| \le ||S_\alpha x||\, |x(\alpha) - y(\alpha)| + |y(\alpha)|\, ||S_\alpha x - S_\alpha y||$$

$$\le ||x||\, ||x - y|| + ||y||\, ||x - y|| \le 2||x - y||.$$

Thus, the operator f satisfies the Lipschitz condition with constant 2, and so it is continuous.

Now let us consider the sequence $\{T_\alpha\}$ constructed for f and prove by induction on α that

$$T_\alpha = \begin{cases} R_{\alpha+1} B, & \text{if } \alpha < \omega, \\ R_\alpha B, & \text{if } \alpha \ge \omega, \end{cases} \tag{4}$$

where $R_\alpha = I - S_\alpha$. In fact, for $\alpha = 0$ (4) becomes the equality $T_0 = R_1 B$, which is readily checked. Now assume that (4) holds for all $\alpha < \alpha_0$, and let us check that it remains valid for $\alpha = \alpha_0$. Suppose first that $\alpha_0 - 1$ exists and $\alpha_0 < \omega$. Then, by hypothesis, $T_{\alpha_0 - 1} = R_{\alpha_0} B$, and so

$$T_{\alpha_0} = \overline{\mathrm{co}}\, f(T_{\alpha_0 - 1}) = \overline{\mathrm{co}}\, f(R_{\alpha_0} B) = \overline{\mathrm{co}}(R_{\alpha_0 + 1} B) = R_{\alpha_0 + 1} B.$$

If $\alpha_0 - 1$ exists and $\alpha_0 > \omega$ (the case $\alpha_0 = \omega$ cannot occur), then $\alpha_0 - 1 \ge \omega$, and an analogous argument shows that $T_{\alpha_0} = R_{\alpha_0} B$.

Now suppose that $\alpha_0 - 1$ does not exist. Then

$$T_{\alpha_0} = \bigcap_{\beta<\alpha_0} T_\beta = \bigcap_{\beta<\alpha_0} R_\beta B = R_{\alpha_0} B,$$

as needed. From (4) it immediately follows that for any $\alpha < \delta$ the set $f(T_\alpha)$ is not relatively compact. At the same time the operator f is ultimately compact, since $f^\infty(B) = T_{\delta+1} = \{0\}$.

Similar examples can be constructed in spaces with "sufficiently nice" properties. Indeed, let δ be an ordinal number of countable power, $\delta \neq \omega$, and, as above, set A $= \{\alpha: 0 \leq \alpha \leq \delta\}$. Fix some bijection $\phi: \mathbf{N} \to$ A, where $\mathbf{N}$ denotes the natural numbers, and define an operator f in the Hilbert space l_2 by the rule

$$(fx)_n = \begin{cases} 0, & \text{if } \phi(n) = 0, \\ \sup\{|x_{\phi(m)}|: \phi(m) < \phi(n)\}x_n, & \text{if } \phi(n) > 0. \end{cases}$$

Then it is readily verified that the properties of f are analogous to those of the operator constructed above.

In the proofs, given below, of Lemmas 1.6.10 and 1.6.11 and of Theorem 1.6.12 we shall consider, for the sake of simplicity, the case of the Hausdorff MNC, and in **1.6.13** we shall indicate which of its properties were actually used.

1.6.10. Lemma. *Let R_n $(n = 1, 2, \dots)$ be subsets of the bounded set Q in the metric space E, and let $\mathcal{U}$ denote the family of all sets A that are representable in the form*

$$A = \bigcup_{n=1}^{\infty} A_n, \tag{5}$$

where A_n is finite and $A_n \subset R_n$ for all n. Then there is an $A^* \in \mathcal{U}$ such that

$$\chi(A^*) = \sup\{\chi(A): A \in \mathcal{U}\}(\equiv s). \tag{6}$$

Proof. Let A^k $(k = 1, 2, \dots)$ be sets in $\mathcal{U}$ such that $\lim_{k\to\infty} \chi(A^k) = s$, and let the representation (5) for A^k be $A^k = \bigcup_{n=1}^\infty A_n^k$. Consider the set $\tilde{A}^k = \bigcup_{n=k}^\infty A_n^k$. Since $\tilde{A}^k$ differs from A^k by only finitely many terms, we have

$$\chi(\tilde{A}^k) = \chi(A^k), \quad \lim_{k\to\infty} \chi(\tilde{A}^k) = s.$$

Now set $A^* = \bigcup_{k=1}^\infty \tilde{A}^k$. The set A^* belongs to the family $\mathcal{U}$, because $A^* = \bigcup_{k=1}^\infty \bigcup_{n=k}^\infty A_n^k = \bigcup_{n=1}^\infty \bigcup_{k=1}^n A_n^k$ and $\bigcup_{k=1}^n A_n^k \subset R_n$ $(n = 1, 2, \dots)$. We claim that (6) holds. In fact, the inequality $\chi(A^*) \leq s$ follows from the inclusion $A^* \in \mathcal{U}$ established above. On the other hand, $\chi(A^*) \geq \chi(\tilde{A}^k)$ for any k, since $A^* \supset \tilde{A}^k$. Consequently,

$$\chi(A^*) \geq \lim_{k\to\infty} \chi(\tilde{A}^k) = s. \textbf{ QED}$$

1.6.11. Lemma. *Let (E, ρ) be a complete metric space, M a closed bounded set in E, $f: M \to E$ a χ-condensing operator, and $R \subset M$. Then*

$$\lim_{n\to\infty} \chi[f^n(R)] = 0. \tag{7}$$

(Notice that the operator f^n is not necessarily defined on the whole set R. In that case by $f^n(R)$ we mean the image under f^n of the part of R on which f^n is defined.)

Proof. Denote $R_n = f^n(R)$ $(n = 1, 2, \ldots)$. Clearly, $R_n \subset f(M)$, and since f is χ-condensing, the set $Q = f(M)$ is bounded. Thus, all conditions of Lemma 1.6.10 are satisfied, and consequently there is a set $A^* \in \mathcal{U}$ such that (6) holds. Let the representation (5) of A^* be $A^* = \bigcup_{n=1}^{\infty} A_n^*$. Next, for each $x \in A_n^*$ pick an element $y \in f^{n-1}(R) = R_{n-1}$ such that $f(y) = x$, and then denote the set of all elements y constructed in this manner by B_{n-1}. Now set $B = \bigcup_{n=1}^{\infty} B_n$. Clearly, $B \in \mathcal{U}$, and so $\chi(B) \leq \chi(A^*)$. Furthermore,

$$f(B) = \bigcup_{n=2}^{\infty} A_n^*,$$

and so

$$\chi[f(B)] = \chi(A^*) \geq \chi(B). \tag{8}$$

Since f is χ-condensing, from this inequality it follows that the set B is totally bounded and hence, thanks to the completeness of E, that $\overline{B}$ is compact. But then $f(\overline{B})$ is also compact, since $\chi[f(B)] = 0$. Finally, (8) implies $\chi(A^*) = 0$. Thus, we showed that any set $A \in \mathcal{U}$ is totally bounded ($\chi(A) \leq \chi(A^*) = 0$).

Suppose now that (7) does not hold. This means that there are a positive ε_0 and an infinite increasing sequence of positive integers n_k $(k = 1, 2, \ldots)$ such that

$$\chi[f^{n_k}(R)] > \varepsilon_0 \tag{9}$$

for any k. Fix an arbitrary element $x_1 \in f^{n_1}(R)$ and choose $x_2 \in f^{n_2}(R)$ such that $\rho(x_1, x_2) \geq \varepsilon_0$. The existence of such an element x_2 follows from the fact that inequality (9) means, in particular, that the set $f^{n_2}(R)$ has no finite ε_0-net in E. For the same reason, $\{x_1, x_2\}$ cannot be an ε_0-net for $f^{n_3}(R)$, hence, there is an $x_3 \in f^{n_3}(R)$ whose distance to

$\{x_1, x_2\}$ is not smaller than ε_0. Continuing in this manner we produce a sequence $\{x_k\}$ which, on the one hand, is necessarily totally bounded (because it can be completed to a set A that belongs to $\mathcal{U}$) and, on the other hand, is not totally bounded since the distance between any of its elements is not smaller than ε_0. We thus arrived at a contradiction, which completes the proof of the lemma. **QED**

1.6.12. Theorem. *Let E be a Banach space, M a closed bounded set in E, and $f: M \to E$ a χ-condensing operator. Then the sets T_α, constructed according to the rules (1), are compact for $\alpha \geq \omega$.*

Proof. It suffices to show that the set

$$T_\omega = \bigcap_{n=0}^{\infty} T_n$$

is totally bounded. In turn, to establish the total boundedness of T_ω it suffices to establish the total boundedness of any of its countable subsets A, which obviously can be written as a union of one-element sets $A_n \subset T_n$. Notice that the conditions of Lemma 1.6.10 are fulfilled here (with T_n in the role of R_n and T_0 in that of Q), and so there is an $A^* \in \mathcal{U}$ with maximal MNC. Thus, our task reduces to proving that $\chi(A^*) = 0$. Write $A^* = \bigcup_{n=1}^{\infty} A_n^*$, where $A_n^* \subset T_n$ and is finite. If $x \in A^*$, then in its $(1/n)$-neighborhood there is an element of the form

$$\tilde{x} = \sum_{i=1}^{m} \lambda_i f(x_i) \quad (\lambda_i \geq 0, \sum_{i=1}^{m} \lambda_i = 1, x_i \in T_{n-1}). \tag{10}$$

Let $\tilde{A}_n^*$ denote the collection of elements $\tilde{x}$, taken one for each $x \in A_n^*$. It is readily seen that the MNC of $\tilde{A}^* = \bigcup_{n=1}^{\infty} \tilde{A}_n^*$ is equal to the MNC of A^*: in fact, the elements of these sets can be arranged in sequences $\{y_n\}, \{z_n\}$ such that $||y_n - z_n|| \to 0$ as $n \to \infty$. Let B_{n-1} denote the collection of elements x_i, constructed for all elements $x \in A_n^*$ in accordance with formula (10). Then $B = \bigcup_{n=0}^{\infty} B_n \in \mathcal{U}$, and so $\chi(B) \leq \chi(A^*)$. On the other hand, since $\operatorname{co} f(B) \supset \tilde{A}^*$ and $\chi[\operatorname{co} f(B)] = \chi[f(B)]$, we get

$$\chi[f(B)] = \chi[\operatorname{co} f(B)] \geq \chi(\tilde{A}^*) = \chi(A^*).$$

Thus, $\chi(B) \leq \chi(A^*) \leq \chi[f(B)]$, which in view of the fact that f is χ-condensing implies $\chi(A^*) = 0$. **QED**

1.6.13. Remark. Let us isolate those properties of the MNC that were used to prove assertions **1.6.10–1.6.12**. In Lemma 1.6.10 we used: a) real-valuedness; b) additive nonsingularity, i.e., the invariance of the MNC with respect to the adjunction of a single element to a (nonempty) set; c) monotonicity. In Lemma 1.6.11, in addition to a)–c), the following properties were required: d) nonnegativity; e) subordination to the Hausdorff MNC, i.e., existence of a constant $c > 0$ such that $\psi(\Omega) \leq c\chi(\Omega)$ for any bounded set Ω; f) regularity. Finally, in the proof of Theorem 1.6.12 we used properties a)–c), f) and also: g) continuity, and h) invariance under passage to the convex hull. In particular, assertions **1.6.10–1.6.12** are valid for the Kuratowski MNC and the MNC β (see **1.3.1**).

1.7. K-OPERATORS

In this section we continue our study of those general properties of condensing operators, the formulation of which does not involve MNCs. The starting point will be the definition of the notion of an ultimately compact operator in the form used in subsection **1.6.4** (f): $\overline{\text{co}}\, f(\Omega \cap M) = \Omega$ must imply the compactness of Ω. It turns out that if one requires that the compactness of Ω be implied by a less restrictive equality, then the properties of the operator improve. This situation corresponds roughly to considering operators that are condensing with respect to nicer MNCs. For example, if the equality in question is replaced by $\overline{\text{co}}\,[T \cup f(\Omega \cap M)] = \Omega$, where T is a one-element set, then one obtains the K_1-operators, for which an analogue of the Schauder principle already holds (see **1.7.3, 1.7.4**), in contrast to the case of ultimately compact operators. For K_2-operators (i.e., in the case where the set T in the above equality consists of two elements) it is possible to construct a fixed-point index theory that is richer compared with the cases of the ultimately compact and the K_1-operators (see Chapter 3).

1.7.1. Definition. Let A be a family of subsets of a Banach space E, and let $M \subset E$. A continuous operator $f: M \to E$ is called an *A-operator* if for any $T \in A$ and any set $\Omega \subset E$ one has the implication:

$$\overline{\text{co}}\,[T \cup f(\Omega \cap M)] = \Omega \;\Rightarrow\; \Omega \text{ is compact.}$$

Let K_n denote the set of all n-element subsets of E (in particular, $K_0 = \{\emptyset\}$), and let K_∞ and K_c designate the set of all finite and respectively compact subsets of E. We let $\mathcal{K}_n$ [resp. $\mathcal{K}_\infty$, $\mathcal{K}_c$] denote the set of all K_n- [resp. K_∞-, K_c-] operators. This section is devoted precisely to the study of the operators in the classes $\mathcal{K}_n, \mathcal{K}_\infty$, and $\mathcal{K}_c$, for which we shall use the common term of "*K-operators*" and the notation $\mathcal{K} = \{\mathcal{K}_n\}_{n=0}^{\infty} \cup \mathcal{K}_\infty \cup \mathcal{K}_c$.

1.7.2. Remarks. (a) Clearly,

$$\mathcal{K}_0 \supset \mathcal{K}_1 \supset \ldots \supset \mathcal{K}_n \supset \ldots \supset \mathcal{K}_\infty \supset \mathcal{K}_c. \tag{1}$$

(b) Earlier we proved (see Lemma 1.6.4, assertion (f)) that the class of K_0-operators coincides with that of ultimately compact operators.

Next we prove a generalization of Schauder's theorem to the class of K_1-operators (cf. Theorem 1.5.11).

1.7.3. Theorem. *Suppose the K_1-operator f maps the nonempty closed convex subset M of the Banach space E into itself. Then f has at least one fixed point in M.*

Proof. Fix some point $x_0 \in M$ and denote by $\mathfrak{M}$ the family of all sets $\Omega \subset M$ with the property that

$$\overline{\mathrm{co}}\,[\{x_0\} \cup f(\Omega)] \subset \Omega.$$

Clearly, $\mathfrak{M}$ is nonempty, since $M \in \mathfrak{M}$.

Set $\Omega_\infty = \bigcap_{\Omega \in \mathfrak{M}} \Omega$. It is readily verified that $x_0 \in \Omega_\infty$, and so $\Omega_\infty \neq \emptyset$. Moreover, Ω_∞ is contained in M and is invariant under f. The assertion of our theorem will follow from Schauder's principle provided we can show that Ω_∞ is convex and compact. To this end we observe that

$$\overline{\mathrm{co}}\,[\{x_0\} \cup f(\Omega_\infty)] = \Omega_\infty.$$

In fact, if $x_1 \in \Omega_\infty$ and $x_1 \notin \overline{\mathrm{co}}\,[\{x_0\} \cup f(\Omega_\infty)]$, then, as is readily checked, $\Omega_\infty \setminus \{x_1\} \in \mathfrak{M}$, which contradicts the definition of Ω_∞. Since f is a K_1-operator, the equality established above implies the compactness of Ω_∞. The convexity of Ω_∞ is an obvious consequence of the same equality. **QED**

The next example shows that the theorem just proved fails for ultimately compact operators; this means, in particular, that the classes $\mathcal{K}_0$ and $\mathcal{K}_1$ are distinct.

1.7.4. Example. In the Banach space c_0 of sequences that converge to zero consider the operator f defined by the formula

$$f(x) = f(x_1, x_2, \ldots) = (1, x_1, x_2, \ldots)$$

(we met f before in **1.6.8**, where it was shown that f is ultimately compact on the unit ball B in c_0). Clearly, f maps B into itself. However, f has no fixed point in B: in fact, it is readily verified that the only possible fixed point for f is the sequence $(1, 1, \ldots)$, which does not lie in c_0.

1.7.5. Theorem. *$\mathcal{K}_n \neq \mathcal{K}_{n+1}$ for any nonnegative integer n.*

Proof. We need to exhibit an operator $f \in \mathcal{K}_n \setminus \mathcal{K}_{n+1}$. We take for E the space l_1 of absolutely summable sequences, and take for the domain M of the operator f the closed ball in l_1 of radius 3 and center zero. We first define f on the vectors $e_1, e_2, \dots$ of the canonical basis:

$$f(e_k) = \begin{cases} e_1, & \text{if } k = 1, \dots, n+1, \\ 2e_{k+1} - e_{\langle k \rangle}, & \text{if } k \geq n+2; \end{cases}$$

here $\langle k \rangle$ denotes (the reminder of the division of $k-2$ by n)+2.

Further, let B_k ($k = 1, 2, \dots$) denote the closed ball of radius ε and center e_k. The positive number $\varepsilon < 1$ will be chosen later. For an arbitrary $x \in M$ we put

$$f(x) = \begin{cases} e_1, & \text{if } x \notin \bigcup_{k=1}^{\infty} B_k, \\ \frac{1}{\varepsilon}\|x - e_k\| e_1 + \frac{1}{\varepsilon}(\varepsilon - \|x - e_k\| f(e_k)), & \text{if } x \in B_k. \end{cases}$$

Notice that the balls B_k are disjoint, since $\varepsilon < 1$. It is not hard to check that the operator f thus defined maps M into itself and is continuous. We claim that f is not a K_{n+1}-operator. In fact, if in the equality

$$\overline{\text{co}}\,[T \cup f(\Omega \cap M)] = \Omega \tag{2}$$

we put

$$\Omega = \overline{\text{co}}\{e_1, e_2, \dots, e_{n+2}, 2e_{n+3} - e_{\langle n+2 \rangle}, 2e_{n+4} - e_{\langle n+3 \rangle}, \dots\} \subset M,$$

then obviously $\Omega \supset \{e_1, e_2, \dots\}$. Consequently,

$$f(\Omega) = \bigcup_{k=n+2}^{\infty} [e_1, 2e_{k+1} - e_{\langle k \rangle}],$$

where $[a, b]$ denotes the segment with endpoints $a, b \in E$. Hence, if one takes for T the $(n+1)$-element set $\{e_2, e_3, \dots, e_{n+2}\}$, then (2) will hold; at the same time, Ω is not compact. Thus, $f \notin \mathcal{K}_{n+1}$.

Now let us show that if (2) holds for a noncompact set Ω and a finite set T, then T contains at least $n+1$ points. This obviously will mean that $f \in \mathcal{K}_n$.

Thus, suppose Ω is noncompact and satisfies (2) for some finite set T. Then Ω intersects infinitely many balls B_k: otherwise, the left-hand side of equality (2) would be the convex hull of a finite set, and hence compact. Let C denote the set of all $k \geq n+2$ for which $B_k \cap \Omega \neq \emptyset$. Then

$$f(\Omega \cap M) \subset \bigcup_{k \in C} [e_1, 2e_{k+1} - e\langle k \rangle] \subset L(D),$$

where D is the set consisting of the vectors e_1 and $2e_{k+1} - e_{\langle k \rangle}$ $(k \in C)$ and $L(D)$ denotes the closed linear span of D. Suppose that we can produce a set $\tilde{T}$ with the following properties:

$$\tilde{T} \subset \Omega \cap M;$$

$$\tilde{T} \text{ consists of } n+1 \text{ points;}$$

$$\tilde{T} \text{ is linearly independent} \tag{3}$$

$$\rho(\tilde{T}, L(D)) > 0$$

(here $\rho(A, B)$ denotes the Hausdorff distance between the sets A and B). Then from (2) it will follow that T consists of at least $n+1$ points. Indeed, (2) and (3) imply the inclusion

$$\tilde{T} \subset L(T \cup D) \subset L(T) + L(D).$$

Let us label the points of the set $\tilde{T}$: $\tilde{T} = \{x_1, \dots, x_{n+1}\}$. Then for any $i = 1, \dots, n+1$,

$$x_i \in L(T) + z_i, \tag{4}$$

where $z_i \in L(D)$. From (3) it follows, in particular, that for any $z \in L(D)$ the set $\tilde{T} \cup \{z\}$ is linearly independent, and then so is the set $\{x_i - z_i : i = 1, \dots, n+1\}$. From (4) one derives the inclusion

$$\{x_i - z_i : i = 1, \dots, n+1\} \subset L(T),$$

which means that T consists of no less than $n+1$ points.

Therefore, in order to prove the theorem it suffices to exhibit a set $\tilde{T}$ with properties (3).

To this end, we first note that the following alternative holds for the set C: either C contains all positive integers starting with some m, and $m-1 \notin C$; or there is no m such that the last assertion is true. In the first case we take as an $(n+1)$-element set $T_1 = \{e_k : k = m, \dots, m+n\}$. In the second case, we take for T_1 an arbitrary $(n+1)$-element subset of the (infinite) set $\{e_k : k \in C, k-1 \notin C\}$. From here on the proofs for the two cases are analogous.

From the coordinate representation of the elements of T_1 and D and the definition of the norm in l_1 it readily follows that the distance from any element e_k of T_1 to the linear span of the set $D \cup T_1 \setminus \{e_k\}$ is not smaller than $1/2$ (in the second case it is even not smaller than 1).

Now to define the sought-for set $\tilde{T}$ proceed as follows: chose one representative $x_k \in B_k \cap \Omega$ for each k such that $e_k \in T_1$, and set $\tilde{T} = \{x_k\}$. Thus, the elements of $\tilde{T}$ are

"ε-perturbations" of the elements of T_1: $||x_k - e_k|| \le \varepsilon$. Since the number of elements of T_1 is known beforehand (it is $n+1$), we can choose $\varepsilon > 0$ in the definition of the operator f so that $\tilde{T}$ will possess properties (3). **QED**

1.7.6. Remarks. (a) The class $\mathcal{K}_\infty$ is different from all classes $\mathcal{K}_n$ ($n = 0, 1, \dots$). In fact, suppose that $\mathcal{K}_n = \mathcal{K}_\infty$ for some n; then in view of the inclusions $\mathcal{K}_n \supset \mathcal{K}_{n+1} \supset \mathcal{K}_\infty$ (see (1)) we would have $\mathcal{K}_n = \mathcal{K}_{n+1}$, which is impossible. It is now clear that the class $\mathcal{K}_c$, too, is different from any of the classes $\mathcal{K}_n$.

(b) The question of whether the classes $\mathcal{K}_\infty$ and $\mathcal{K}_c$ coincide or not is open.

We next turn to the investigation of the relationships between condensing and K-operators.

1.7.7. Theorem. *Let M be a subset of a Banach space E and let $f: M \to E$ be an operator that is condensing with respect to an* MNC *ψ which is assumed to be monotone and invariant under the operation of adjunction of compact sets (for example, ψ is semi-additive and regular). Then $f \in \mathcal{K}_c$.*

Proof. Suppose that for a compact set T and a set Ω one has

$$\overline{\mathrm{co}}\,[T \cup f(\Omega \cap M)] = \Omega. \tag{5}$$

Then

$$\psi(\Omega) = \psi\{\overline{\mathrm{co}}\,[T \cup f(\Omega \cap M)]\} = \psi[T \cup f(\Omega \cap M)] = \psi[f(\Omega \cap M)].$$

In view of the monotonicity of ψ, the last equality gives

$$\psi(\Omega \cap M) \le \psi(\Omega) = \psi[f(\Omega \cap M)],$$

and so $\Omega \cap M$ is relatively compact. Now the continuity of f implies that $f(\Omega \cap M)$ is relatively compact, and in view of (5) this means that Ω is compact. **QED**

1.7.8. Remark. Clearly, if in the preceding theorem we replace the requirement that ψ be invariant under adjunction of compact sets by that of semi-additive nonsingularity, i.e., that ψ be invariant under the adjunction of one-element (and hence of finite) sets, then the conclusion is that f is a K_∞-operator.

1.7.9. Theorem. *Let $M \subset E$ and let $f: M \to E$ be condensing with respect to a semi-additive, translation-invariant (i.e., $\psi(\Omega + x) = \psi(\Omega)$)* MNC *$\psi$. Then $f \in \mathcal{K}_\infty$.*

Proof. Suppose that for some finite set $T = \{x_1, \dots, x_n\}$ and a set Ω one has

$$\overline{\text{co}}\,[T \cup f(\Omega \cap M)] = \Omega. \tag{6}$$

If the set $f(\Omega \cap M)$ is empty then, obviously, Ω is compact, since the left-hand side in (6) is the convex hull of a finite set. In the opposite case, let $x_0 \in f(\Omega \cap M)$. Since ψ is invariant under translations, $\psi(\{x_0\}) = \psi(\{x_i\})$ for $i = 1, 2, \dots, n$, and by the semi-additivity of ψ,

$$\psi(T) = \psi(\{x_1, \dots, x_n\}) = \max\{\psi(\{x_1\}), \dots, \psi(\{x_n\})\} = \psi(\{x_0\}). \tag{7}$$

Furthermore, in view of the monotonicity of ψ (which is a consequence of its semi-additivity),

$$\psi(\{x_0\}) \le \psi[f(\Omega \cap M)]. \tag{8}$$

Combining (7) and (8) and using the semi-additivity of ψ once more, we obtain from (6) that $\psi[f(\Omega \cap M)] = \psi(\Omega)$. To complete the proof one argues exactly as in the final part of the proof of Theorem 1.7.7. **QED**

The following subsections are devoted to the question whether Theorem 1.6.6 and the assertions of **1.7.7–1.7.9** admit converses, i.e., to the construction of an MNC with respect to which the K-operators under consideration are condensing. First we describe one way of constructing such an MNC, and then we prove that the K-operators are condensing with respect to that measure and study its properties.

1.7.10. Construction of a suitable MNC. Let E be an arbitrary Banach space. In the set 2^E of all subsets of E we introduce an equivalence relation R as follows: $(\Omega_1, \Omega_2) \in R$ whenever there are compact sets T_1 and T_2 such that

$$\Omega_2 \subset \overline{\text{co}}\,(T_1 \cup \Omega_1), \quad \Omega_1 \subset \overline{\text{co}}\,(T_2 \cup \Omega_2).$$

The symmetry, reflexivity, and transitivity of R are readily checked. Let Q denote the quotient set $2^E/R$. The equivalence class of the set Ω will be denoted by $\tilde{\Omega}$.

In Q we introduce the following partial order "$\le$": $\tilde{\Omega}_1 \le \tilde{\Omega}_2$ whenever there exist representatives $\Omega_1' \in \tilde{\Omega}_1$ and $\Omega_2' \in \tilde{\Omega}_2$ such that $\Omega_1' \subset \Omega_2'$. Let us check that "$\le$" is indeed an order relation. Reflexivity is obvious. Transitivity is established as follows. Let $\tilde{\Omega}_1 \le \tilde{\Omega}_2$ and $\tilde{\Omega}_2 \le \tilde{\Omega}_3$. This means that there are $\Omega_1' \in \tilde{\Omega}_1, \Omega_2', \Omega_2'' \in \tilde{\Omega}_2$, and $\Omega_3'' \in \tilde{\Omega}_3$ such that $\Omega_1' \subset \Omega_2'$ and $\Omega_2'' \subset \Omega_3''$. Since $(\Omega_2', \Omega_2'') \in R$, there is a compact set T such that $\Omega_2' \subset \overline{\text{co}}\,(T \cup \Omega_2'')$. But then

$$\Omega_1' \subset \Omega_2' \subset \overline{\text{co}}\,(T \cup \Omega_2'') \subset \overline{\text{co}}\,(T \cup \Omega_3'').$$

It remains to observe that $\overline{\text{co}}(T \cup \Omega_3'') \in \tilde{\Omega}_3$.

Finally, let us check that the relation "$\leq$" is anti-symmetric. Suppose $\tilde{\Omega}_1 \leq \tilde{\Omega}_2$ and $\tilde{\Omega}_2 \leq \tilde{\Omega}_1$. This means that there are $\Omega_1', \Omega_1'' \in \tilde{\Omega}_1$ and $\Omega_2', \Omega_2'' \in \tilde{\Omega}_2$ such that $\Omega_1' \subset \Omega_2'$ and $\Omega_2'' \subset \Omega_1''$. Since Ω_1' and Ω_1'' belong to the same equivalence class, there is a compact set T_1 such that $\Omega_1'' \subset \overline{\text{co}}(T_1 \cup \Omega_1')$. Similarly, there is a compact set T_2 such that $\Omega_2' \subset \overline{\text{co}}(T_2 \cup \Omega_2'')$. This implies the simultaneous inclusions

$$\Omega_1'' \subset \overline{\text{co}}(T_1 \cup \Omega_1') \subset \overline{\text{co}}(T_1 \cup \Omega_2'),$$

$$\Omega_2' \subset \overline{\text{co}}(T_2 \cup \Omega_2'') \subset \overline{\text{co}}(T_2 \cup \Omega_1''),$$

i.e., $(\Omega_1'', \Omega_2') \in R$, which means precisely that $\tilde{\Omega}_1 = \tilde{\Omega}_2$.

In the partially ordered set $(Q, \leq)$, for any two elements $\tilde{\Omega}_1$ and $\tilde{\Omega}_2$ one can indicate their supremum. Specifically, for the element $\tilde{\Omega}_3 = \max\{\tilde{\Omega}_1, \tilde{\Omega}_2\}$ we take a class that contains the union of two representatives of the classes $\tilde{\Omega}_1$ and $\tilde{\Omega}_2$. It is readily verified that $\tilde{\Omega}_3$ does not depend on the choice of those representatives. To show that $\tilde{\Omega}_3$ is indeed the supremum, suppose $\tilde{\Omega}_4$ is such that $\tilde{\Omega}_1 \leq \tilde{\Omega}_4$ and $\tilde{\Omega}_2 \leq \tilde{\Omega}_4$. This means that there are sets $\Omega_1' \in \tilde{\Omega}_1, \Omega_2' \in \tilde{\Omega}_2$ and $\Omega_4', \Omega_4'' \in \tilde{\Omega}_4$, such that $\Omega_1' \subset \Omega_4'$ and $\Omega_2' \subset \Omega_4''$. Hence, $\Omega_1' \cup \Omega_2' \subset \Omega_4' \cup \Omega_4''$. But, as is readily seen, $\Omega_4' \cup \Omega_4'' \in \tilde{\Omega}_4$, and so $\tilde{\Omega}_3 \leq \tilde{\Omega}_4$.

Now define a function ψ on 2^E with values in Q by the rule $\psi(\Omega) = \tilde{\Omega}$. We claim that ψ is is an MNC in the sense of the general definition. In fact, the closure of the convex hull of any set clearly belongs to the same equivalence class as the set itself, and so $\psi(\Omega) = \psi(\overline{\text{co}}\,\Omega)$.

It is a trivial matter to check that ψ enjoys the properties of semi-additivity (and hence that of monotonicity) and invariance under the operation of adjunction of an arbitrary compact set. Moreover, ψ is obviously regular in the sense that $\psi(\Omega) = \tilde{0}$ if and only if Ω is relatively compact (here $\tilde{0}$ designates the class containing all relatively compact sets).

1.7.11. Theorem. *There exists an* MNC ψ *that is semi-additive and invariant under adjunction of compact sets, such that any* K_c*-operator is* ψ*-condensing.*

Proof. The MNC ψ constructed above has the desired properties. In fact, suppose

$$\psi(\Omega) \leq \psi[f(\Omega)] \quad (\Omega \subset M = D(f)).$$

This means that in the class $\tilde{\Omega}$ there is a set Ω', and in the class containing $f(\Omega)$—a set Ω'', such that $\Omega' \subset \Omega''$. Since $(\Omega, \Omega') \in R$, there is a compact set T_1 such that $\Omega \subset \overline{\text{co}}(T_1 \cup \Omega')$. Similarly, there is a compact set T_2 such that $\Omega'' \subset \overline{\text{co}}(T_2 \cup f(\Omega))$. This gives

$$\Omega \subset \overline{\text{co}}(T_1 \cup \Omega') \subset \overline{\text{co}}(T_1 \cup \Omega'') \subset \overline{\text{co}}(T_1 \cup \overline{\text{co}}(T_2 \cup f(\Omega)))$$

$$\subset \overline{\text{co}}\,(T_1 \cup T_2 \cup f(\Omega)) = \overline{\text{co}}\,(T \cup f(\Omega)),$$

where $T = T_1 \cup T_2$ is compact. Thus, we showed that there is a compact set T such that

$$\Omega \subset \overline{\text{co}}\,(T \cup f(\Omega)). \tag{9}$$

Now let us show that Ω is relatively compact, which will complete the proof of the theorem. To this end we exhibit a set Ω_∞ which contains Ω and has the property that $\overline{\text{co}}\,(T \cup f(\Omega_\infty \cup M)) = \Omega$. Then, since f is a K_c-operator, the set Ω_∞ will be compact, implying that Ω is relatively compact. To construct Ω_∞ we use a transfinite induction process ressembling closely the one described in **1.6.1**. We define the transfinite sequence $\{\Omega_\alpha\}$ by the rules:

$$\begin{aligned} \Omega_0 &= \overline{\text{co}}(T \cup f(\Omega)), \\ \Omega_\alpha &= \overline{\text{co}}(T \cup f(\Omega_{\alpha-1} \cap M)), \quad \text{if } \alpha - 1 \text{ exists}, \\ \Omega_\alpha &= \overline{\text{co}}\left(\bigcup_{\beta<\alpha} \Omega_\beta\right), \quad \text{if } \alpha - 1 \text{ does not exist}. \end{aligned}$$

Next, we check by transfinite induction on α that the following two assertions hold true:

1) the sequence $\{\Omega_\alpha\}$ is nondecreasing with respect to inclusion, i.e., $\beta \leq \alpha$ implies $\Omega_\beta \subset \Omega_\alpha$;

2) $\Omega_\alpha \subset \overline{\text{co}}\,(T \cup f(\Omega_\alpha \cap M))$ for all α.

(Notice that the second assertion is an obvious corollary of the first; for us, however, it will be more convenient to prove them simultaneously.)

For $\alpha = 0$ the first assertion is plain. Also, since, by (9), $\Omega = \Omega \cap M \subset \Omega_0 \cap M$, we have

$$\Omega_0 = \overline{\text{co}}\,(T \cup f(\Omega)) \subset \overline{\text{co}}\,(T \cup f(\Omega_0 \cap M)),$$

i.e., both 1) and 2) are true for $\alpha = 0$. Suppose they are true for all $\alpha \leq \alpha_0$; let us prove that they are also true for $\alpha = \alpha_0$. First, if $\alpha_0 - 1$ exists, then by assertion 2) for $\alpha = \alpha_0 - 1$,

$$\Omega_{\alpha_0-1} \subset \overline{\text{co}}\,(T \cup f(\Omega_{\alpha_0-1} \cap M)) = \Omega_{\alpha_0}.$$

Now if $\beta < \alpha_0$, then $\beta \leq \alpha_0 - 1$, and so

$$\Omega_\beta \subset \Omega_{\alpha_0-1} \subset \Omega_{\alpha_0},$$

i.e., 1) holds for $\alpha = \alpha_0$. Further,

$$\Omega_{\alpha_0} = \overline{\text{co}}\,(T \cup f(\Omega_{\alpha_0-1}) \cap M)) \subset \overline{\text{co}}\,(T \cup f(\Omega_{\alpha_0} \cap M)),$$

because $\Omega_{\alpha_0-1} \subseteq \Omega_{\alpha_0}$ (as we just proved). Now suppose $\alpha_0 - 1$ does not exist. Then, clearly, by the definition of Ω_{α_0}, $\Omega_\beta \subset \Omega_{\alpha_0}$ for all $\beta < \alpha_0$. In particular, this implies that

$$\overline{\text{co}}\,(T \cup f(\Omega_\beta \cap M)) \subset \overline{\text{co}}\,(T \cup f(\Omega_{\alpha_0} \cap M)).$$

Using the inductive hypothesis, we conclude that

$$\Omega_\beta \subset \overline{\text{co}}\,(T \cup f(\Omega_\beta \cap M)) \subset \overline{\text{co}}\,(T \cup f(\Omega_{\alpha_0} \cap M))$$

for any $\beta < \alpha_0$. Consequently,

$$\Omega_{\alpha_0} = \overline{\text{co}}\Big(\bigcup_{\beta<\alpha_0} \Omega_\beta \Big) \subset \overline{\text{co}}(T \cup f(\Omega_{\alpha_0} \cap M)),$$

as needed.

The sequence $\{\Omega_\alpha\}$ stabilizes starting with some index δ: $\Omega_\alpha = \Omega_\delta$ for all $\alpha \geq \delta$. Set $\Omega_\infty = \Omega_\delta$. Then clearly

$$\Omega_\infty = \overline{\text{co}}\,(T \cup f(\Omega_\infty \cap M));$$

moreover, since $\{\Omega_\alpha\}$ is nondecreasing, $\Omega_\infty \supset \Omega$. **QED**

1.7.12. Remark. In the last part of the proof we essentially showed that an operator $f: M \to E$ is a K_c-operator if and only if, for any set Ω and any compact set T, the inclusion

$$\Omega \subset \overline{\text{co}}\,(T \cup f(\Omega \cap M))$$

implies Ω is relatively compact.

By analogy with Theorem 1.7.1 one can establish the following results.

1.7.13. Theorem. *There exists an* MNC ψ_∞ *that is semi-additive and invariant with respect to the adjunction of one-element sets, such that any* K_∞*-operator is* ψ_∞*-condensing.*

1.7.14. Theorem. *There exists a semi-additive* MNC ψ_0 *such that any* K_0*-compact (i.e., ultimately compact) operator is* ψ_0*-condensing.*

The proofs of these theorems proceed according to the same scheme as that of Theorem 1.7.11; the only thing that differs is the equivalence relation R on 2^E used in the definition of the sought-for MNC.

Specifically, in the case of the K_∞-operators two sets Ω_1 and Ω_2 are regarded as equivalent if there exists finite sets $T_1, T_2 \in 2^E$ such that

$$\Omega_1 \subset \overline{\text{co}}\,(T_2 \cup \Omega_2), \quad \Omega_2 \subset \overline{\text{co}}\,(T_1 \cup \Omega_1),$$

whereas in the case of the K_0-operators they are considered as equivalent if

$$\Omega_1 \subset \overline{\text{co}}\Omega_2, \quad \Omega_2 \subset \overline{\text{co}}\,\Omega_1.$$

1.7.15. Remark. It can observed that the properties of the MNCs with respect to which one or another given class of K–operators is condensing improve as the considered class in the chain $\mathcal{K}_0 \to \mathcal{K}_\infty \to \mathcal{K}_c$ gets narrower and narrower. At the same time, one cannot expect to succeed in constructing, for an arbitrary n ($1 \leq n \leq \infty$), an MNC ψ_n which, compared with ψ_0, will possess in addition one of the properties of the MNC ψ_∞, namely, invariance with respect to the adjunction of one-element sets, and such that the operators in the class $\mathcal{K}_n$ will be ψ_n-condensing. In fact, according to Remark 1.7.8, this would imply that the classes $\mathcal{K}_n, \mathcal{K}_{n+1}, \dots$ coincide, which contradicts Theorem 1.7.5.

1.8. SURVEY OF THE LITERATURE

1.8.1. The Hausdorff and Kuratowski MNCs. The definition of the MNC α and Theorem 1.1.8 are based on the works of K. Kuratowski [95, 96], where some applications to topology are also indicated. The MNC χ is introduced following the works of L. S. Gol′denshteĭn, I. Gohberg, and A. S. Markus [55] and of L. S. Gol′denshteĭn and A. S. Markus [56]. Our exposition of the properties of α and χ follows works of A. Ambrosetti, Yu. G. Borisovich, J. Daneš, G. Darbo, A. Furi, R. D. Nussbaum, W. V. Petryshyn, B. N. Sadovskiĭ, Yu. I. Sapronov, A. Vignoli, and others (see the list of references). Theorem 1.1.5 is taken from [29, 155]. The assertion in **1.1.6** for the Hausdorff MNC is presented according to [55], while for the Kuratowski MNC we follow the papers of A. Furi and A. Vignoli [49] and R. D. Nussbaum [116].

The results of **1.1.9–1.1.11** rely on the papers [55, 56], where it is also proved that when E is a Banach space with a basis $\{e_1, e_2, \dots\}$ the following inequalities hold:

$$a \inf_n \sup_{x \in \Omega} ||(I - P_n)x|| \leq \chi(\Omega) \leq \inf_n \sup_{x \in \Omega} ||(I - P_n)x||,$$

where P_n is the projector onto the linear span of the vectors $\{e_1, e_2, \dots, e_n\}$ and

$$a = \overline{\lim_{n \to \infty}} \, ||I - P_n||.$$

Formula (1) in **1.1.9** is taken from V. G. Kurbatov's paper [99]. Concerning the Kuratowski and Hausdorff MNCs in uniform spaces the reader is referred to the papers by B. N. Sadovskiĭ [155, 160].

The papers of R. D. Nussbaum [117] and A. Ambrosetti [13] contain a study of the Kuratowski MNC in the space $C(K, X)$ of continuous functions on a compact metric space

(K, d) with values in a compact metric space (X, ρ). Let Ω be a bounded set in $C(K, X)$. For arbitrary $\delta > 0$ set

$$\omega(\delta, \Omega) = \sup\{\rho[f(k_1), f(k_2)]\colon k_1, k_2 \in K,\ d(k_1, k_2) < \delta,\ f \in \Omega\}.$$

Let

$$a = \lim_{\delta \to 0} \omega(\delta, \Omega), \quad b = \sup_{k \in K} \alpha_X\{f(k)\colon f \in \Omega\};$$

here α_X denotes the Kuratowski MNC in X. The following bounds hold:

$$\max\{a/2, b\} \le \alpha(\Omega) \le 2a + b.$$

In particular, if the set $\{f(k)\colon f \in \Omega,\ k \in K\}$ is relatively compact, then $a/2 \le \alpha(\Omega) \le 2a$.

The Hausdorff MNC in the space $C([0,1], E)$, where E is a Banach space with a basis, can be computed, according to N. A. Erzakova's work [47], by the formula

$$\chi(\Omega) = \lim_{n \to \infty} \lim_{\delta \to 0} \sup_{f \in \Omega} \omega(f, \delta, n)$$

where

$$\omega(f, \delta, n) = \max_{\delta \le \xi \le 1-\delta} \min_{y \in E} \max_{\xi - \delta \le t \le \xi + \delta} \|f(t) - P_n y\|$$

and P_n is the canonical projector onto the linear span of the first n vectors in the basis of E.

In a metric space the Hausdorff MNC is also invariant under the passage to the convex hull, if the convex hull is understood in the sense of some convex structure on the space in question (see L. A. Talman's paper [172]).

A description of the elementary properties of the inner Hausdorff MNC can be found in R. D. Nussbaum's paper [119]. Let E be a normed space and let P_α ($\alpha \in \mathrm{A}$, A a directed set) be a family of compact operators in E such that $P_\alpha x \underset{\alpha}{\to} x$ for any fixed $x \in E$. For arbitrary $\Omega \subset E, \alpha \in A$, and $\delta > 0$ define the set Ω_α^δ by

$$\Omega_\alpha^\delta = \{x \in \Omega\colon \|x - P_\beta x\| < \delta \text{ for all } \beta \ge \alpha\}.$$

Then (see N. A. Erzakova's paper [45])

$$\chi_{\mathrm{i}}(\Omega) = \lim_{\delta \to 0} \limsup_{\alpha} \sup_{x \in \Omega} \inf_{y \in \Omega_\alpha^\delta} \|x - y\|.$$

In the case $E = l_p$ this formula can be written in the simpler form

$$\chi_{\mathrm{i}}(\Omega) = \overline{\lim_{n \to \infty}} \sup_{x \in \Omega} \inf_{y \in \Omega} [\|P_n x - y\|^p + \|(I - P_n)x\|^p]^{1/p},$$

where P_n denotes the canonical projector onto the linear span of the first n elements of the basis. The last formula can also be generalized to the case of normed spaces with a basis for which

$$||x||^p = ||P_n x||^p + ||x - P_n x||^p \quad (x \in E).$$

1.8.2. The general definition of an MNC. Definition 1.2.1 is taken from B. N. Sadovskiĭ's papers [155, 160]. Examples 1.2.5, 1.2.7 are borrowed from the same place, and subsection **1.2.6**—from the book [15]. In subsections 1.2.8 and 1.2.9, concerned with measures of weak noncompactness, we follow F. S. De Blasi's paper [31], in which there is also given a proof of the following assertion: *any decreasing sequence of weakly closed subsets of a Banach space whose measure of weak noncompactness tends to zero has nonempty intersection.*

An analogue of the Hausdorff MNC is the following MNC in a locally convex space (C. J. Himmelberg, J. R. Porter, and F. S. Van Vleck [61]). Let $\mathcal{B}$ be a fundamental system of convex neighborhoods of zero in the locally convex space E. For any bounded set $\Omega \subset E$ let $Q(\Omega)$ denote the set of all $V \in \mathcal{B}$ such that $\Omega \subset S + V$ for some totally bounded set S. The set where the function Q assumes its values considered to be partially ordered by inclusion. Q is obviously "anti-monotone": $\Omega_1 \subset \Omega_2$ implies $Q(\Omega_1) \geq Q(\Omega_2)$, as well as invariant under passage to the convex hull. Moreover, Q is regular in the sense that $Q(\Omega) = \mathcal{B}$ if and only if Ω is totally bounded.

In the space $L_p[0,1]$ $(p \geq 1)$ the following function is an MNC:

$$\mu(\Omega) = \lim_{\text{mes}\, D \to 0} \sup_{x \in \Omega} ||P_D x||,$$

where P_D is the operator of multiplication by the indicator function of the set D (see [47]). This quantity measures to what extent the norms of the functions from Ω are not equi-absolutely continuous. The MNC μ enjoys all the properties of MNCs listed in **1.1.4**, except for regularity: $\mu(\Omega) = 0$ for any totally bounded set Ω, but there exist sets $\Omega \subset L_p[0,1]$ that are not totally bounded and for which nevertheless $\mu(\Omega) = 0$ [47]. We shall return to the study of the function μ in Chapter 4.

An interesting question is what can be said about the structure of the space of MNCs on a Banach space? Following the paper of V. A. Bondarenko [17], we call an MNC ϕ on E *universal* if for any real-valued MNC ψ on E there exists a function $f: \mathbf{R} \to \mathbf{R}$ such that $\psi(\Omega) = f[\phi(\Omega)]$ for all $\Omega \in 2^E$. As one can readily see, the problem of the existence of a universal MNC is equivalent to that of whether it is possible to partition 2^E into disjoint classes such that any MNC is constant in each such class. In the case of real-valued MNCs it is required, in addition, that the power of the set of these classes does not exceed the power of the continuum. The following assertions hold true (see [17]).

Theorem. *In any separable Banach space there exists a universal* MNC.

Theorem. *In any separable Banach space, in the set of all real-valued* MNC*s that are invariant under adjunction of one-element sets there exists a universal* MNC.

A number of authors have considered other systems of axioms that isolate objects similar to MNCs. Some of these will be described in **1.8.8–1.8.11**.

1.8.3. The MNC β. The definition of the MNC β given here is taken from the papers of L. S. Gol′denstheĭn, I. Gohberg, and A. S. Markus [55] and of L. S. Gol′denshteĭn and A. S. Markus [56] (see also V. Istrăţescu [64] and J. Daneš [27]). Theorems 1.3.3 and 1.3.4 are taken from N. A. Erzakova's paper [43].

In the spaces l_p $(1 \leq p < \infty)$ the MNCs β and χ are connected by the relation

$$\beta(\Omega) = 2^{1/p}\chi(\Omega);$$

in the space l_∞ they are not proportional (see [43]). In the spaces L_p $(1 \leq p < \infty)$,

$$\beta(\Omega) \geq 2^{1/p}\chi(\Omega),$$

and equality holds if Ω is compact in the sense of convergence in measure (see N. A. Erzakova's work [42]). The MNCs β and χ in L_p are not proportional for $p = 2$, and are proportional for $p \neq 2$.

1.8.4. Sequential MNCs. The exposition in Section 1.4 follows the papers of B. N. Sadovskiĭ [155, 160].

1.8.5. Condensing operators. The definitions of (k, α)-bounded, (k, χ)-bounded, and χ-condensing operators follow papers by G. Darbo [29], L. S. Gol′denshteĭn, I. Gohberg and A. S. Markus [55], L. S. Gol′denshteĭn and A. S. Markus [56], and B. N. Sadovskiĭ [154]; the general definitions of condensing operators, operators condensing in the proper sense, and condensing families of operators are taken from B. N. Sadovskiĭ's papers [155, 160]. Example 1.5.3, constructed by V. Sviridov, is discussed in [162]. Theorem 1.5.7 appears, in different versions, in papers of B. N. Sadovskiĭ [155], R. D. Nussbaum [116], J. L. R. Webb [180], A. E. Rodkina and B. N. Sadovskiĭ [150]. The theorem on the Fréchet derivative of a condensing operator can be found in papers of J. Daneš [26] and R. D. Nussbaum [118].

Among the results that serve as criteria for an operator to be condensing we should mention the following result (J. R. L. Webb [180], A. A. Kalmykov [69]).

Theorem. *Let E be a Banach space, $f: E \to E$ a continuous map, and $\Omega \subset E$. Then $\alpha[f(\Omega)] \leq k$ if and only if for any $\varepsilon > 0$ there exists a finite-dimensional operator $S_\varepsilon: \Omega \to E$ such that $\|f(x) - S_\varepsilon(x)\| \leq k + \varepsilon$ for all $x \in \Omega$.*

The following simple assertion (see R. D. Nussbaum's paper [116]) can also serve to test if an operator is condensing.

Theorem. *Let (X, d) be a metric space, $\{C_i: i = 1, \ldots, n\}$ an open cover of X, and $\{\lambda_i: i = 1, \ldots, n\}$ a partition of unity subordinate to this cover. Further, let E be a Banach space and let $f_i: C_i \to E$ be (k, α)-bounded maps. Then the map $x \mapsto f(x) = \sum_{i=1}^{n} \lambda_i(x) f_i(x)$ is (k, α)-bounded.*

In R. W. Leggett's papers [101, 102] a study is made of classes of condensing-type operators that are invariant under equivalent renormalizations of the underlying space. Let $\tau(f) = \inf\{q > 0: f \text{ is } (q, \alpha)\text{-bounded}\}$, where $f: E \to E$ and $(E, \|\cdot\|)$ is a Banach space. The operator f is said to be *topologically strictly α-condensing* if one can find a norm $\|\cdot\|_1$ on E that is equivalent to the original norm $\|\cdot\|$, such that $\tau_1(f) < 1$, where $\tau_1(f)$ is the number $\tau(f)$ computed in the space $(E, \|\cdot\|_1)$.

Theorem. *In order for the operator f to be topologically strictly α-condensing it is necessary, and in the case where f is linear also sufficient, that* $\lim_{n\to\infty}[\tau(f^n)]^{1/n} < 1$.

Theorem 1.5.11 and the assertions in **1.5.12** are taken from B. N. Sadovskiĭ's papers [154, 155, 160]. They generalize the principle of M. A. Krasnosel′skiĭ for operators of the type "contraction plus compact" and of G. Darbo [29] for (k, α)-operators with $k \leq 1$. A modified version of Theorem 1.5.11 for the case of χ_{i}-condensing maps can be found in the paper of A. S. Potapov and B. N. Sadovskiĭ [138] (we should emphasize that the inner Hausdorff MNC χ_{i} is not an MNC in the sense of the general definition).

The connection between χ-condensing and contractive operators was studied by V. S. Kozyakin [83].

Theorem. *Let (M, ρ) be a complete metric space, and let $f: M \to M$ be a χ-condensing operator that satisfies the following three requirements:*

a) *there exists a compact set $K \subset M$ such that $f(A) = A$ implies that $A \subset K$;*

b) $\sup_n \rho(f^n(x), x) < \infty$ *for all $x \in K$;*

c) *there exists a neighborhood* U *of* K *such that* $\sup_n \operatorname{diam} f^n(U) < \infty$.

Then there exist a complete metric space T, *a point* $t \in T$, *and a continuous map* $g: M \to T$ *such that*

1) $g(x) = t$ *for all* $x \in K$;

2) g *maps* $M \setminus K$ *homeomorphically onto* $T \setminus \{t\}$;

3) *the map* $g \circ f \circ g^{-1}$ *is a contraction on* T.

In [83] there are also given necessary and sufficient conditions for an operator that acts in a complete metric space and is condensing with respect to a regular monotone MNC which is invariant under the adjunction of one-element sets, to be contractive in an equivalent metric.

1.8.6. Ultimately compact operators. A countable sequence of the type described in **1.6.1** was used by G. Darbo in [29]. In our exposition of the results of Section 1.6 we followed B. N. Sadovskiĭ's works [154, 155, 158, 160].

1.8.7. K**-operators.** The exposition in Section 1.7 follows the papers of A. S. Potapov [133, 134] and of A. S. Potapov and B. N. Sadovskiĭ [137]. Mappings of locally convex spaces that are similar to the K-operators were considered by E. A. Lifshits and B. N. Sadovskiĭ [104].

Below we describe a number of constructions that single out various systems of MNCs as well as classes of operators of condensing and ultimately compact type.

1.8.8. Distinguishing maps. In this and the next subsections we follow the paper of Yu. G. Borisovich and Yu. I. Sapronov [22]. We confine ourselves to the Banach space setting, though the results decribed here are also valid in locally convex spaces.

Thus, let E be a Banach space, and let $\mathrm{CL}(E)$ denote the set of all closed subsets of E, equipped with the exponential topology (see [96]). A family $\Phi = \{\Phi_\lambda\} \subset \mathrm{CL}(E)$ is called a Φ-*system* if, first, the intersection of any subfamily of sets from Φ belongs to Φ and, second, for any $\Omega \in \mathrm{CL}(E)$ there is a $\Phi_\lambda \in \Phi$ such that $\Omega \subset \Phi_\lambda$.

Let K be a cone in the real linear space X, which defines in X a partial order "$\leq$" [85]. A map $\phi: \mathrm{CL}(E) \to K$ is said to be *distinguishing* if it satisfies the following four conditions:

1) ϕ is monotone: $\phi(\Omega_1) \leq \phi(\Omega_2)$ whenever $\Omega_1 \subset \Omega_2$;

2) $\phi(\Omega \cup R) \leq \phi(\Omega)$ whenever $\phi(R) = 0$;

3) $\phi(\bigcap_{\Phi_\lambda \supset \Omega} \Phi_\lambda) = \phi(\Omega)$ for all Ω;

4) ϕ is nonsingular: $\phi(x) = 0$ for all $x \in E$.

The *kernel* $\operatorname{Ker}\phi$ of the distinguishing map ϕ is defined to be the set $\{\Omega \in \mathrm{CL}(E): \phi(\Omega) = 0\}$.

If Φ consists of all closed convex subsets of E, then it is readily verified that any distinguishing map is a monotone nonsingular MNC (defined only on closed subsets).

If for an Φ-system one takes the family of all subsets of E with the property that, together with any point, they contain the ray emanating from the origin that passes through that point, then the construction described above allows one to define the so called *projective measures of noncompactness,* i.e., MNCs ϕ with the property that $\phi(t\Omega) = \phi(\Omega)$ for all $t > 0$. A study of the projective MNCs and of the operators that are condensing with respect to such measures can be found in the paper of Yu. G. Borisovich and Yu. I. Sapronov referred to above and in A. A. Kalmykov's paper [70].

In the case where $\Phi = 2^E$ and $\operatorname{Ker}\phi$ consists only of convex sets, the distinguishing map ϕ serves as a "measure of nonconvexity" of sets. One concrete recipe for constructing measures of nonconvexity can be found in a paper by J. Eisenfeld and V. Lakshmikantham [40].

1.8.9. Operators that are compatible with a distinguishing map. An operator $f: E \to E$ is said to be *compatible with the distinguishing map* ϕ if, first, $\operatorname{Ker}|\phi$ is invariant under f: $\phi[\overline{f(\Omega)}] = 0$ for all $\Omega \in \operatorname{Ker}\phi$, and, second, $\phi[\overline{f(\Omega)}] \not\geq \phi(\Omega)$ for all $\Omega \notin \operatorname{Ker}\phi$.

An operator $f: \overline{U} \to E$ is said to *admit Φ-restriction to an invariant set Φ_λ* if $f(U \cap \Phi_\lambda) \subset \Phi_\lambda$.

The basic property of the operators compatible with a distinguishing map is given by the following

Theorem. *Any operator $f: \overline{U} \to E$ that is compatible with a distinguishing map ϕ admits a Φ-restriction with the following properties:*

1) $\Phi_\lambda \cap \overline{U} \neq \emptyset$;

2) $\Phi_\lambda \in \operatorname{Ker}\phi$;

3) *for any prescribed set $R \in \operatorname{Ker}\phi$, the set Φ_λ contains R.*

Using this theorem one can construct compact invariant sets of operators that are compatible with a distinguishing map.

The notion of family of operators compatible with a distinguishing map is defined in a similar manner, and an analogue of the preceding theorem holds true for such families.

1.8.10. MNCs with a kernel. In this subsection we follow the book of J. Banaš and K. Goebel [15] and describe a certain subclass of the class of all MNCs (in the sense of the general definition) on a Banach space, which satisfy a certain set of additional axioms.

A nonempty family $\mathcal{P}$ of bounded subsets of a Banach space E is called a *kernel (of* MNC*)* if

1) $\Omega \in \mathcal{P}$ implies $\overline{\Omega} \in \mathcal{P}$ and $\operatorname{co} \Omega \in \mathcal{P}$;

2) $\Omega \in \mathcal{P}, \Omega_1 \subset \Omega$, and $\Omega_1 \neq \emptyset$ imply $\Omega_1 \in \mathcal{P}$;

3) $\Omega_1, \Omega_2 \in \mathcal{P}$ and $\lambda \in [0,1]$ imply $\lambda\Omega_1 + (1-\lambda)\Omega_2 \in \mathcal{P}$;

4) the set of all closed subsets belonging to $\mathcal{P}$ is closed in the set of all bounded closed subsets of E in the Hausdorff metric.

An MNC $\psi: 2^E \to [0, \infty)$ is called a *measure of noncompactness with kernel* $\mathcal{P}$ if it vanishes only on the sets in $\mathcal{P}$, is monotone, and possesses the following two properties:

1) $\psi[\lambda\Omega_1 + (1-\lambda)\Omega_2] \leq \lambda\psi(\Omega_1) + (1-\lambda)\psi(\Omega_2)$ for all $\lambda \in [0,1]$;

2) any decreasing (with respect to inclusion) sequence $\{\Omega_n\}$ of closed bounded subsets of E such that $\psi(\Omega_n) \to 0$ as $n \to \infty$ has a nonempty intersection.

The simplest nontrivial example of an MNC with the kernel $\mathcal{P}$ consisting of the bounded subsets of a subspace E_1 of E is the function defined as follows: $\psi(\Omega) = \chi(\Omega) + \rho(\Omega, E_1)$, where ρ is the Hausdorff metric.

In the book [15] the reader can find a study of various properties of MNCs with a kernel, fixed-point theorems for operators that are condensing with respect to such measures, and a study of MNCs with a kernel in concrete spaces.

1.8.11. Measures of compactness. Here we follow the works of G. S. Jones [67] and F. S. De Blasi [30]. We describe a method of constructing functions of MNC-type in metric spaces. Thus, let (M, d) be a complete bounded metric space, and let $\mathcal{N}$ be a family of compact subsets of M. A function $\psi: 2^M \to [0, \infty)$ is called a *measure of compactness* [30] if it is monotone, invariant under passage to the closure, vanishes on the sets in $\mathcal{N}$, and if $\psi(\{a\} \cup A) = \psi(A)$ for all $A \in 2^M$ and all $a \in M$ such that $\psi(\{a\}) = 0$. A monotone (with respect to inclusion) map $h: 2^M \to 2^M$ called a *ψ-closure* [67] if

1) $\overline{h}(\Omega) = h(\Omega)$;

2) $h^2 = h$;

3) $\Omega \subset h(\Omega)$;

4) $\psi[h(\Omega)] = \psi(\Omega)$.

Theorem. *Suppose the map $f: M \to M$ is continuous and ψ-condensing (i.e., $\psi(f(\Omega)) < \psi(\Omega)$ whenever $\psi(\Omega) \neq 0$). Suppose further that f is reducible on $h(\mathcal{N})$ in the following sense: any subset of $h(\mathcal{N})$ that has more than one element and is invariant under f contains a proper subset invariant under f. Then f has at least one fixed point.*

In [67] one can find a description of measures of compactness that are singled-out by a somewhat different system of axioms. Also, in [30, 67] one can find methods for constructing nontrivial measures of compactness and a study of some of their properties.

1.8.12. Locally condensing maps. A continuous map $f: M \to E$, where M is an open subset of the Banach space E, is said to be *locally strictly ψ-condensing* (see R. D. Nussbaum's paper [116]; here ψ is an MNC on E) if for any point $x \in M$ there exist a neighborhood V_x of x and a number $k_x < 1$ such that the restriction $f|_{V_x}$ is (k_x, ψ)-condensing. Any locally strictly α-condensing map with compact fixed-point set F is ultimately compact on some neighborhood of F, and in the sequence $\{T_\alpha\}$ (see **1.6.1**) a compact set occurs already at the first transfinite step [116].

CHAPTER 2

THE LINEAR THEORY

It is well known (see, for example, [34]) that if A is a compact linear operator acting in a Banach space E, then for $\lambda = 0$ the Fredholm alternative holds for the equation

$$\lambda x - Ax = 0. \tag{1}$$

Furthermore, any nonzero point λ of the spectrum of A is an eigenvalue of finite multiplicity; the spectrum of A is an at most countable set, the only possible limit point of which is zero.

For any bounded linear operator A there is a smallest number R such that for $|\lambda| > R$ the Fredholm alternative holds for equation (1) and λ can only be an isolated eigenvalue of finite multiplicity. In this chapter we obtain expressions for the computation and estimation of R in terms that are connected with MNCs. In the same terms we give conditions for the continuity of parts of the spectrum that lie inside the disc of radius R, for strongly continuous families of operators.

The approach discussed here is based on associating with each Banach space E a special Banach space E^+ such that the norms on E^+ are in a natural correspondence with certain MNCs in E. Under this association, to each bounded linear operator $C\colon E_1 \to E_2$ will correspond an operator $C^+\colon E_1^+ \to E_2^+$ such that C^+ is invertible if and only if C is Fredholm. All this enables us to give conditions for the invertibility of the operator C^+, and hence for the Fredholmness of C, in terms of MNCs.

Throughout this chapter $L(E_1, E_2)$ will denote the space of all bounded linear operators acting from E_1 to E_2. For the case $E_1 = E_2$ we put $L(E_1, E_2) = L(E_1)$.

2.1. FREDHOLM OPERATORS

In this section we consider an example (which, as will turn out later, is quite general) and we give necessary definitions and facts, a detailed exposition of which can be found, for example, in [54].

2.1.1. An example of Fredholm operator. Let E be a complex Banach space and let $B, C \in L(E)$ be such that $\|B\| \leq k$ and C is compact. Consider the operator $A = B + C$. Clearly, for $|\lambda| > k$ the solvability of the equation

$$\lambda x - Ax = y \tag{1}$$

is equivalent to the solvability of the equation

$$x - (\lambda I - B)^{-1}Cx = (\lambda I - B)^{-1}y. \tag{2}$$

The operator $(\lambda I - B)^{-1}C$ is compact. Hence, the Fredholm alternative holds for equation (2), and 1 can only be an eigenvalue of finite multiplicity of $(\lambda I - B)^{-1}C$. It follows that the Fredholm alternative holds also for equation (1), and λ can only be an eigenvalue of finite multiplicity. From the above arguments we deduce the following properties of the operator $F = \lambda I - A$:

a) $\alpha = \dim N(F) < \infty$, where $N(F) = \{x \in E\colon Fx = 0\}$;

b) $\overline{FE} = FE$ (where the overline denotes closure);

c) $\beta = \operatorname{codim} FE < \infty$;

d) the number $\operatorname{ind} F = \beta - \alpha$, called the *index of the operator* F, is equal to zero.

2.1.2. Definition of a Φ-point [54]. If $A \in L(E)$ and $\lambda \in \mathbf{C}$ are such that the operator $F = \lambda I - A$ has properties a)–c), then λ is called a *Φ-point* [or a *Fredholm point*] of the operator A. The set of all Φ-points of A is denoted by $\Phi(A)$.

2.1.3. Theorem on the structure of the set $\Phi(A)$ [54]. *The set $\Phi(A)$ of an operator $A \in L(E)$ is open and, as such, is the union of finitely or countably many (connected) components.*

The index of the operator $\lambda I - A$ is constant on any of the components of $\Phi(A)$. If the Banach space E is infinite-dimensional, then for any operator $A \in L(E)$ there exists at least one point $\lambda \notin \Phi(A)$.

2.1.4. Theorem on the invertibility of the operator $\lambda I - A$ [54]). *If the component $G \subset \Phi(A)$ contains at least one point λ for which the operator $\lambda I - A$ is invertible, then $\lambda I - A$ is invertible at all points of G, except possibly for some isolated points.*

Now let us consider the case of an operator $F \in L(E_1, E_2)$.

2.1.5. Definition of Φ_-- and Φ_+-operators [54]. F is said to be a *Fredholm operator* (or a *Φ-operator*) if it has the properties a)–d) given in **2.1.1**, where E is replaced by E_1. If F has only properties a) and b), then it is called a *Φ_+-operator.*

2.1.6. Theorem on perturbation of a Φ-operator [54]. *For any Φ-operator $F: E_1 \to E_2$ there is a $\rho > 0$ such that if $B \in L(E_1, E_2)$ and $\|B\| < \rho$, then $F + B$ is a Φ-operator and* $\operatorname{ind}(F + B) = \operatorname{ind} F$.

2.2. THE "+"-OPERATION AND NORMAL MEASURES OF NONCOMPACTNESS

In this section we describe the quotient space of the space of bounded sequences by the subspace of totally bounded sequences and the connection of the norms in this quotient space with MNCs.

2.2.1. Basic definitions. Let E be a Banach space with norm $\|\cdot\|$. We let BE denote the space of all bounded sequences $X = (x_1, \dots, x_n, \dots)$, $x_n \in E$, with the natural coordinatewise linear operations and the norm $\|X\| = \sup\{\|x_n\|: n = 1, 2, \dots\}$. Clearly, the space BE is also Banach. Let KE denote the closed subspace of BE consisting of the relatively compact sequences, i.e., sequences for which the set of their elements is relatively compact. Finally, let E^+ denote the quotient space BE/KE with the natural linear operations and norm, i.e., for $\mathcal{X} \in E^+$,

$$\|\mathcal{X}\| = \inf\{\|X\|: X \in \mathcal{X}\}. \tag{1}$$

Let us show that

$$\|\mathcal{X}\| = \chi(X), \quad X \in \mathcal{X} \tag{2}$$

(from here on X denotes, depending on the context, a sequence and also the set of its elements).

Let $\|\mathcal{X}\| = d$. Then for any $\varepsilon > 0$ one can find an $X' \in \mathcal{X}$ such that $\|X'\| \leq d + \varepsilon$. Clearly, for any $X \in \mathcal{X}$ the totally bounded sequence $Y = X - X'$ is a $(d + \varepsilon)$-net of the set of elements of X, and so, in view of the arbitrariness of ε,

$$\chi(X) \leq d = \|\mathcal{X}\|. \tag{3}$$

Let us prove the opposite inequality. Let Q be a q-net of the set of elements of X. Consider the sequence Z, constructed according to the following recipe: z_n is an element of Q whose distance to some $x_n \in X$ is smaller than q (if there are several such, then one can take any of them). Clearly, $Z \in KE$, $X - Z \in \mathcal{X}$ and $||X - Z|| \leq q$. Consequently,

$$\chi(X) \geq \|\mathcal{X}\|,$$

which in conjunction with (3) completes the proof.

Thus, the Hausdorff MNC defines a norm on E^+ via formula (2). Generalizing this observation, one arrives at the following definition.

2.2.2. Definition of a normal MNC. A *normal measure of noncompactness* in the Banach space E is defined to be a seminorm ψ on BE with the property that $\psi(X) = 0$ if and only if $X \in KE$.

2.2.3. Lemma. *Every normal* MNC ψ *in a Banach space* E *defines a norm* ψ^+ *on the space* E^+ *by the formula*

$$\psi^+(\mathcal{X}) = \psi(X),\ X \in \mathcal{X}.$$

Conversely, every norm ψ^+ *on the space* E^+ *generates a normal* MNC ψ *in* E *via the same formula, read from right to left.*

Proof. Clearly, it suffices to verify that ψ^+ is well defined in the sense that $X, Y \in \mathcal{X}$ implies $\psi(X) = \psi(Y)$. Thus, suppose $X - Y \in KE$. Then

$$\psi(Y) = \psi(Y) - \psi(Y - X) \leq \psi(X) \leq \psi(X - Y) + \psi(Y) = \psi(Y),$$

i.e., $\psi(Y) = \psi(X)$. **QED**

2.2.4. Remark. From now on we shall assume that the following condition is satisfied: the space E^+ with the norm ψ^+ is complete.

2.2.5. Remarks. (a) A straightforward corollary of Theorems 1.1.4 and 1.3.3 is that the MNCs α, β, χ, considered on bounded sequences, are normal MNCs.

(b) Under some small supplementary restrictions, any normal MNC ψ generates in the Banach space in question an MNC $\tilde{\psi}$, defined on all bounded subsets by the rule

$$\tilde{\psi}(\Omega) = \sup\{\psi(X): X \in B\Omega\} \tag{4}$$

(here $B\Omega$ designates the set of sequences of elements of Ω). This fact will not be used further, and we omit its precise statement and proof. Rather, we merely notice that, by Theorem 1.4.5,

$$\frac{1}{2}\chi(\Omega) \leq \tilde{\chi}(\Omega) \leq \chi(\Omega), \tag{5}$$

and consequently, in view of the inequalities given in **1.1.7** and **1.3.2**,

$$\frac{1}{4}\alpha(\Omega) \leq \tilde{\alpha}(\Omega) \leq 2\alpha(\Omega), \tag{6}$$

and

$$\frac{1}{4}\beta(\Omega) \leq \tilde{\beta}(\Omega) \leq 2\beta(\Omega). \tag{7}$$

A generalization of the preceding remarks leads to the following notions.

2.2.6. Definition of the notions of M- and M_χ-normality. An MNC ϕ, defined on the bounded subsets of a Banach space E, is said to be *M-normal* if the function $\overline{\phi}$, defined as

$$\overline{\phi}(X) = \phi(X), \quad X \in BE, \tag{8}$$

is a normal MNC and the MNC $\tilde{\overline{\phi}}$ defined by formula (4) is equivalent to ϕ:

$$c_1\phi(\Omega) \leq \tilde{\overline{\phi}}(\Omega) \leq c_2\phi(\Omega)$$

for any bounded $\Omega \in E$.

When ϕ is equivalent to χ we say that ϕ is *M_χ-normal.* In this case the norm $\overline{\phi}^+$ on E^+ is equivalent to the quotient norm (and to the norm $\overline{\chi}^+$).

A sufficient set of conditions for the M_χ-normality of ϕ is that ϕ be monotone, algebraically semi-additive, semi-homogeneous, and equivalent to the MNC χ.

2.3. FREDHOLMNESS CRITERIA FOR OPERATORS

In this section we describe the construction of an operator $A^+: E_1^+ \to E_2^+$ for each given operator $A: E_1 \to E_2$ and we establish a criterion for the Fredholmness of A in terms of properties of A^+ and normal MNCs.

If E_1 and E_2 are Banach spaces, then any operator $C \in L(E_1, E_2)$ induces an operator (denoted with the same letter) that acts from BE_1 to BE_2 according to the rule: $CX =$

$(Cx_1, Cx_2, \dots)$ for $X = (x_1, x_2, \dots)$. Clearly, $C \in L(BE_1, BE_2)$, and $C(KE_1) \subset KE_2$. Hence, one can define an operator $C^+ \in L(E_1^+, E_2^+)$ by the formula

$$C^+\mathcal{X} = CX + KE_2 \quad (\mathcal{X} = X + KE_1).$$

The operator C^+ possesses a number of obvious properties, listed in the following lemma.

2.3.1. Lemma. *Let E_1, E_2, E_3 be Banach spaces, let $C_1, C_2: E_1 \to E_2, C_3: E_2 \to C_3$ be bounded linear operators, and let $a \in \mathbf{C}$. Then*

a) $(C_1 + C_2)^+ = C_1^+ + C_2^+$;

b) $(aC_1)^+ = aC_1^+$;

c) $(C_3C_2)^+ = C_3^+C_2^+$.

2.3.2. Theorem (criterion for membership in the class Φ_+). *$C \in L(E_1, E_2)$ is a Φ_+-operator* (see **2.1.5**) *if and only if C^+ is injective.*

Proof. Suppose C is a Φ_+-operator. Since the kernel N of C is finite-dimensional, the space E_1 admits a direct sum decomposition $E_1 = N \oplus E_1'$, and C maps the Banach space E_1' bijectively, linearly and continuously onto the normed linear space $CE_1 \subset E_2$. Next, since CE_1 is closed in E_2 by hypothesis, it is itself a Banach space, and we can apply Banach's theorem on the inverse operator, according to which the restriction $\hat{C}$ of the operator C to the subspace E_1' has a continuos inverse $\hat{C}^{-1}: CE_1 \to E_1'$. Now let $\mathcal{X} \in E_1^+$ be such that $C^+\mathcal{X} = 0$. This means that for any $X \in \mathcal{X}$ the sequence $Y = CX$ lies in KE_2. Write X as $X = X_1 + X_2$ with $X_1 \in BE_1'$ and $X_2 \in BN \subset KE_1$. Then $Y = \hat{C}X_1$, and so $X_1 = \hat{C}^{-1}Y \in KE_1$. Consequently, $X = X_1 + X_2 \in KE_1$, i.e., $\mathcal{X} = 0$. Thus, if C is a Φ_+-operator, then C^+ is injective.

Conversely, suppose C^+ is injective. If the kernel N of C were finite-dimensional, then using the Riesz lemma (see [107]) one could find a bounded sequence $X = (x_1, x_2, \dots) \in N$ such that $X \notin KE_1$. Setting $\mathcal{X} = X + KE_1$, we would have $\mathcal{X} \neq 0$ and $C^+\mathcal{X} = 0$, contradicting the assumption on C^+. Thus, $\dim N < \infty$.

Now let us prove that $\overline{CE_1} = CE_1$. To this end we write, as above, $E_1 = N \oplus E_1'$. We claim that $\hat{C}^{-1}: CE_1 \to E_1'$ is a bounded operator. Indeed, suppose this is not the case. Then there exists a sequence of nonzero elements $x_n \in E_1'$ such that $\|x_n\| \to \infty$ as $n \to \infty$ and $\|\hat{C}x_n\| \le H$ $(n = 1, 2, \dots)$. Then the elements $z_n = x_n/\|x_n\|$ lie on the unit sphere of the space E_1' and $y_n = \hat{C}z_n \to 0$ as $n \to \infty$. In conjunction with the injectivity of the operator C^+, this implies that the sequence $Z = (z_1, z_2, \dots)$ lies in KE_1 (otherwise C^+ would send the nonzero element $\mathcal{Z} = Z + KE_1$ into zero). If z is an arbitrary limit point of the sequence Z, we conclude that $\|z\| = 1$ and $\hat{C}z = 0$, which contradicts the invertibility of $\hat{C}$.

Now let us return to the proof that the subspace CE_1 is closed. Let $y \in \overline{CE_1}$, i.e., there exists a sequence $y_n \in CE_1 = \hat{C}E_1'$ such that $y_n \to y$. By the foregoing arguments, the sequence $\hat{C}^{-1}y_n$ converges to some element $x \in E_1'$. But then $y = \hat{C}x \in CE_1$. **QED**

2.3.3. Theorem (Fredholmness criterion). *The operator $C \in L(E_1, E_2)$ is Fredholm if and only if C^+ is bijective (i.e., it is injective and $C^+E_1^+ = E_2^+$).*

Proof. Suppose C is Fredholm. Then, by the preceding theorem, C^+ is injective and so it remains to verify that it is surjective, i.e., that for any $Y \in BE_2$ one can find an $X \in BE_1$ and a $Z \in KE_2$ such that $CX = Y + Z$. Since the codimension of CE_1 in E_2 is finite, the space E_2 can be decomposed as $E_2 = CE_1 \oplus E_2'$ (where $\dim E_2' < \infty$). But then one can write $Y = Y_1 + Y_2$, where $Y_1 \in B(CE_1)$ and $Y_2 \in BE_2' \subset KE_2$. Setting $Z = -Y_2$, we are led to the following problem: given $Y_1 \in B(CE_1)$, show that there exists an $X \in BE_1$ such that $CX = Y$. Represent the space E_1 as $E_1 = N \oplus E_1'$ (where N is again the kernel of C) and set $X = \hat{C}^{-1}Y$, where $\hat{C}$ denotes the restriction of C to E_1'. Then obviously $CX = Y_1$.

We proved that the Fredholmness of C implies the bijectivity of C^+. Now suppose C^+ is bijective. Then, by the preceding theorem, C is a Φ_+-operator, and it remains to verify that $\dim(E_2/CE_1) < \infty$. Suppose this is not the case. Pick a sequence $\{U_n : n = 1, 2, \ldots\}$ in the space E_2/CE_1 such that $\|U_n\| = 1$ and $\|U_n - U_m\| \geq 1/2$ for $n, m = 1, 2, \ldots, n \neq m$ (such a sequence can be constructed with the help of the Riesz lemma). Next, construct a sequence $Y = (y_1, y_2, \ldots)$ such that $y_n \in U_n$, $n = 1, 2, \ldots$. Since C^+ is surjective, there exist sequences $X = (x_1, x_2, \ldots) \in BE_1$ and $Z = (z_1, z_2, \ldots) \in KE_2$ such that $CX = Y + Z$. With no loss of generality one can asssume that Z converges (otherwise one can pass to a subsequence). Further, since $y_n + z_n \in CE_1$, one concludes that $z_m - z_m$ belongs to $U_n - U_m$. In fact,

$$z_m - z_n = (y_n - y_m) - (y_n + z_n) + (y_m + z_m);$$

here the first term lies in $U_n - U_m$, while the last two lie in CE_1. But then

$$\|U_n - U_m\| \leq \|z_m - z_n\| \to 0 \quad \text{when } n, m \to \infty,$$

which contradicts the choice of U_n. **QED**

Using theorems 2.3.2 and 2.3.3 we can now obtain conditions for the Fredholmness of an operator C, or for its membership in the class of Φ_+-operators, that are formulated in terms of normal MNCs.

2.3.4. Definition. Suppose that in the Banach spaces E_1 and E_2 there are given normal MNCs ψ_1 and ψ_2, respectively. An operator $F: E_1 \to E_2$ is said to be (k, ψ_1, ψ_2)-*bounded above* [resp. *below*] if $X \in BE_1$ implies $FX \in BE_2$ and $\psi_2(FX) \leq k\psi_1(X)$ [resp. $\psi_2(FX) \geq k\psi_1(X)$]. The operator F is said to be (ψ_1, ψ_2)-*bounded above* [resp. *below*] if it is (k, ψ_1, ψ_2)-bounded above [resp. below] for some $k > 0$.

2.3.5. Definition of the (ψ_1, ψ_2)-norm. The number $\|F\|^{\psi_1,\psi_2}$, defined as

$$\|F\|^{\psi_1,\psi_2} = \sup\{k: F \text{ is } (k, \psi_1, \psi_2)\text{-bounded above}\}, \tag{1}$$

is called the (ψ_1, ψ_2)-*norm* of the operator F.

The (ψ_1, ψ_2)-norms of operators are sometimes referred to as their *measures of noncompactness.*

It is readily verified that

$$\|F\|^{\psi_1,\psi_2} = \sup\{\psi_2(FX): X \in BE_1, \psi_1(X) = 1\} \tag{2}$$

and $\|F\|^{\psi_1,\psi_2} = \|F^+\|_{E_1^+ \to E_2^+}$.

2.3.6. Theorem (sufficient condition of membership in the class Φ_+). *Suppose that in the Banach spaces E_1, E_2 there are given* MNC*s ψ_1, ψ_2 and that the operator $C \in L(E_1, E_2)$ is (ψ_1, ψ_2)-bounded below. Then C is a Φ_+-operator.*

Proof. By Theorem 2.3.2, it suffices to verify that the operator C^+ is injective. Since C is (ψ_1, ψ_2)-bounded below, for any $\mathcal{X} = X + KE_1 \in E_1^+$ we have

$$\psi_2(C^+\mathcal{X}) = \psi_2(CX) \geq k\psi_1(X) = k\psi_1^+(\mathcal{X}),$$

where $k > 0$. This implies the injectivity of C^+. **QED**

2.3.7. Theorem (sufficient condition of Fredholmness). *Let $C = D + A$, where $D, A \in L(E_1, E_2)$ and D is Fredholm. Suppose that in the space E_1 there is given an* MNC *ψ. Finally, suppose that there is a $k < 1$ such that*

$$\psi(AX) \leq k\psi(DX)$$

for all $X \in BE_1$. Then C is a Fredholm operator and its index is equal to the index of D.

Proof. By Theorem 2.3.3, the operator D^+ is invertible, and so one can represent C^+ as

$$C^+ = D^+ + A^+ = [I^+ + A^+(D^+)^{-1}]D^+.$$

If we can show that the norm of the operator $U = A^+(D^+)^{-1}$, which acts in the space (E_2^+, ψ^+), is smaller than one, then both assertions of the theorem will be established. In fact, firstly, that will mean that C^+ is invertible, i.e., C is Fredholm. Next, this assertion remains valid for every operator $C_\lambda = D + \lambda A$ $(0 \leq \lambda \leq 1)$, which in conjunction with Theorem 2.1.6 readily implies that $\operatorname{ind} C = \operatorname{ind} C_1 = \operatorname{ind} C_0 = \operatorname{ind} D$.

Thus, let us estimate the norm of the operator U in (E_2^+, ψ^+). Let $\mathcal{Y} \in E_2^+$; set $\mathcal{X} = (D^+)^{-1}\mathcal{Y}$ and pick $X \in \mathcal{X}$. Then

$$\psi^+(U\mathcal{Y}) = \psi^+(A^+\mathcal{X}) = \psi(AX) \leq k\psi(DX) = k\psi^+(D^+\mathcal{X}) = k\psi^+(\mathcal{Y}).$$

This inequality shows that $||U|| \leq k \leq 1$. **QED**

2.4. THE (ψ_1, ψ_2)-NORMS OF AN OPERATOR

In a number of situations estimates of the (ψ_1, ψ_2)-norms of operators $A \in L(E_1, E_2)$ can be obtained by computing other numerical characteristics, which usually generate in a simple manner an equivalent norm for operators $A^+ \in L(E_1^+, E_2^+)$. The present section is devoted to the description of such characteristics.

Suppose that in the spaces E_1 and E_2 there are given M-normal MNCs ϕ_1 and ϕ_2, respectively (see **2.2.6**), and let $A \in L(E_1, E_2)$. Denote by K the set of all positive numbers k such that the operator S is (k, ϕ_1, ϕ_2)-bounded and set

$$||A||^{(\phi_1, \phi_2)} = \inf K.$$

2.4.1. Theorem. *There exists a constant κ such that*

$$||A||^{(\phi_1, \phi_2)} \leq \kappa ||A||^{\overline{\phi}_1, \overline{\phi}_2},$$

where $\overline{\phi}_1, \overline{\phi}_2$ are constructed from ϕ_1, ϕ_2 by means of formula (8) *in* **2.2.6** *and $||A||^{\overline{\phi}_1, \overline{\phi}_2}$ is defined by means of formula* (1) *in* **2.3.5**.

Proof. Since ϕ_1 and ϕ_2 are M-normal, there exist constants $c_{11}, c_{12}, c_{21}, c_{22} > 0$ such that

$$c_{11}\phi_1(\Omega) \leq \tilde{\overline{\phi}}_1(\Omega) \leq c_{12}\phi_1(\Omega),$$

$$c_{21}'\phi_2(\Omega) \leq \tilde{\overline{\phi}}_2(\Omega) \leq c_{22}\phi_2(\Omega),$$

where $\tilde{\bar{\phi}}_1$ and $\tilde{\bar{\phi}}_2$ are constructed from $\bar{\phi}_1$ and $\bar{\phi}_2$ by means of formula (4) in **2.2.5**. It is easy to check that

$$\tilde{\bar{\phi}}_2(A\Omega) \leq \|A\|^{\bar{\phi}_1,\bar{\phi}_2}\tilde{\bar{\phi}}_1(\Omega).$$

Consequently,

$$\phi_2(A\Omega) \leq c_{12}^{-1}\tilde{\bar{\phi}}_2(A\Omega) \leq c_{21}^{-1}\|A\|^{\bar{\phi}_1,\bar{\phi}_2}\tilde{\bar{\phi}}_1(\Omega) \leq c_{21}^{-1}c_{12}\|A\|^{\bar{\phi}_1,\bar{\phi}_2}\phi_1(\Omega),$$

i.e.,

$$\|A\|^{(\bar{\phi}_1,\bar{\phi}_2)} \leq c_{21}^{-1}c_{12}\|A\|^{\bar{\phi}_1,\bar{\phi}_2}. \quad \textbf{QED}$$

Clearly, one also has the inequality

$$\|A\|^{\bar{\phi}_1,\bar{\phi}_2} \leq \|A\|^{(\phi_1,\phi_2)}.$$

Thus,

$$\|A\|^{\bar{\phi}_1,\bar{\phi}_2} \leq \|A\|^{(\phi_1,\phi_2)} \leq \kappa\|A\|^{\bar{\phi}_1,\bar{\phi}_2}. \tag{1}$$

It is readily seen that if in addition to the hypotheses of Theorem 2.4.1 one requires that $\|\cdot\|^{(\phi_1,\phi_2)}$ be a seminorm on $L(E_1, E_2)$, then $\|\cdot\|^{(\phi_1,\phi_2)}$ generates a norm on the set of all operators A^+ with $A \in L(E_1, E_2)$. Moreover, that norm is equivalent to the norm of A^+ as an operator acting from $(E_1^+, \overline{\phi}_1^+)$ into $(E_2^+, \overline{\phi}_2^+)$. In order that $\|\cdot\|^{(\phi_1,\phi_2)}$ be a seminorm it suffices, for example, that ϕ_2 be a monotone, algebraically semi-additive, semi-homogeneous, regular MNC.

Now let us consider the case where $E_1 = E_2 = E$ and ϕ_1, ϕ_2 are the Hausdorff MNCs; for brevity we denote $\|\cdot\|^{(\chi,\chi)} = \|\cdot\|^{(\chi)}$.

2.4.2. Theorem (a formula for the computation of $\|A\|^{(\chi)}$). *Let $A \in L(E)$. Then*

$$\|A\|^{(\chi)} = \chi(AS) = \chi(AB), \tag{2}$$

where S and B denote the unit sphere and respectively the unit ball in E.

Proof. The equality $\chi(AS) = \chi(AB)$ is obvious. Next, the inequality $\|A\|^{(\chi)} \geq \chi(AB)$ holds because $\chi(B) = 1$. Let us prove the opposite inequality. Let Ω be an arbitrary bounded set and let $\{y_i\}_{i=1}^n$ be a finite d-net of Ω, i.e., $\Omega \subset \bigcup_{i=1}^n B(y_i, d)$. Then $A\Omega \subset \bigcup_{i=1}^n AB(y_i, d)$. Using the monotonicity, semi-additivity, and translation-invariance of the Hausdorff MNC, we obtain

$$\chi(AB) \leq \chi\{AB(0, d)\}.$$

From the linearity of A and the semi-homogeneity of χ it follows that $\chi(A\Omega) \leq d\,\chi(AB)$, whence $\|A\|^{(\chi)} \leq \chi(AB)$, as needed. **QED**

2.4.3. Remark. Formula (2) implies the obvious inequality

$$\|A\|^{(\chi)} \leq \|A\|. \tag{3}$$

In some cases formula (2) permits an easy computation of the norm $\|A\|^{(\chi)}$. Let us give some examples of such computations.

2.4.4. Example (computation of $\|A\|^{(\chi)}$ in the space l_1). Let $A : l_1 \to l_1$ be a continuous operator. Then in the standard basis $e_i = \{\delta_{ik}\}_{k=1}^{\infty}$ (where δ_{ik} is the Kronecker symbol) A is represented by a matrix $(a_{ij})_{i,j=1}^{\infty}$. Let us show that

$$\|A\|^{(\chi)} = \lim_{n\to\infty} \sup_k \sum_{j=n}^{\infty} |a_{jk}|. \tag{4}$$

Applying the expression for the Hausdorff MNC in l_1 given in **1.1.9**, we get

$$\|A\|^{(\chi)} = \lim_{n\to\infty} \sup_{x\in S} \sum_{j=n}^{\infty} \left| \sum_{k=1}^{\infty} a_{jk}\xi_k \right|,$$

where $x = (\xi_1, \xi_2, \dots)$ and $\|x\| = \sum_{k=1}^{\infty} |\xi_k| = 1$. Since

$$\sup_{x\in S} \sum_{j=n}^{\infty} \left| \sum_{k=1}^{\infty} a_{jk}\xi_k \right| \leq \sup_{x\in S} \sum_{j=n}^{\infty} \sum_{k=1}^{\infty} |a_{jk}|\,|\xi_k| = \sup_{x\in S} \sum_{k=1}^{\infty} \sum_{j=n}^{\infty} |a_{jk}|\,|\xi_k| \leq \sup_k \sum_{j=n}^{\infty} |a_{jk}|,$$

we have

$$\|A\|^{(\chi)} \leq \lim_{n\to\infty} \sup_k \sum_{j=n}^{\infty} |a_{jk}|.$$

Now notice that the set AS contains all sequences y_k of the form $y_k = (a_{1k}, a_{2k}, \dots)$, which are the images of the unit vectors e_i. But, by formula (1) in **1.1.9**,

$$\chi(\{y_k\}_{k=1}^{\infty}) = \lim_{n\to\infty} \sup_k \sum_{j=n}^{\infty} |a_{jk}|.$$

Since $\chi(AS) \geq \chi(\{y_k\}_{k=1}^{\infty})$, we obtain the needed opposite inequality

$$\|A\|^{(\chi)} \geq \lim_{n\to\infty} \sup_k \sum_{j=n}^{\infty} |a_{jk}|,$$

which completes the proof of (4). **QED**

2.4.5. Corollary. *In order for an operator A, given by a matrix $(a_{ij})_{j,k=1}^{\infty}$, to be compact as an operator $l_1 \to l_1$ it is necessary and sufficient that*

$$\lim_{n\to\infty} \sup_k \sum_{j=n}^{\infty} |a_{jk}| = 0.$$

2.4.6. Example (computation of $\|A\|^{(\chi)}$ in the space c_0). Now let the operator A, given in the standard basis e_i (see **2.4.4**) by the matrix $(a_{ij})_{j,k=1}^{\infty}$, act from c_0 to c_0. We claim that

$$\|A\|^{(\chi)} = \overline{\lim_{j\to\infty}} \sum_{k=1}^{\infty} |a_{jk}|. \tag{5}$$

In fact, from formula (1) in **1.1.9** it follows that

$$\|A\|^{(\chi)} = \lim_{n\to\infty} \sup_{x\in S} \max_{j\ge n} \left| \sum_{k=1}^{\infty} a_{jk}\xi_k \right|,$$

where $x = (\xi_1, \xi_2, \dots) \in S$. Hence,

$$\|A\|^{(\chi)} \le \lim_{n\to\infty} \sup_{j\ge n} \sum_{k=1}^{\infty} |a_{jk}| = \overline{\lim_{j\to\infty}} \sum_{k=1}^{\infty} |a_{jk}|.$$

The opposite inequality can be established as follows. Pick sequences of positive integers $\{j_n\}_{n=1}^{\infty}$ and $\{m_n\}_{n=1}^{\infty}$ such that $j_n \ge n$ $(n = 1, 2, \dots)$ and

$$\lim_{n\to\infty} \sum_{k=1}^{\infty} |a_{j_n k}| = \overline{\lim_{j\to\infty}} \sum_{k=1}^{\infty} |a_{jk}|.$$

Let $x_n = \{\xi_k^n\}_{k=1}^{\infty} \in c_0$ be a sequence of vectors such that $|\xi_k^n| = 1$, $a_{j_n k}\xi_k^n = |a_{j_n k}|$ $(k = 1, \dots, m_n)$, and $\xi_k^n = 0$ for $k > m_k$.

Consider the set $M = \{Ax_n\}_{n=1}^{\infty}$. Since $\|x_n\| = 1$, we have, by **1.1.4**,

$$\|A\|^{(\chi)} \ge \chi(M) = \lim_{n\to\infty} \sup_p \max_{j\ge n} \left| \sum_{k=1}^{\infty} a_{jk}\xi_k^p \right|$$

$$\ge \lim_{n\to\infty} \left| \sum_{k=1}^{\infty} a_{j_n k}\xi_k^n \right| = \lim_{n\to\infty} \sum_{k=1}^{m_n} |a_{j_n k}| = \overline{\lim_{j\to\infty}} |a_{jk}|,$$

which completes the proof of (5).

2.4.7. Corollary. *In order for an operator A, given by a matrix $(a_{jk})_{j,k=1}^{\infty}$, to be compact as an operator $c_0 \to c_0$ it is necessary and sufficient that* $\lim_{j\to\infty} \sum_{k=1}^{\infty} |a_{jk}| = 0$.

2.4.8. Example (computation of $\|A\|^{(\chi)}$ in the space $C[0,1]$). Consider the operator $A: C[0,1] \to C[0,1]$ which acts according to the rule $(Ax)(t) = a(t)x(t)$, where $a : [0,1] \to \mathbf{R}$ is a continuous function. We claim that

$$\|A\|^{(\chi)} = \max_{t\in[0,1]} |a(t)| \;(= \|A\|). \tag{6}$$

In fact, since $\|A\| = \max_{t\in[0,1]} |a(t)|$, inequality (3) implies that

$$\|A\|^{(\chi)} \le \max_{t\in[0,1]} |a(t)|.$$

Let us prove the opposite inequality. Let t_0 be a point at which $|a(t)|$ attains its maximum and let $\{t_n\} \subset [0,1]$ be a monotone sequence that converges to t_0. Consider a sequence of continuous functions $\{x_n\}$ such that $\|x_n\| = 1$ and $x_n(t_n) = -x_n(t_{n-1}) = 1$. Clearly, the oscillation of the function Ax_n on the segment $[t_n, t_{n+1}]$ is not smaller than $|a(t_n)+a(t_{n+1})|$. Set $M = \{Ax_n\}_{n=1}^{\infty}$. Since $\|x_n\| = 1$, formula (2) in **1.1.10** yields

$$\|A\|^{(\chi)} \ge \chi(M) = \frac{1}{2} \lim_{\delta\to 0} \sup_n \max_{|t_1-t_2|\le\delta;\, t_1,t_2\in[0,1]} |a(t_1)x_n(t_1) - a(t_2)x_n(t_2)|$$

$$\ge \frac{1}{2} \lim_{n\to\infty} |a(t_n) + a(t_{n+1})| = |a(t_0)| = \max_{t\in[0,1]} |a(t),$$

which completes the proof of formula (6).

2.4.9. Remark. Since in examples **2.4.4**, **2.4.6**, and **2.4.8** we constructed sequences $\{x_n\}$ such that $\|A\|^{(\chi)} = \chi(\{Ax_n\})$ and $\chi(\{x_n\}) = 1$, we can say that the formulas for $\|A\|^{(\chi)}$ in l_1, c_0, and $C[0,1]$ define the $\overline{\chi}$-norm of the reduced operators in the indicated spaces ($\overline{\chi}$ is defined by formula (8) in **2.2.6**).

2.4.10. Other equivalent norms. Here we give a number of recipes for defining equivalent norms for operators $A^+ \in L(E_1^+, E_2^+)$.

As the first of such norm we consider $\|A\|^{(\alpha)} = \frac{1}{2}\alpha(AS)$, where S is the unit sphere in E_1. For the second norm we take $\|A\|^{(\beta)} = \beta(AS)$. Clearly, in view of the properties of the MNCs α and β, $\|\cdot\|^{(\alpha)}$ and $\|\cdot\|^{(\beta)}$ are seminorms on the set of operators $A \in L(E_1, E_2)$. From the inequalities established in **1.1.7** and **1.3.2** it follows that

$$\frac{1}{2}\|A\|^{(\chi)} \le \|A\|^{(\alpha)} \le \|A\|^{(\chi)} \tag{7}$$

and

$$\frac{1}{2}\|A\|^{(\beta)} \leq \|A\|^{(\chi)} \leq \|A\|^{(\beta)}. \tag{8}$$

This shows that the seminorms $\|A\|^{(\alpha)}$ and $\|A\|^{(\beta)}$ are equivalent to $\|A\|^{(\chi)}$.

Let us consider another norm, denoted $\|A\|_{(\lambda)}$ and defined as

$$\|A\|_{(\lambda)} = \inf_{\operatorname{codim} L<\infty} \|A|_L\|,$$

where $A|_L$ denotes the restriction of the operator A to the subspace L. Clearly, $\|A\|_{(\lambda)}$ is a seminorm. In the next section (see **2.5.5**) we will show that

$$\frac{1}{2}\|A\|^{(\chi)} \leq \|A\|_{(\lambda)} \leq 2\|A\|^{(\chi)}.$$

As an alternative characteristic of the "degree of noncompactness" of an operator $A \in L(E_1, E_2)$ one can take the distance from A to the subspace $V(E_1, E_2)$ of compact operators from E_1 to E_2. That distance, denoted here by $\|A\|_T$, is given by the formula

$$\|A\|_T = \inf\{\|A - K\| : K \in V(E_1, E_2)\}.$$

Clearly, $\|\cdot\|_T$ is a norm on the set of operators $A^+ \in L(E_1^+, E_2^+)$. It is also clear that $\|A\|^{(\chi)} \leq \|A\|_T$. Now suppose that in E_2 there is a basis $\{e_j\}_{j=1}^{\infty}$ and that the limit $b = \underline{\lim}_{n\to\infty}\|Q_n\|$ exists, where Q_n denotes the projector onto the closed linear span of the set $\{e_j\}_{j=n}^{\infty}$ parallel to the linear span of the set $\{e_j\}_{j=1}^{n-1}$. We claim that in this case $\|A\|_T \leq b\|A\|^{(\chi)}$. In fact,

$$\|A\|_T \leq \inf_n \|Q_n A\| = \inf_n \sup_{x\in S} \|Q_n Ax\| \leq \inf_n \left\{ \sup_{x\in S}(\|Q_n Ax - Q_n y_k(x)\| + \|Q_n y_k(x)\|) \right\},$$

where $y_k(x)$ designates the point of the $[\chi(AS)+\varepsilon]$-net $\{y_k\}_{k=1}^{N}$ of the set AS, the distance from which to Ax is smaller than $\chi(AS)+\varepsilon$. Thus,

$$\|A\|_T \leq \inf_n\{\|Q_n\|[\chi(AS)+\varepsilon] + \|Q_n y_k(x)\|\} \leq \underline{\lim}_{n\to\infty} \|Q_n\|[\chi(AS)+\varepsilon],$$

which completes the proof.

If the constant $b = \underline{\lim}_{n\to\infty}\|Q_n\| = 1$, as is the case, for example, if $E_2 = l_p$ with $p \geq 1$, or if $E_2 = c_0$, then

$$\|A\|^{(\chi)} = \|A\|_T. \tag{9}$$

As the example given below shows, in the general case formula (9) is not valid.

2.4.11. An example where $\|A\|^{(\chi)}$ does not coincide with $\|A\|_T$. Set $E = l_p \times c$ (the elements of E are pairs (x, y), $x \in l_p$, $y \in c$, with the norm $\|(x,y)\| = (\|x\|^2 + \|y\|^2)^{1/2}$) and consider the operator $A \in L(E)$ defined by

$$A(x,y) = (0, x). \tag{10}$$

Let us prove that $\|A\|_T = 1$. Since $\|A\| = 1$ and E has a basis, it suffices to show that $\|A - K\| \geq 1$ for all finite-rank operators K. Let $\tilde{e}_j = (e_j, 0)$, where $e_j = \{\delta_{jk}\}_{k=1}^{\infty}$ ($j = 1, 2, \ldots$; δ_{jk} is the Kronecker symbol); then $\|A\tilde{e}_j\| = 1$. The needed inequality $\|A - K\| \geq 1$ will follow if we show that $\lim_{j\to\infty} \|K\tilde{e}_j\| = 0$. It suffices to check this last relation for an arbitrary rank-one operator K, i.e., when $K\tilde{e}_j = f(e_j)z$, where $z \in E$, $f = \{f_j\}_{j=1}^{\infty}$, and $f \in l_q$ ($p^{-1} + q^{-1} = 1$). But in this case $\|K\tilde{e}_j\| = |f_j|\,\|z\| \to 0$ when $j \to \infty$.

Now let us estimate $\|A\|^{(\chi)}$. As is readily seen, $\|A\|^{(\chi)} = \chi(M)$, where $M \subset c$ denotes the set consisting of all sequences $y = \{y_j\}_{j=1}^{\infty}$ such that

$$\sum_{j=1}^{\infty} |y_j|^p \leq 1. \tag{11}$$

If condition (11) is satisfied, then, as is readily checked, the value of $\sup_{j,k} |y_j - y_k|$ is maximal only when two coordinates of y are different from zero and of opposite sign, and so

$$\sup_{j,k} |y_j - y_k| \leq \max_{0 \leq t \leq 1} [1 + (1 - t^p)^{1/p}] = 2^{1-1/p}.$$

Thus, for each $y \in M$ one can choose a stationary sequence $\kappa(y) = \{\kappa_j(y)\}_{j=1}^{\infty}$ such that $\sup_j |y_j - \kappa_j(y)| \leq 2^{-1/p}$. The set of stationary sequences $\{\kappa(y) : y \in M\}$ is a compact $2^{-1/p}$-net of M. Hence,

$$\|A\|^{(\chi)} \leq 2^{-1/p} < 1 = \|A\|_T.$$

2.5. THE MEASURE OF NONCOMPACTNESS OF THE CONJUGATE OPERATOR

It is know (see, for example [34]) that the norm of the conjugate (adjoint) of an operator coincides with the norm of the original operator, and that the conjugate of a compact operator is compact. It is therefore natural to expect that the MNC of a conjugate operator will coincide with the MNC of the original operator. However, as the examples given below will show, this is not true in general. The present section is devoted to establishing various relations between the MNCs of an operator and of its conjugate.

2.5.1. Theorem on the χ- and α-norms of the conjugate operator. *Let $A \in L(E_1, E_2)$. Then*

$$\frac{1}{2}\|A\|^{(\chi)} \leq \|A^*\|^{(\chi)} \leq 2\|A\|^{(\chi)} \tag{1}$$

and

$$\frac{1}{2}\|A\|^{(\alpha)} \leq \|A^*\|^{(\alpha)} \leq 2\|A\|^{(\alpha)} \tag{2}$$

(see **2.4.1** and **2.4.10** for notations).

Proof. Set $q = \|A\|^{(\chi)}$. Let S and S^* denote the unit spheres in the spaces E_1 and E_2^*, respectively, and let ε be an arbitrary positive number. Now let $\{y_j\}_{j=1}^m$ be some $(q + \varepsilon/3)$-net of the set AS and $a = \max_{1 \leq j \leq m} \|y_j\|$. Divide the segment $[-a, a]$ into n disjoint parts $\Lambda_1, \ldots, \Lambda_n$ of diameters smaller than $\varepsilon/3$. For any collection $k_1, \ldots, k_m$ of m positive integers smaller than n consider the (possibly empty) set $M(k_1, \ldots, k_m)$ consisting of the functionals $g \in S^*$ with the property that $g(y_j) \in \Lambda_{k_j}$ for $j = 1, \ldots, m$. Notice that $A^*S^* = \bigcup_{(k_1,\ldots,k_m)} A^*M(k_1, \ldots, k_m)$. Hence, in order to estimate $\alpha(A^*S^*)$ it suffices to estimate the diameter of the set $A^*M(k_1, \ldots, k_m)$. If $\phi_1, \phi_2 \in A^*M(k_1, \ldots, k_m)$, then there exist $g_1, g_2 \in M(k_1, \ldots, k_m)$ such that $\phi_1 = A^*g_1$, $\phi_2 = A^*g_2$. Given an arbitrary $x \in S$, pick elements y_p, $p = 1, \ldots, m$, such that $\|Ax - y_p\| \leq q+\varepsilon/3$. Then

$$|\phi_1(x) - \phi_2(x)| = |g_1(Ax) - g_2(Ax)|$$

$$\leq |g_1(Ax) - g_1(y_p)| + |g_1(y_p) - g_2(y_p)| + |g_2(y_p) - g_2(Ax)| \leq 2q+\varepsilon,$$

i.e., $\|\phi_1 - \phi_2\| \leq 2q+\varepsilon$, and consequently $\alpha(A^*S^*) \leq 2\|A\|^{(\chi)}$, or (in the notations of **2.4.10**) $\|A^*\|^{(\alpha)} \leq \|A\|^{(\chi)}$. Now using inequality (7) in **2.4.10** one obtains

$$\|A^*\|^{(\chi)} \leq 2\|A\|^{(\chi)} \tag{3}$$

and

$$\|A^*\|^{(\alpha)} \leq 2\|A\|^{(\alpha)}. \tag{4}$$

Next, replacing A by A^* in the preceding argument and denoting the unit sphere in E_1^{**} by S^{**} it is not hard to show that $\alpha(A^{**}S^{**}) \leq 2\|A^*\|^{(\chi)}$. The inclusion $A^{**}S^{**} \supset AS$ implies that $\alpha(AS) \leq 2\|A^*\|^{(\chi)}$. Therefore,

$$\|A\|^{(\alpha)} \leq \|A^*\|^{(\chi)},$$

or, by inequalities (7) in **2.4.10**,

$$\frac{1}{2}\|A\|^{(\chi)} \leq \|A^*\|^{(\chi)} \tag{5}$$

and

$$\frac{1}{2}\|A\|^{(\alpha)} \leq \|A^*\|^{(\alpha)}. \tag{6}$$

Combining (3), (5) and (4), (6) we obtain the claimed inequalities (1) and (2). **QED**

2.5.2. Theorem on the λ- and χ-norms. *Let $A \in L(E_1, E_2)$. Then*

$$\|A\|_{(\lambda)} = \|A^*\|^{(\chi)}. \tag{7}$$

(See **2.4.1** and **2.4.10** for notations.)

Proof. First let us establish the inequality

$$\|A\|_{(\lambda)} \leq \|A^*\|^{(\chi)}.$$

Let $q > \|A^*\|^{(\chi)}$. Denote the unit ball in E_2^* by B_2^* and let $\{f_i: f_i \in E_1^*, i = 1, \dots, m\}$ be a finite q-net of the set $A^*B_2^*$. Denote $L = \bigcap_{i=1}^m \operatorname{Ker} f_i$. Clearly, $\operatorname{codim} L < \infty$. We claim that $\|A|_L\| \leq q$. In fact, given an arbitrary $x_0 \in L$, pick a functional $f_0 \in E_2^*$ such that $\|f_0\| = 1$ and $\|Ax_0\| = f_0(Ax_0)$. By the definition of the conjugate operator, $f_0(Ax_0) = (A^*f_0)(x_0)$. But for A^*f_0 there is an f_i such that $A^*f_0 = f_i + \overline{f}$ and $\|\overline{f}\| \leq q$. Consequently,

$$\|Ax_0\| = (A^*f_0)(x_0) = f_i(x_0) + \overline{f}(x_0) \leq q\|x_0\|.$$

Thus, $\|A|_L\| \leq q$, which proves the needed inequality.

To prove the opposite inequality, pick some subspace L such that $\operatorname{codim} L < \infty$ and let $\|A|_L\| = q$, $f \in E_2^*, \|f\| = 1$. Consider A^*f as a functional on L. Then

$$\|(A^*f)|_L\| = \sup_{x \in L} |(A^*f)x|/\|x\| = \sup_{x \in L} |f(Ax)|/\|x\| \leq \|f\| \cdot \|A|_L\| = q.$$

By the Hahn-Banach theorem (see [107]), the functional $(A^*f)|_L$ can be extended to the entire space E_1 preserving its norm. Let f_1 be such an extension. Clearly, $A^*f = f_1 + f_2$, where f_2 belongs to $L^\perp = \{f: f \in E_1^*, \operatorname{Ker} f \supset L\}$ and $\|f_2\| \leq \|A^*\| + q$. Consequently, $A^*B_2^* \subset qB_1^* + M$, where B_1^* is the unit ball in E_1^* and M is some bounded precompact (since $\dim L^\perp < \infty$) subset of $L^\perp$. The last inclusion implies that $\|A\|_{(\lambda)} \geq \|A^*\|^{(\chi)}$ as needed. **QED**

2.5.3. Theorem on the χ-norm of the operator A^{**}. *For any operator $A \in L(E_1, E_2)$,*

$$\|A^{**}\|^{(\chi)} \leq \|A\|^{(\chi)}. \tag{8}$$

Proof. Let B_1 and B_2 denote the unit balls in the spaces E_1 and E_2, respectively. Take $q > \|A\|^{(\chi)}$ and let $\{y_i\}$ be a finite q-net of the set AB_1. Then $AB_1 \subset \bigcup_i B^i$, where $B^i = y_i + qB_2$. The ball B_j is weakly* dense in the unit ball B_j^{**} of the space E_j^{**}, $j = 1, 2$, and the operator A^{**} is weakly continuous (see [153]). Let the overline stand for weak* closure in the respective ambient space. Then

$$A^{**}B_1^{**} \subset \overline{AB_1} \subset \overline{\bigcup_i B^i} = \bigcup_i (B^i)^{**},$$

where $(B^i)^{**} = qB_2^{**} + x_i$. Inequality (8) follows from the above inclusion. **QED**

2.5.4. Corollary. *For any operator $A \in L(E_1, E_2)$,*

$$\|A\|^{(\chi)} \geq \|A^*\|_{(\lambda)} \tag{9}$$

and

$$\|A\|_{(\lambda)} = \|A^{**}\|_{(\lambda)}. \tag{10}$$

Proof. By Theorem 2.5.2, $\|A^*\|_{(\lambda)} = \|A^{**}\|^{(\chi)}$, and so (9) is a consequence (8). Since $E_j \subset E_j^{**}$, $j = 1, 2$, one has $\|A\|_{(\lambda)} \leq \|A^{**}\|_{(\lambda)}$. Now applying Theorem 2.5.2 and inequality (9), we conclude that

$$\|A\|_{(\lambda)} = \|A^*\|^{(\chi)} \geq \|A^{**}\|_{(\lambda)},$$

as needed. **QED**

2.5.5. Corollary. *For any operator $A \in L(E_1, E_2)$,*

$$\frac{1}{2}\|A\|^{(\chi)} \leq \|A\|_{(\lambda)} \leq 2\|A\|^{(\chi)} \tag{11}$$

and

$$\frac{1}{2}\|A\|_{(\lambda)} \leq \|A^*\|_{(\lambda)} \leq 2\|A\|_{(\lambda)}. \tag{12}$$

Proof. Inequalities (11) are obvious consequences of inequalities (1) and Theorem 2.5.2. To prove (12) it suffices to observe that inequalities (1) imply the following inequalities for the operator A^*:

$$\frac{1}{2}\|A^*\|^{(\chi)} \leq \|A^{**}\|^{(\chi)} \leq 2\|A^*\|^{(\chi)},$$

and then to apply Theorem 2.5.2. **QED**

2.5.6. Remarks on the sharpness of the constants. Let us give some examples which show that the constants in the inequalities (1) and (12) are sharp.

Consider the space E and the operator A defined in Example 2.4.11. As one can easily see, in this case $E^* = l_1 \times l_q$ (with $p^{-1}+q^{-1} = 1$) and A^* acts according to the rule

$$A^*(x, y) = (0, Ux),$$

where $x = (\xi_1, \xi_2, \dots)$ and $Ux = (\xi_2, \xi_3, \dots)$. It is readily verified that $\|A^*\|^{(\chi)} = \chi(M)$, where $M \subset l_q$ consists of all sequences $y = \{\eta_i\}_{i=1}^{\infty}$ such that $\sum_{i=1}^{\infty} |\eta_i| \leq 1$. Applying formula (1) of **1.1.9** for the computation of the Hausdorff MNC in the spaces l_p, we get $\chi(M) = 1$. As shown in Example 2.4.10, $\|A\|^{(\chi)} \leq 2^{-1/p}$. Consequently,

$$\|A^*\|^{(\chi)} \geq 2^{1/p}\|A\|^{(\chi)}. \tag{13}$$

Since p is an arbitrary real number in $(1, \infty)$, inequality (13) shows that the coefficient 2 in the inequalities (1) cannot be replaced by a smaller number. Now consider the operator $B = A^*$. Then $B^* = A^{**}$ and $\|B\|^{(\chi)} = 1$, $\|B^*\|^{(\chi)} \leq \|A\|^{(\chi)} \leq 2^{-1/p}$. Hence, $2^{-1/p}\|B\|^{(\chi)} \geq \|B^*\|^{(\chi)}$. Therefore, the coefficient 1/2 in (1) cannot be replaced by a larger number. The sharpness of the constants in the inequalities (1) is thus established.

It follows from Theorem 2.5.2 that

$$\|A\|_{(\lambda)} = \|A^*\|^{(\chi)} = 1, \quad \|A^*\|_{(\lambda)} = \|A^{**}\|^{(\chi)} \leq 2^{-1/p},$$

whence

$$2^{-1/p}\|A\|_{(\lambda)} \geq \|A^*\|_{(\lambda)}. \tag{14}$$

Since here p is an arbitrary number larger than one, the factor 1/2 in inequalities (12) cannot be made larger. Considering now the operator $B = A^*$ and using (10), we obtain

$$\|B^*\|_{(\lambda)} \geq 2^{1/p}\|B\|_{(\lambda)},$$

and therefore the coefficient 2 in inequality (12) cannot be made smaller.

2.5.7. The norm $\|A^*\|_T$. Since the conjugate of a compact operator is compact (see [34]), one has the obvious inequality

$$\|A^*\|_T \leq \|A\|_T$$

(concerning the notations, see **2.4.10**). Hence, in the case of reflexive spaces E_1, E_2 one has

$$\|A^*\|_T = \|A\|_T. \tag{15}$$

Let us give an example which shows that in the general case equality (15) does not hold.

Consider the operator A, given by formula (10) of **2.4.11**, as an operator acting in the space $E = l_1 \times c_0$. Clearly, $\|A\| = 1$. Using formula (1) of **1.1.9** it is readily checked that $\|A\|^{(\chi)} = 1$. Hence, $\|A\|_T = 1$. It is easy to see that $E^* = l_1 \times m$ and A^* acts as

$$A^*(x, y) = (0, x).$$

Arguing as in the analysis of Example 2.4.11, one can show that $\|A^*\|^{(\chi)} \leq 1/2$. From (1) it follows that

$$\|A^*\|^{(\chi)} \geq \frac{1}{2}\|A\|^{(\chi)},$$

and so $\|A^*\|^{(\chi)} = 1/2$. Now let K denote the rank-one operator acting in E^* according to the rule

$$K(x, y) = \left(0, z_0 \sum_{j=1}^{\infty} \xi_j\right),$$

where $x = (\xi_1, \xi_2, \dots)$ and $z_0 = (1/2, 1/2, \dots)$. Clearly,

$$\|(A^* - K)(x, y)\| = \left\| \left(0, \left\{\xi_k - \frac{1}{2}\sum_{j=1}^{\infty} \xi_j\right\}_{k=1}^{\infty}\right) \right\|$$

$$= \sup_k \left|\xi_k - \frac{1}{2}\sum_{j=1}^{\infty} \xi_j\right| \leq \frac{1}{2}\sum_{j=1}^{\infty} |\xi_j| \leq \frac{1}{2}\|(x, y)\|.$$

Consequently, $\|A^* - K\| \leq \frac{1}{2}$, and since $\|A^*\|_T \geq \|A^*\|^{(\chi)} = \frac{1}{2}$, we conclude that

$$\|A^*\|_T = \frac{1}{2} < 1 = \|A\|_T.$$

2.5.8. Remark. Since relation (9) of **2.4.10** holds for $E_2 = l_p$ $(p \geq 1)$, application of equality (15) to the case $E_1 = l_p$, $E_2 = l_q$ $(p, q > 1)$ gives

$$\|A^*\|^{(\chi)} = \|A\|^{(\chi)}. \tag{16}$$

Using formulas (4) of **2.4.4** and (5) of **2.4.6** it is not difficult to show that (16) also remains valid when $E_1 = E_2 = c_0$.

2.5.9. The Hilbert space case. Let $A \in L(H)$, where H is a Hilbert space. First, let us show that in this case

$$\|A^*\|_{(\lambda)} = \|A\|_{(\lambda)}. \tag{17}$$

Let L be a subspace of finite codimension in H and M its orthogonal complement. Denote by P_L and P_M the orthogonal projectors onto L and M. Clearly,

$$\|A|_L\| = \|AP_L\| = \|P_L A^*\|. \tag{18}$$

Consider the subspace $K = \mathrm{Ker}(P_M A^* P_L)$. Since K is the kernel of a finite-rank map, $\operatorname{codim} K < \infty$. The subspace $P_L K$ also has finite codimension. If $x \in K$, then $A^* P_L x \in L$. Hence, $A^* P_L|_K = P_L A^* P_L$, and consequently

$$\|P_L A^*\| \geq \|P_L A^* P_L\| = \|A^* P_L|_K\| = \|A^*|_{P_L K}\|.$$

In conjunction with (18) this yields the inequality

$$\|A\|_{(\lambda)} \geq \|A^*\|_{(\lambda)},$$

the application of which to the operator A^* completes the proof.

It follows from relation (17) that

$$\|A^*\|^{(\chi)} = \|A\|^{(\chi)}$$

for any operator $A \in L(H)$. In fact, by Theorem 2.5.2 and relation (17),

$$\|A^*\|^{(\chi)} = \|A\|_{(\lambda)} = \|A^*\|_{(\lambda)} = \|A^{**}\|^{(\chi)} = \|A\|^{(\chi)}.$$

2.6. THE FREDHOLM SPECTRUM OF A BOUNDED LINEAR OPERATOR

In this section, relying on the Fredholmness criterion for bounded linear operators established in Section 2.3, we study the set of all complex numbers λ for which the operator $\lambda I - A$ is not Fredholm. The boundaries of the annulus that contains this set can be easily calculated in terms of the so-called lower and upper ψ-norms of the operator A, where ψ is some normal MNC. In order to examine simultaneously the real and the complex cases, we need the notions of the complexification of a space and of an operator (see [78]).

2.6.1. Definition of the complexification. The *complexification of the real linear space* E is defined to be the complex linear space E_c whose elements are the pairs (x, y), $x, y \in E$, with the linear operations defined as follows:

$$(x_1, y_2)+(x_2, y_2) = (x_1+x_2, y_1+y_2),$$

$$(\mu+i\nu)(x, y) = (\mu x - \nu y, \mu y+\nu x).$$

If E is a Banach space, then E_c is equipped with the norm

$$\|(x, y)\| = \max_\theta \|x \cos\theta+y \sin\theta\|.$$

Between the real subspace $\hat{E} = \{(x, 0): x \in E\}$ of E_c and the space E one has the isometric isomorphism $(x, 0) \leftrightarrow x$. By convention, the complexification of a complex space E is taken to be E itself.

The *complexification of a linear operator* A acting in the real linear space E is defined to be the linear operator $A_c: E_c \to E_c$ acting as

$$A_c(x, y) = (Ax, Ay).$$

If E is Banach, then $\|A_c\| = \|A\|$. A linear operator $B : E_c \to E_c$ is said to be *real* if $B\hat{E} \subset \hat{E}$. In particular, the complexification A_c of any linear operator that acts in a real space is a real operator. By convention, the complexification of a linear operator A that acts in a complex space is taken to be A itself.

2.6.2. The spectrum of an operator. Let A be a linear operator in a Banach space E. The *spectrum* of A is defined to be the set $\sigma(A)$ of all $\lambda \in \mathbf{C}$ such that the operator $\lambda I_c - A_c$ does not have a bounded inverse defined on the whole space E_c.

The spectrum of any bounded operator acting in a (nonzero) Banach space is a nonempty bounded closed set (see [34]).

2.6.3. The spectral radius. Recall (see [34]) that the *spectral radius* of an operator $A \in L(E)$ is defined to be the number

$$R(A) = \sup\{|\lambda|: \lambda \in \sigma(A)\}.$$

For the spectral radius one has the following formula of I. M. Gel′fand (see [34]):

$$R(A) = \lim_{n\to\infty} \|A^n\|^{1/n}, \tag{1}$$

and also the inequality $R(A) \leq \|A\|$.

For any operator $A \in L(E)$ and any $\varepsilon > 0$ one can find a norm $\|\cdot\|_*$ on E (equivalent to the original norm) such that

$$R(A) \leq \|A\|_* \leq R(A)+\varepsilon.$$

For our purposes it will be convenient to introduce the notions of the lower norm and inner spectral radius of an operator $A \in L(E)$.

2.6.4. Definition. The *lower norm* of the operator $A \in L(E)$ is defined to be the number

$$\|A\|_\bullet = \begin{cases} \|A^{-1}\|^{-1}, & \text{if } A \text{ is bijective} \\ 0, & \text{otherwise.} \end{cases} \tag{2}$$

The *inner spectral radius* of A is

$$r(A) = \inf\{|\lambda| : \lambda \in \sigma(A)\}.$$

2.6.5. Lemma. *The following assertions hold true for any operator $A \in L(E)$:*
(a) $\|Ax\| \geq \|A\|_\bullet \|x\|$ *for all $x \in E$;*
(b) $r(A) > 0$ *if and only if A is bijective, in which case* $r(A) = [R(A^{-1})]^{-1}$*;*
c) $r(A) = \lim_{n\to\infty} \|A^n\|_\bullet^{1/n}$.

Proof. (a) If $\|A\|_\bullet = 0$, the assertion is obvious. If now $\|A\|_\bullet \neq 0$, then A has a bounded inverse A^{-1}, and so

$$\|A\|_\bullet \|x\| = \|A^{-1}\|^{-1}\|A^{-1}Ax\| \leq \|A^{-1}\|^{-1}\|A^{-1}\|_\bullet \|Ax\| = \|Ax\|.$$

(b) Since the spectrum $\sigma(A)$ is closed, $r(A) > 0$ if and only $0 \notin \sigma(A)$, i.e., if and only if the operator $0I_c - A_c = -A_c$ is bijective. The latter obviously holds if and only if A is bijective. Next, it is readily seen that if A is bijective, then $\lambda \in \sigma(A)$ if and only if $\lambda^{-1} \in \sigma(A^{-1})$, and so

$$r(A) = \inf\{|\lambda| : \lambda \in \sigma(A)\} = [\sup\{|\lambda| : \lambda \in \sigma(A)\}]^{-1}$$

$$= [\sup\{|\mu| : \mu \in \sigma(A^{-1})\}]^{-1} = [R(A^{-1})]^{-1}.$$

(c) If A is not bijective, then clearly the same is true for all the powers A^n, and so $\|A^n\|_\bullet = 0$ for all n and the right-hand side of the equality to be proven is equal to zero. Obviously, in this case $r(A) = 0$. If now A is bijective, then using the equality just established and formula (1) we obtain

$$r(A) = [R(A^{-1})]^{-1} = \big(\lim_{n\to\infty} \|(A^{-1})^n\|^{1/n}\big)^{-1}$$

$$= \lim_{n\to\infty} \big(\|(A^{-1})^n\|^{-1}\big)^{1/n} = \lim_{n\to\infty} \|A^n\|^{1/n}. \quad \textbf{QED}$$

2.6.6. Definition of the lower ψ-norm. Let E be a Banach space with a normal MNC ψ and let $A \in L(E)$. The number

$$\|A\|_\psi = \inf\{\psi(AX): X \in BE, \psi(X) = 1\}$$

is called the *lower ψ-norm* of the operator A.

The ψ-norm $\|A\|^\psi$ of the operator A, defined in **2.3.5**, will be also referred to as the *upper ψ-norm* of A.

The upper ψ-norm may be equal to infinity, and the lower ψ-norm may be equal to zero. To handle these situations, we make the folowing conventions: $0 \cdot \infty = 0$, and, if a is a positive number, $a/0 = \infty$, $a/\infty = 0$, $a \cdot \infty = \infty$, $a + \infty = \infty$.

Some easily verifiable properties of the upper and lower ψ-norms of operators are listed in the following lemma.

2.6.7. Lemma. *Let A and A_1 be linear operators acting in a Banach space E with a normal* MNC ψ. *Then:*

(a) *A is (ψ,ψ)-bounded above if and only if $\|A\|^\psi < \infty$; in this case A is $(\|A\|^\psi, \psi, \psi)$-bounded above* (see **2.3.4**)*;*

(b) *A is (ψ,ψ)-bounded below if and only if $\|A\|_\psi > 0$; in this case A is $(\|A\|_\psi, \psi, \psi)$-bounded below;*

(c) *$\|A\|^\psi = \|A^+\|^{\psi^+}$, where $\|A^+\|^{\psi^+}$ is the upper norm of the operator A^+ in the Banach space (E^+, ψ^+);*

(d) *$\|A\|_\psi = \|A^+\|_{\psi^+} = \big[\|(A^+)^{-1}\|^{\psi^+}\big]^{-1}$, where $\|A^+\|_{\psi^+}$ is the lower norm of the operator A^+ in the Banach space (E^+, ψ^+);*

(e) *$\|A + A_1\|^\psi \le \|A\|^\psi + \|A_1\|^\psi$;*

(f) *$\|\lambda A\|^\psi = |\lambda|\, \|A\|^\psi$ for all $\lambda \in \mathbf{C}$;*

(g) *$\|\lambda A\|_\psi = |\lambda|\, \|A\|_\psi$ for all $\lambda \in \mathbf{C}$;*

(h) *$\|AA_1\|^\psi \le \|A\|^\psi \|A_1\|^\psi$;*

(i) $\|AA_1\|_\psi \geq \|A\|_\psi \|A_1\|_\psi$;

(j) $\|A\|^\psi = 0$ *if and only if* A *maps any bounded set* $\Omega \subset E$ *into a relatively compact set.*

2.6.8. Definition. Let E be a real or complex Banach space and let $A \in L(E)$. Recall that a point $\lambda \in \mathbf{C}$ is called a *Φ-point* of the operator A if the operator $\lambda I_c - A_c$ is Fredholm. The set of all Φ-points of A is denoted by $\Phi(A)$. The complement of $\Phi(A)$ in $\mathbf{C}$ is called the *Fredholm spectrum* of the operator A and is denoted by $\sigma_\Phi(A)$. The *outer* and *inner Fredholm radii* of the operator A are defined to be the numbers

$$R_\Phi(A) = \sup\{|\lambda| : \lambda \in \sigma_\Phi(A)\}$$

(with the convention that $\sup \emptyset = 0$), and respectively

$$r_\Phi(A) = \inf\{|\lambda| : \lambda \in \sigma_\Phi(A)\}.$$

Our immediate goal is to study the properties of the Fredholm spectrum $\sigma_\Phi(A)$ with the help of Theorem 2.3.3. To this end we need the following two lemmas concerning properties of the operators $(A_c)^+$, $(A^+)_c$, and A_c.

2.6.9. Lemma. *There exists an isomorphism* $\mathfrak{J} : (E_c)^+ \to (E^+)_c$ *such that*

$$\mathfrak{J}^{-1}(A^+)_c \mathfrak{J} = (A_c)^+$$

for any operator $A \in L(E)$.

Proof. Let $\mathcal{Z} \in (E_c)^+$, $\mathcal{Z} = Z + KE_c$, where $Z = ((x_1, y_1), (x_2, y_2), \ldots)$. Set $\mathfrak{J}\mathcal{Z} = (\mathcal{X}, \mathcal{Y})$, where $\mathcal{X} = X + KE$, $\mathcal{Y} = Y + KE$, $X = (x_1, x_2, \ldots), Y = (y_1, y_2, \ldots)$. It is readily verified that the mapping $\mathfrak{J}$ is well defined, linear, and bijective. Also, it is a straightforward matter to check that

$$\mathfrak{J}^{-1}(A^+)_c \mathfrak{J}\mathcal{Z} = ((Ax_1, Ay_1), (Ax_2, Ay_2), \ldots) + KE_c = (A_c)^+ \mathcal{Z}. \quad \textbf{QED}$$

2.6.10. Lemma. *Let* E *be a Banach space and* $A \in L(E)$. *Then*

(a) *the complexification* E_{1c} *of any subspace* E_1 *of* E *is a subspace of* E_c, *and if* $\dim E_{1c} < \infty$ *or* $\dim E_1 < \infty$, *then* $\dim E_{1c} = \dim E_1$;

(b) *any subspace* E_2 *of* E_c *is the complexification of some subspace* E_1 *of* E;

(c) *the operator* A_c *is Fredholm if and only* A *is Fredholm; in this case,* $\operatorname{ind} A_c = \operatorname{ind} A$.

Proof. (a) Clearly, E_{1c} is a subspace of E_c. Next, it is readily verified that if the vectors x_i, $i = 1, 2, \ldots, n$ form a basis of the space E_1, then the vectors $(x_i, 0)$ form a basis of the (complex) space E_{1c}. Hence, $\dim E_1 < \infty$ implies $\dim E_{1c} = \dim E_1$.

Suppose now that $\dim E_{1c} < \infty$ and z_i, $i = 1, 2, \ldots, n$ is a basis of E_{1c}. Then obviously $\dim E_1 \leq 2n$, and, by what we proved above, $\dim E_{1c} = \dim E_1$.

(b) Set

$$\operatorname{Re} E_2 = \{x \in E \colon \exists y \in E\ [(x, y) \in E_2]\},$$

$$\operatorname{Im} E_2 = \{y \in E \colon \exists x \in E\ [(x, y) \in E_2]\}.$$

Clearly, $\operatorname{Re} E_2$ and $\operatorname{Im} E_2$ are subspaces of E. We claim that $\operatorname{Re} E_2 = \operatorname{Im} E_2$. In fact, if $x \in \operatorname{Re} E_2$, i.e., $(x, y) \in E_2$ for some y, then $i(x, y) = (-y, x) \in E_2$, and hence $x \in \operatorname{Im} E_2$. Conversely, if $x \in \operatorname{Im} E_2$, i.e., $(y, x) \in E_2$ for some y, then $-i(y, x) = (x, -y) \in E_2$, i.e., $x \in \operatorname{Re} E_2$. The proved equality implies that $E_2 = E_{1c}$, where $E_1 = \operatorname{Re} E_2 = \operatorname{Im} E_2$.

(c) Let $N(A)$ [resp. $N(A_c)$] be the kernel of the operator A [resp. A_c]. Obviously, $N(A_c) = [N(A)]_c$. From this and assertion (a) it follows that if one of the subspaces $N(A), N(A_c)$ is finite-dimensional, then so is the other, and then $\dim N(A_c) = \dim N(A)$.

To show that the range AE is closed if and only if $A_c E_c$ is so, it suffices to remark, first, that $A_c E_c = (AE)_c$ and, second, that the convergence of a sequence (x_n, y_n) to (x, y) in E_c is equivalent to the convergence of x_n to x and of y_n to y in E.

Finally, let us show that $\operatorname{codim} A_c E_c < \infty$ if and only if $\operatorname{codim} AE < \infty$, in which case $\operatorname{codim} A_c E_c = \operatorname{codim} AE$. This will complete the proof of the last assertion of the lemma. We have

$$\operatorname{codim} A_c E_c < \infty \iff E_c = (AE)_c \oplus E_2 \ (\dim E_2 < \infty)$$

$$\iff E = AE \oplus E_1 \ (E_{1c} = E_2) \iff \operatorname{codim} AE = \dim E_1 = \dim E_2. \quad \textbf{QED}$$

2.6.11. Theorem on the Fredhom spectrum. *The following assertions hold true for any operator $A \in L(E)$:*

(a) $\sigma_\Phi(A) = \sigma(A^+)$;

(b) *$\sigma_\Phi(A)$ is a closed bounded set, and $\sigma_\Phi(A) = \emptyset$ if and only if* $\dim E < \infty$;

(c) $\sigma_\Phi(A) \subset \sigma(A)$;

(d) *$r_\Phi(A) \leq R_\Phi(A)$ if and only if E is infinite-dimensional;*

(e) *$r_\Phi(A) > 0$ if and only if A is Fredholm;*

(f) *if $\lambda \in \mathbf{C}$ and $|\lambda| > R_\Phi(A)$, then $\lambda I - A$ is a Fredholm operator of index zero;*

(g) *if $\lambda \in \mathbf{C}$ and $|\lambda| < r_\Phi(A)$, then $\lambda I - A$ is a Fredholm operator and* $\operatorname{ind}(\lambda I - A) = \operatorname{ind} A$;

(h) *if the operator A is bounded above with respect to the normal* MNC ψ, *then*

$$R_\Phi(A) = \lim_{n\to\infty} (\|A^n\|^\psi)^{1/n};$$

(i) *if the operator A is bounded below with respect to the normal* MNC ψ, *then*

$$r_\Phi(A) = \lim_{n\to\infty} (\|A^n\|_\psi)^{1/n};$$

(j) $R_\Phi(A) = \inf\{\|A\|^\psi : \psi$ is a normal MNC in $E\}$;

(k) $r_\Phi(A) = \sup\{\|A\|_\psi : \psi$ is a normal MNC in $E\}$;

(l) *if $E = E_1 \oplus E_2$ and the subspaces E_1 and E_2 are invariant under the operator A, then*

$$\sigma_\Phi(A) = \sigma_\Phi(A_1) \cup \sigma_\Phi(A_2),$$

where A_i is the restriction of A to E_i, $i = 1, 2$.

Proof. By definition, $\lambda \in \sigma_\Phi(A)$ if and only if the operator $\lambda I_c - A_c$ is not Fredholm, i.e., by Theorem 2.3.3, if and only if the operator $(\lambda I_c - A_c)^+$ is not bijective. Using Lemmas 2.3.1 and 2.6.9, we further deduce that $\lambda \in \sigma_\Phi(A)$ if and only if the operator $\mathfrak{I}^{-1}[\lambda(I^+)_c - (A^+)_c]\mathfrak{I}$ is not bijective, which in turn is true if and only if $\lambda \in \sigma(A^+)$.

(b) This assertion is a straightforward consequence of (a) and of the fact that $E^+ = \{0\}$ if and only if $\dim E < \infty$.

(c) If $\lambda \notin \sigma(A)$, then the operator $\lambda I_c - A_c$ is bijective, and hence Fredholm; but then $\lambda \notin \sigma_\Phi(A)$.

(d) As established in (b), $\sigma_\Phi(A) \neq \emptyset$ if and only if E is infinite-dimensional. If $\lambda_0 \in \sigma(A)$, then clearly $r_\Phi(A) \leq |\lambda_0| \leq R_\Phi(A)$. If, however, $\dim E < \infty$, then $r_\Phi(A) = \infty$ and $R_\Phi(A) = 0$.

(e) Since the Fredholm spectrum is closed, $r_\Phi(A) > 0$ if and only if $0 \notin \sigma_\Phi(A)$, i.e., if and only if the operator A_c if Fredholm; the latter in turn holds if and only if A itself is Fredholm (see **2.6.10**).

(f) Clearly, all complex numbers λ satisfying the inequality $|\lambda| > R_\Phi(A)$ lie in the same connected component of the set $\Phi(A)$, and in that component there are points that do not belong to $\sigma(A)$ (because the spectrum is bounded). But then it follows from **2.1.2** that for any such λ the operator $\lambda I_c - A_c$ is Fredholm of index zero. The needed assertion now follows upon applying Lemma 2.6.10.

(g) If $|\lambda| < r_\Phi(A)$, then $t\lambda \notin \sigma_\Phi(A)$ for all $t \in [0, 1]$. Consequently, the operator $t\lambda I_c - A_c$ is Fredholm; moreover, by **2.1.2**, its index does not depend on t, i.e., in particular, $\operatorname{ind}(\lambda I_c - A_c) = \operatorname{ind}(-A_c) = \operatorname{ind} A_c$. Applying again Lemma 2.6.10, we obtain the needed assertion.

Notice that (a) implies the relations $R_\Phi(A) = R(A^+)$ and $r_\Phi(A) = r(A^+)$. Consequently, assertions (h) to (k) follow immediately from the corresponding properties of spectral radii and the relations $\|A\|^\psi = \|A^+\|^{\psi^+}$, $\|A\|_\psi = \|A^+\|_{\psi^+}$.

(l) In view of (a), it suffices to verify that $\sigma(A^+) = \sigma(A_1^+) \cup \sigma(A_2^+)$. Clearly, the space E^+ is representable as $E^+ = E_1' \oplus E_2'$, where E_i' is the subspace consisting of the elements $X+KE$ with $X \in BE_i$, $i = 1, 2$. Next, since $AE_i \subset E_i$, one also has $A^+E_i' \subset E_i'$, $i = 1, 2$. Hence, $\sigma(A^+) = \sigma(A_1') \cup \sigma(A_2')$. It remains to observe that E_i^+ is isomorphic to E_i' (via $X+KE_i \longleftrightarrow X+KE$), and the operator A_i^+ is similar to A_i', so that $\sigma(A_i^+) = \sigma(A_i')$, $i = 1, 2$. **QED**

2.6.12. A representation theorem for bounded linear operators. *For any operator $A \in L(E)$ and any $\varepsilon > 0$ one can find a representation $A = A_1 + A_2$ such that A_2 is finite-rank and $R(A_1) < R_\Phi(A)+\varepsilon$.*

Proof. In the complex plane **C** consider the disc K of radius $R_\Phi(A)$ centered at zero. The exterior of K lies in a single connected component of the set $\Phi(A)$, in which there are regular points of the operator A. By **2.1.4**, in the exterior of K there are only isolated points of the spectrum of A. Hence, one can find a circle Γ of radius r ($r < R_\Phi(A)+\varepsilon$) centered at zero , such that on Γ there are no points of $\sigma(A)$. Orienting Γ anticlockwise, consider the operator

$$P_\Gamma = \frac{1}{2\pi i} \int_\Gamma (\lambda I_c - A_c)^{-1} d\lambda.$$

Since the exterior of the disc of radius r centered at zero may contain only finitely many points of the spectrum, all of which lie in $\Phi(A)$, the projector $I_c - P_\Gamma$ maps E_c onto a finite-dimensional subspace (see [54, Theorems 4.1, 4.2]). Clearly,

$$A_c = A_c P_\Gamma + A_c(I_c - P_\Gamma).$$

If the space E is complex (and hence $E_c = E, A_c = A$), the proof ends here, since the operator $A_c(I_c - P_\Gamma)$ has finite rank, and the spectral radius of $A_c P_\Gamma$, equal to the spectral radius of the restriction of A_c to $P_\Gamma E_c$, is smaller than $R_\Phi(A)+\varepsilon$.

If now E is real, it suffices to show that P_Γ is a real operator (see **2.6.1**). Indeed, in that case the projector $I_c - P_\Gamma$ is also real, and upon denoting the restrictions of $A_c P_\Gamma$ and $A_c(I_c - P_\Gamma)$ to E (more precisely, to $\hat{E}$) by A_1 and A_2, respectively, we obtain the sought-for representation.

Put $\lambda = re^{i\theta}$. Then

$$P_\Gamma = \frac{r}{2\pi} \int_0^{2\pi} (re^{i\theta} I_c - A_c)^{-1} e^{i\theta} d\theta$$

$$= \frac{r}{2\pi}\int_0^{\pi}(rI_c - e^{-i\theta}A_c)^{-1}d\theta + \frac{r}{2\pi}\int_{\pi}^{2\pi}(rI_c - e^{i\theta}A_c)^{-1}d\theta,$$

or

$$P_\Gamma = \frac{r}{2\pi}\int_0^{\pi}[(rI_c - e^{-i\theta}A_c)^{-1} + (rI_c - e^{i\theta}A_c)^{-1}]d\theta. \tag{3}$$

Let us show that the integrand is a real operator, which will imply that P_Γ is real, as needed. Let $(rI_c - e^{-i\theta}A_c)^{-1}(x,0) = (u,v)$, i.e., $(rI_c - e^{-i\theta}A_c)(u,v) = (x,0)$. The last equation is equivalent to the following system:

$$ru - (\cos\theta)Au - (\sin\theta)Av = x,$$

$$rv - (\cos\theta)Av - (\sin\theta)Au = 0.$$

Notice that if in this system one replaces v by $-v$ and θ by $-\theta$, then both equations are preserved. This means that $(rI_c - e^{i\theta}A_c)(u,-v) = (x,0)$, i.e., $(rI_c - e^{i\theta}A_c)^{-1}(x,0) = (u,-v)$. Thus, we showed that on applying the two operator terms appearing in the integrand of (3) to the real vector $(x,0)$ one obtains the complex-conjugate vectors (u,v) and $(u,-v)$; the whole sum is therefore equal to the real vector $(2u,0)$. This shows that the operator P_Γ is real and completes the proof of the theorem. **QED**

2.6.13. Corollary. *Suppose that in the Banach space E there is given a normal* MNC ψ. *Let $A \in L(E)$ and $\|A\|^\psi < 1$. Then A admits a representation*

$$A = A_1 + A_2,$$

where A_2 is a finite-rank operator and the spectral radius of A_1 is smaller than one.

This assertion provides a complete characterization of the linear condensing maps; indeed, conversely, if A admits a representation of the indicated type, then in an equivalent norm A_1 is contractive, and then A is condensing with respect to the corresponding Hausdorff MNC.

2.7. NORMAL MEASURES OF NONCOMPACTNESS AND PERTURBATION THEORY FOR LINEAR OPERATORS

If a sequence of bounded linear operators A_n in an infinite-dimensional Banach space converges strongly to a bounded operator A_∞, then, in general, one cannot guarantee that for sufficiently large n the spectra of the operators A_n and A_∞ are close. However, there are well-known cases (see [79, Chapter VIII, Sec. 4]) where for some eigenvalues λ_∞ of A_∞ any operator A_n with sufficently large n has in a neighborhood of λ_∞ a set of eigenvalues $\{\lambda_n^1, \dots, \lambda_n^{k_n}\}$, the total multiplicity (i.e., the sum of the dimensions of the corresponding eigenspaces) of which is equal to the multiplicity of λ_∞. Such eigenvalues are termed *stable* (see [79]) and, as shown in [175] and [79], the stability of an eigenvalue λ_∞ is usually connected with the strong convergence of the resolvents of the operators A_n in a neighborhood of λ_∞.

In the present section we formulate, in terms of normal MNCs, necessary and sufficient conditions for the closeness of the parts of the spectra of the operators A_n and A_∞ lying outside some disc, and of the Riesz projectors corresponding to these parts of the spectra; these conditions also guarantee the strong convergence of the resolvents.

2.7.1. The main conditions. We begin with a situation that is frequently encountered in applications. Suppose given linear operators $A(\xi)$, acting in a complex Banach space E, and depending on a parameter ξ that takes values in a metric space Ξ. Let ξ_∞ be a limit point of the space Ξ and suppose that $A(\xi)$ is strongly continuous at ξ_∞. Further, suppose that A, regarded as a function of the two variables x and ξ, is jointly condensing (see **1.5.5**) with constant q with respect to a monotone, algebraically semi-additive, semi-homogeneous, regular MNC ϕ. Suppose $\xi_n \to \xi_\infty$ as $n \to \infty$. Let us investigate the properties of the sequence of operators $A_n = A(\xi_n)$, $n = 1, 2, \dots, \infty$.

Recall that the MNC ϕ defines a normal MNC $\overline{\phi}$ in BE by the rule $\overline{\phi}(X) = \phi(\{x_n\})$, where $\{x_n\}$ denotes the set of elements of the sequence X. In view of the monotonicity of ϕ, for any subsequence $\{x_{n_k}\}$ of X, regarded as an independent sequence Y, one has $\overline{\phi}(Y) \le \overline{\phi}(X)$. The normal MNCs with this last property will be called *monotone*.

Another consequence of the monotonicity of ϕ is that

$$\overline{\phi}(A_kX) = \phi(\{A_kx_n\}) \le \phi(\bigcup_k \{A_kx_n\}) \le q\phi(\{x_n\}) = q\overline{\phi}(X),$$

i.e., $\|A_k\|^{\overline{\phi}} \le q$ for all $k = 1, 2, \dots, \infty$. Hence, by the strong continuity of $A(\xi)$ in ξ and Theorem 2.6.11, the sequence of operators A_n satisfies the following condition.

(A) *The operators A_n converge strongly to a bounded linear operator A_∞ as $n \to \infty$, and outside a closed disc K of radius q centered at zero each A_n has only isolated points of the spectrum, each of which can only be an eigenvalue of finite multiplicity.*

The sequence of operators A_n enjoys one supplementary property, which we isolate as an independent definition.

2.7.2. Jointly condensing sequences of operators. We say that the sequence of bounded linear operators A_n is *jointly condensing with constant* q with respect to the normal MNC ψ if for any bounded sequence $X = \{x_n\}$,

$$\psi(Y) \leq q\psi(X), \tag{1}$$

where $Y = \{A_n x_n\}$.

Under the conditions considered in **2.7.1** inequality (1) is obviously satisfied, since

$$\overline{\phi}(Y) = \phi(\{A_n x_n\}) \leq \phi[A(\Xi)X] \leq q\phi(X) = q\overline{\phi}(X).$$

The monotonicity of ϕ implies an additional property of the sequence $\{A_n\}$: any of its subsequences $\{A_{n_k}\}$, regarded as an independent sequence, is jointly condensing, with the same constant as $\{A_n\}$, with respect to the normal MNC $\overline{\phi}$. Sequences of operators that enjoy this last property will be said to be *regularly jointly condensing.*

In the case where $q = 0$ we say that the sequence $\{A_n\}$ is *jointly compact.* Clearly, in order that $\{A_n\}$ be jointly compact it is necessary and sufficient that the sequence $\{A_n x_n\}$ be totally bounded for any bounded sequence $\{x_n\}$.

Next we consider an example of jointly compact sequence that will be important in the ensuing analysis.

2.7.3. Example. *Let $\{P_n\}$ be a sequence of finite-dimensional projectors in a Banach space E and let P_n be strongly convergent to a finite-dimensional projector P_∞. Then in order that $\{P_n\}$ be jointly compact it is necessary and sufficient that* $\dim P_n = \dim P_\infty$ *for all sufficiently large n.*

Indeed, let $E_n = P_n E$, $n = 1, 2, \ldots, \infty$. First, notice that if the vectors $e_1, \ldots, e_p \in E_\infty$ are linearly independent, then so are the vectors $P_n e_1, \ldots, P_n e_p$ for n sufficiently large. In fact, assuming the contrary, one can find a sequence $n_k \to \infty$ and collections $\overline{\alpha}_k = (\alpha_k^1, \alpha_k^2, \ldots, \alpha_k^p)$, such that $\|\overline{\alpha}_k\| = \max_{1 \leq i \leq p} |\alpha_k^i| = 1$ and

$$\alpha_k^1 P_{n_k} e_1 + \alpha_k^2 P_{n_k} e_2 + \ldots + \alpha_k^p P_{n_k} e_p = 0.$$

With no loss of generality we may assume that $\overline{\alpha}_k \to \overline{\alpha}_\infty$. Then $\|\overline{\alpha}_\infty\| = 1$ and $\sum_{i=1}^p \alpha_\infty^i e_i = 0$, which contradicts the linear independence of the vectors $e_1, \ldots, e_p$. Thus, we showed that the strong convergence of the projectors implies the inequality

$$\dim P_n \geq \dim P_\infty$$

for all sufficiently large n.

Let us prove the necessity of our condition. Consider the subspaces $E'_n = P_n E_\infty$. We claim that $E_n = E'_n$ for all sufficiently large n. In fact, assuming the contrary, there is a sequence of numbers $n_k \to \infty$ such that $E'_{n_k} \neq E_{n_k}$. Then there are vectors $x_{n_k} \in E_{n_k}$ such that $\|x_{n_k}\| = 1$ and

$$\inf\{\|x_{n_k} - y\|: y \in E'_{n_k}\} \geq \frac{1}{2}. \tag{2}$$

Since $\{x_{n_k}\} = \{P_{n_k} x_{n_k}\}$, the sequence $\{x_{n_k}\}$ is totally bounded. With no loss of generality we may assume that $x_{n_k} \to x_\infty$ and hence $P_{n_k} x_{n_k} \to P_\infty x_\infty$. Therefore,

$$\|x_{n_k} - P_{n_k} P_\infty x_\infty\| \to 0,$$

and since $P_{n_k} P_\infty x_\infty \in E'_{n_k}$, this contradicts inequality (2). Thus, $\dim E_n \leq \dim E_\infty$, as needed.

We now turn to the sufficiency part. Let $\{x_n\}$ be a bounded sequence of elements of E and $y_n = P_n x_n$. If $e_1, \dots, e_p, \dots$ is a basis in E_∞, then, as we remarked above, for all sufficiently large n the vectors $P_n e_1, \dots, P_n e_p, \dots$ are linearly independent and hence, by the assumption that $\dim P_n = \dim P_\infty$, they also form a basis in the space E_n. Therefore, for such n's, $y_n = \sum_{i=1}^p \alpha_n^i P_n e_i$. We claim that the numbers $\max_{1 \leq i \leq p} |\alpha_n^i| = \|\overline{\alpha}_n\|$ are uniformly bounded in n. In fact, assuming the contrary, there are sequences $n_k \to \infty$, α_{n_k} and $\beta_{n_k} = \overline{\alpha}_{n_k} / \|\overline{\alpha}_{n_k}\|$, and $y_{n_k} = P_{n_k} x_{n_k} / \|\overline{\alpha}_{n_k}\|$, such that $\|\overline{\alpha}_{n_k}\| \to \infty$, $\|\beta_{n_k}\| = 1$, $y_{n_k} \to 0$, and

$$y_{n_k} = \sum_{i=1}^{p} \beta_{n_k}^i P_{n_k} e_i. \tag{3}$$

With no loss of generality we may assume that $\overline{\beta}_{n_k} \to \overline{\beta}_\infty$, and then $\|\overline{\beta}_\infty\| = 1$. Letting $n_k \to \infty$ in (3) we obtain $\sum_{i=1}^\infty \beta_\infty^i e_i = 0$, which contradicts the linear independence of the vectors $e_1, \dots, e_p$. To complete the proof of the sufficiency of our condition it remains to observe that the set

$$\{y: y = \sum_{i=1}^{p} \alpha^i P_n e_i,\ \|\overline{\alpha}\| = \max_i |\alpha^i| \leq r < \infty\}$$

is totally bounded.

We next turn directly to the perturbation theory for regularly jointly condensing sequences of operators.

2.7.4. Theorem on the spectrum. *Suppose that the sequence of bounded operators A_n, acting in a complex Banach space E, satisfies condition* (A) *of* **2.7.2** *and is regularly jointly condensing with constant q with respect to a monotone normal* MNC ψ. *Then:*

(a) *If the operator* A_∞ *has no eigenvalues in some closed set* $Z \subset \mathbf{C} \setminus K(0,q)$ *(where* $K(0,q)$ *denotes the closed disc of radius* q *centered at zero), then for sufficiently large* n, $\sigma(A_n) \cap Z = \emptyset$.

(b) *If the operator* A_∞ *has an eigenvalue* $\lambda_\infty \in \mathbf{C} \setminus K(0,q)$, *then for any* $\mu > 0$ *there is an* n_0 *such that for* $n \geq n_0$ *the operator* A_n *has at least one eigenvalue* λ_n *satisfying* $|\lambda_n - \lambda_\infty| < \mu$.

(c) *If the condition of assertion* (b) *is satisfied, and if* $K(\lambda_\infty, r)$ *is any closed disc such that*

$$K(\lambda_\infty, r) \cap K(0,q) = \emptyset, \quad K(\lambda_\infty, r) \cap \sigma(A_\infty) = \{\lambda_\infty\},$$

then for sufficiently large n *the Riesz projector* P_n *for* A_n *corresponding to the boundary of* $K(\lambda_\infty, r)$ *is defined. Moreover, the sequence* $\{P_n\}$ *is jointly compact and* $\dim P_n = \dim P_\infty$ *for all suficiently large* n.

Before embarking on the proof of assertion (a), let us establish the following lemma.

2.7.5. Lemma. *Suppose the assumptions of Theorem 2.7.4 are in force and the bounded sequence* $\{x_p\}$ *is such that*

$$\lambda_p x_p = A_{n_p} x_p + y_p, \tag{4}$$

where $y_p \to y_0$, *the* λ_p*'s belong to a bounded closed set* $Z_1 \subset \mathbf{C} \setminus K(0,q)$, *and* $\{A_{n_p}\}$ *is a subsequence of* $\{A_n\}$. *Then the sequence* $\{x_n\}$ *it totally bounded and there are a* $\lambda' \in Z_1$ *and a vector* x_0 *such that*

$$\lambda' x_0 = A_\infty x_0 + y_0; \tag{5}$$

moreover, λ' *is a limit point of the sequence* $\{\lambda_p\}$ *and* x_0 *is a limit point of the sequence* $\{x_p\}$.

Proof. Let us show that from any subsequence $X = \{x_{p_k}\}$ of $\{x_p\}$ one can extract a convergent subsequence. Set $r = \inf\{|\lambda| : \lambda \in Z_1\}$. Then $r > q$. With no loss of generality we may assume that the sequence $\{\lambda_{p_k}\}$ converges to some $\lambda' \in Z_1$. The sequence $\{A_{n_{p_k}}\}$ is jointly condensing with constant q with respect to the MNC ψ. Let us estimate $\psi(X)$. We have the chain of inequalities

$$\psi(X) \leq (|\lambda'|/r)\psi(X) = \psi(\{\lambda_{p_k} x_{p_k}\})/r = \psi(\{A_{n_{p_k}} x_{p_k} + y_{p_k}\})/r$$

$$\leq \psi(\{A_{n_{p_k}} x_{p_k}\})/r \leq (q/r)\psi(X).$$

This implies that $\psi(X) = 0$, i.e., the sequence X is totally bounded, and so one can extract from it a convergent subsequence, for which we preserve the notation $\{x_{p_k}\}$. Then from

(4) we obtain

$$\lambda_{p_k} x_{p_k} = A_{n_{p_k}} x_{p_k} + y_{p-k}. \tag{6}$$

If x_0 denotes the limit of $\{x_{p_k}\}$, then letting $k \to \infty$ in (6) we obtain (5). **QED**

2.7.6. Proof of assertion (a). In view of the requirements on the operators A_n, each of them has in the set Z only finitely many points of the spectrum, and those points are eigenvalues of finite multiplicity. Suppose that (a) is not true. Then one can find sequences λ_p, n_p, and x_p, $p = 1, 2, \dots$, such that $\lambda_p \in Z$, $n_p \to \infty$, $\|x_p\| = 1$, and $\lambda_p x_p = A_{n_p} x_p$. Now from the uniform boundedness principle (see, e.g., [34]) it follows that the norms $\|A_{n_p}\|$ are bounded by a common constant. This implies that the sequence $\{\lambda_p\}$ is bounded. But then, by Lemma 2.7.5, the operator A_∞ has an eigenvalue in Z, which contradicts the assumption of assertion (a).

We precede the proof of assertion (b) by two more lemmas.

2.7.7. Lemma. *Suppose that the assumptions of Theorem 2.7.4, item* (a) *are satisfied, and the set Z is bounded. Then for sufficiently large n and for $\lambda \in Z$ the operators $(\lambda I - A_n)^{-1}$ are defined, the norms $\|(\lambda I - A_n)^{-1}\|$ are bounded by a common constant, and*

$$(\lambda I - A_n)^{-1} y \to (\lambda I - A_\infty)^{-1} y \quad (n \to \infty,\ y \in E)$$

uniformly in $\lambda \in Z$.

Proof. The first part of the lemma follows from the assertion (a) just established. Let us prove the second part. Assuming that it is not true, one can find sequences λ_p, x_p, y_p, n_p $(p = 1, 2, \dots)$ such that $\lambda_p \in Z$, $\|x_p\| = 1$, $n_p \to \infty$,

$$y_p = (\lambda_p I - A_{n_p})^{-1} x_p, \tag{7}$$

and $\|y_p\| \to \infty$ when $p \to \infty$. Then (7) yields

$$\lambda_p z_p = A_{n_p} z_p + \|y_p\|^{-1} x_p.$$

But then from Lemma 2.7.5 it follows that there are $\lambda' \in Z$ and z_0, $\|z_0\| = 1$, such that $\lambda' z_0 = A_\infty z_0$, which contradicts the condition of item (a). Now let us show that for any $y \in E$, $\|y\| = 1$ and any $\lambda \in Z$ we have

$$(\lambda I - A_n)^{-1} y \to (\lambda I - A_\infty)^{-1} y$$

when $n \to \infty$, uniformly in λ. Suppose this is not the case. Then there are $\mu_0 > 0, n_p$, and $\lambda_p \in Z$ such that

$$\|(\lambda_p I - A_{n_p})^{-1} y - (\lambda_p I - A_\infty)^{-1} y\| \geq \mu_0, \quad p = 1, 2, \ldots. \tag{8}$$

With no loss of generality we may assume that $\{\lambda_p\}$ converges to some point $\lambda' \in Z$. Then from (8) we deduce that there is an N such that

$$\|(\lambda_p I - A_{n_p})^{-1} y - (\lambda' I - A_\infty)^{-1} y\| \geq \mu_0/2 \tag{9}$$

for all $p \geq N$. Set

$$x_p = (\lambda_p I - A_{n_p})^{-1} y. \tag{10}$$

From what we proved above it follows that the norms $\|x_p\|$ are uniformly bounded. One can rewrite (10) as $\lambda_p x_p = A_{n_p} x_p + y$, and then Lemma 2.7.5 yields a subsequence $\{x_{p_k}\}$ of $\{x_p\}$ that converges to a vector x_0 which satisfies $\lambda' x_0 = A_\infty x_0 + y$. Thus, $x_0 = (\lambda' I - A_\infty)^{-1} y$, i.e.,

$$(\lambda_{p_k} I - A_{n_{p_k}})^{-1} y \to (\lambda' I - A_\infty)^{-1} y, \quad k \to \infty,$$

which contradicts inequality (9), the latter being valid for all $p \geq N$. **QED**

2.7.8. Lemma. *Suppose the simple rectifiable contour Γ lies outside the disc $K(0, q)$ and $\Gamma \cap \sigma(A_\infty) = \emptyset$. Then for sufficiently large n the Riesz projector P_n corresponding to the operator A_n and the contour Γ is defined and $P_n x \to P_\infty x$ when $n \to \infty$, for all $x \in E$.*

Proof. The first part follows from the first part of Lemma 2.7.7. From the second part of Lemma 2.7.7 it follows that

$$\int_\Gamma (\lambda I - A_n)^{-1} x d\lambda \longrightarrow \int_\Gamma (\lambda I - A_\infty)^{-1} x d\lambda, \quad n \to \infty,$$

and consequently $P_n x \to P_\infty x$ when $n \to \infty$, for all $x \in E$. **QED**

2.7.9. Proof of assertion (b). Assuming the contrary, there exist μ_0 and n_p, $p = 1, 2, \ldots$, such that the operators A_{n_p} have no eigenvalues in the disc $K(\lambda_\infty, \mu_0)$. Let Γ_1 denote the boundary of $K(\lambda_\infty, \mu_0)$. Consider the projectors P_{n_p} corresponding to the operators A_{n_p} and the contour Γ_1.

Since inside Γ_1 there is a point of $\sigma(A_\infty)$, we obviously have $P_\infty \neq 0$, i.e., there is an $x \neq 0$ in $P_\infty E$. But then $P_\infty x = x$, whereas $P_p x = 0$ $(p = 1, 2, \ldots)$ because inside Γ_1 there are no points of $\sigma(A_{n_p})$. Consequently, $P_p x$ does not converge to $P_\infty x$, which contradicts Lemma 2.7.8. This proves (b).

2.7.10. Proof of assertion (c). The existence and the strong convergence of the projectors P_n to P_∞ were established in Lemma 2.7.8. It remains to show that for any bounded sequence $\{x_n\}$ the sequence $\{P_n x_n\}$ is totally bounded. Let $0 < \delta < r$. Then

$$P_\infty = \frac{1}{2\pi i}\int_{|\lambda-\lambda_\infty|=r}(\lambda I - A_\infty)^{-1}d\lambda = \frac{1}{2\pi i}\int_{|\lambda-\lambda_\infty|=\delta}(\lambda I - A_\infty)^{-1}d\lambda.$$

By item (a), for n sufficiently large the closed annulus $\delta \le |\lambda - \lambda_\infty| \le r$ contains no points of $\sigma(A_n)$, and

$$P_n = \frac{1}{2\pi i}\int_{|\lambda-\lambda_\infty|=r}(\lambda I - A_n)^{-1}d\lambda = \frac{1}{2\pi i}\int_{|\lambda-\lambda_\infty|=\delta}(\lambda I - A_n)^{-1}d\lambda.$$

This shows that in the definitions of the projectors P_∞ and P_n one can take the number $\delta > 0$ arbitrarily small (but fixed).

Since the sequence $\{A_n\}$ is jointly condensing with constant q with respect to the MNC ψ, we have

$$\psi(\{(\lambda_\infty I - A_n)x_n\}) \ge |\lambda_\infty|\psi(\{x_n\}) - \psi(\{A_n x_n\}) \ge (|\lambda_\infty| - q)\psi(\{x_n\}).$$

Applying this inequality m times and denoting $\gamma = |\lambda_\infty| - q$, we obtain

$$\psi(\{(\lambda_\infty I - A_n)^m x_n\}) \ge \gamma^m \psi(\{x_n\}). \tag{11}$$

An easy induction on m yields the relations

$$(\lambda I - A_n)^m P_n = \frac{1}{2\pi i}\int_{|\lambda-\lambda_\infty|=\delta}(\lambda_\infty - \lambda)^m(\lambda I - A_n)^{-1}d\lambda. \tag{12}$$

In fact, here is the proof of (12) for $m = 1$ (the induction step from m to $m+1$ is carried out following the same scheme):

$$(\lambda_\infty I - A_n)P_n = \frac{1}{2\pi i}\int_{|\lambda-\lambda_\infty|=\delta}[(\lambda_\infty - \lambda)^m(\lambda I - A_n)^{-1} + (\lambda I - A_n)(\lambda I - A_n)^{-1}]d\lambda$$

$$= \frac{1}{2\pi i}\int_{|\lambda-\lambda_\infty|=\delta}(\lambda_\infty - \lambda)(\lambda I - A_n)^{-1}d\lambda,$$

since the integral of the holomorphic function $(\lambda I - A_n)(\lambda I - A_n)^{-1} = I$ is equal to zero.

Now pick an arbitrary bounded sequence $\{x_n\}$ and consider the sequence $\{z_n\}$ obtained from the set

$$\{y_{nm}\colon y_{nm} = \frac{1}{2\pi i}\int_{|\lambda-\lambda_\infty|=\delta}(\lambda_\infty - \lambda)^m \delta^{-m}(\lambda I - A_n)^{-1}x_n d\lambda\} \tag{13}$$

by relabeling its elements: $y_{nm} = z_k$ according to the rule described by the following table:

$$\begin{array}{ccccc}
 & 1\nearrow & 3\nearrow & 6\nearrow & 10\nearrow \\
 & y_{11} & y_{12} & y_{13} & y_{14}\;\cdots \\
\nearrow & & & & \\
 & 2\nearrow & 5\nearrow & 9\nearrow & 14\nearrow \\
 & y_{21} & y_{22} & y_{23} & y_{24}\;\cdots \\
\nearrow & & & & \\
 & 4\nearrow & 8\nearrow & 13\nearrow & 19\nearrow \\
 & y_{31} & y_{32} & y_{33} & y_{34}\;\cdots \\
\nearrow & & & & \\
 & 7\nearrow & 12\nearrow & 18\nearrow & 25\nearrow \\
 & y_{41} & y_{42} & y_{43} & y_{44}\;\cdots \\
\nearrow & \nearrow & \nearrow & \nearrow & \\
 & \cdots & \cdots & \cdots & \cdots
\end{array}$$

Clearly, for any fixed m the sequence $\{y_{nm}\}$ is a subsequence of $\{z_n\}$.

We claim that $\{z_n\}$ is bounded. In fact, $|(\lambda_\infty - \lambda)^m \delta^{-m}| = 1$, and, by Lemma 2.7.7, the norms $\|(\lambda I - A_n)^{-1} x_n\|$ are uniformly bounded. Hence, the norms $\|y_{nm}\|$ are also uniformly bounded.

By the remark made above and the monotonicity of the MNC ψ,

$$\psi(\{y_{nm}\}) \leq \psi(\{z_k\}). \tag{14}$$

Now we turn directly to the proof of the joint compactness of the sequence of projectors P_n. For any bounded sequence $\{x_n\}$ the sequence $\{P_n x_n\}$ is also bounded thanks to the strong continuity of the P_n's. By (12) and (13),

$$(\lambda_\infty I - A_n)^m P_n x_n = \delta^m y_{nm}. \tag{15}$$

Relations (11), (15), and (14) imply the inequalities

$$\psi(\{P_n x_n\}) \leq \gamma^{-m} \psi(\{(\lambda_\infty I - A_n)^m P_n x_n\}) \leq (\delta/\gamma)^m \psi(\{z_k\}). \tag{16}$$

Now choose $\delta < \gamma$. Then inequalities (16) imply $\psi(\{P_n x_n\}) = 0$, i.e., the sequence $\{P_n x_n\}$ is totally bounded.

The last assertion in (c) follows from **2.7.3**. This completes the proof of Theorem 2.7.4. **QED**

2.7.11. The case of a real vector space. *Suppose the operators A_n, which act in the real Banach space E, satisfy condition* (A) *of* **2.7.1** *and are regularly jointly condensing with constant q with respect to the monotone normal* MNC ψ. *Then the conclusions of Theorem 2.7.4 hold true for the complexifications* A_{nc} (see **2.6.1**).

To prove this is suffices to show that if the operators A_n are regularly jointly condensing with constant q with respect to the monotone normal MNC ψ, then their complexifications A_{nc} are regularly jointly condensing, with the same constant q, with respect to the monotone normal MNC ψ_c defined by the formula

$$\psi_c(Z) = \max\{\psi(X), \psi(Y)\}, \tag{17}$$

where $Z \in BE_c$, $Z = \{(x_1, y_1), (x_2, y_2), \dots\}$, and $X, Y \in BE$, $X = \{x_1, x_2, \dots\}$, $Y = \{y_1, y_2, \dots\}$.

In fact, the monotonicity of ψ_c is plain. Now let $z_n = (x_n, y_n)$ and $\{z_n\} \in BE_c$. Then

$$\psi_c(\{A_n z_n\}) = \max\{\psi(\{A_n x_n\}), \psi(\{A_n y_n\})\} \leq q \max\{\psi(\{x_n\}), \psi(\{y_n\})\} = q\psi_c(\{z_n\}).$$

Thus, the sequence $\{A_n\}$ is jointly condensing. The proof of this fact for an arbitrary subsequence $\{A_{n_k}\}$ is carried out in analogous manner.

2.7.12. Remark. Under the assumptions of Theorem 2.7.4, instead of a sequence of operators that is regularly jointly condensing with constant q with respect to a monotone normal MNC ψ one can consider a sequence of operators $\{A_n\}$ that satisfies the following requirement:

(K) *For any $\delta > 0$ and any subsequence $\{A_{n_k}\}$ of $\{A_n\}$ there is an M_χ-normal* MNC *ϕ with respect to which $\{A_{n_k}\}$ is jointly condensing with constant $q+\delta$.*

In the present case the proof given in **2.7.5–2.7.10** must be slightly modified. We mention here only the changes that must be operated in the proof of the total boundedness of the sequence $\{P_n x_n\}$, given in **2.7.10**. Notice that, by formula (12) and Lemma 2.7.7, one has the estimate

$$\|(\lambda_\infty I - A_n)^m P_n\| \leq c\delta^{m+1},$$

where c is some constant.

If the points $\{y_1, \dots, y_l\}$ form an ε-net for the sequence $\{x_n\}$, then by the strong continuity of the operators A_n and of the projectors P_n, the set $\{(\lambda_\infty I - A_n)^m P_n x_p \colon n = 1, 2, \dots;\ p = 1, \dots, l\}$ is a totally bounded $(c\delta^{m+1}\varepsilon)$-net for the sequence $\{(\lambda_\infty I - A_n)^m P_n x_n\}$. Hence, by inequality (11),

$$\phi(\{P_n x_n\}) \leq \gamma^{-m} \phi(\{(\lambda_\infty I - A_n)^m x_n\})$$

$$\leq c_1 \gamma^{-m} \chi(\{(\lambda_\infty I - A_n)^m x_n\}) \leq c_1 \gamma^{-m} c\delta^{m+1} \chi(\{x_n\}), \tag{18}$$

where c_1 is a constant such that $\phi(X) \le c_1\chi(X)$ for any sequence X (the existence of such a constant is guaranteed by the M_χ-normality of ϕ). Now choosing $\delta < \gamma$ and taking into account that (18) holds for all m, we conclude that $\phi(\{P_n x_n\}) = 0$.

Condition (K) is in some sense also necessary. For a more precise formulation we need some notations.

2.7.13. Notations. Let the operators A_n satisfy condition (A) of **2.7.1**. Then there exists an arbitrarily small $\varepsilon > 0$ such that

$$\Gamma_\varepsilon \cap \sigma(A_n) = \emptyset, \quad n = 1, 2, \dots, \infty; \tag{19}$$

here Γ_ε denotes the boundary of the disc $K(0, q+\varepsilon)$.

Orient the contour Γ_ε anticlockwise and define the Riesz projector, as usual, by the formula

$$Q_n(\varepsilon) = \frac{1}{2\pi i}\int_{\Gamma_\varepsilon} (\lambda I - A_n)^{-1} d\lambda.$$

Now set $P_n(\varepsilon) = I - Q_n(\varepsilon)$ and $A_n(\varepsilon) = A_n Q_n(\varepsilon)$.

2.7.14. Theorem. *In order for the sequence of projectors $P_n(\varepsilon)$ to be jointly compact and have the property that the resolvents $(\lambda I - A_n(\varepsilon))^{-1}$ $(\lambda \in \mathbf{C} \setminus K(0, q+\varepsilon))$ converge strongly to $(\lambda I - A_\infty(\varepsilon))^{-1}$ for any $\varepsilon > 0$ for which* (19) *holds, it is necessary and sufficient that the sequence $\{A_n\}$ satisfy condition* (K).

Proof. Necessity. We describe the construction of a normal MNC with respect to which the sequence $\{A_n\}$ will be jointly condensing with constant $q+\delta$. The construction for an arbitrary subsequence $\{A_{n_k}\}$ is analogous—it suffices to regard the subsequence as an independent sequence.

Thus, given $\delta > 0$, choose $\varepsilon < \delta$ such that (19) holds for the contour Γ_ε, and consider the operator $\mathfrak{A} : BE \to BE$ (recall that BE denotes the space of all bounded sequences of elements of E) that acts according to the rule: if $X = \{x_n\}$, then

$$\mathfrak{A}X = \{A_n(\varepsilon)x_n\}. \tag{20}$$

Since the operators $P_n(\varepsilon)$ and A_n are strongly continuous, the uniform boundedness principle guarantees that $\mathfrak{A}$ is bounded in BE.

Now let us show that $\mathfrak{A}$ induces an operator $\mathfrak{A}^+$ in E^+. To this end it suffices to verify that $\mathfrak{A}$ maps KE into itself (recall that KE is the subspace of totally bounded sequences in BE and $E^+ = BE/KE$). Thus, let $Y = \{y_n\}$ and $Y \in KE$. Consider the sequence $\mathfrak{A}Y = \{A_n(\varepsilon)y_n\}$. Let $\{A_{n_k}(\varepsilon)y_{n_k}\}$ be some subsequence of $\mathfrak{A}Y$. Extract from $\{y_{n_k}\}$ a convergent subsequence $\{y_{n_{k_p}}\}$. Then by the strong convergence of A_n to A_∞, the sequence $\{A_{n_{k_p}}(\varepsilon)y_{n_{k_p}}\}$ converges, i.e., $\mathfrak{A}Y$ is totally bounded.

Let us check that $|\lambda| \le q+\varepsilon$ for any $\lambda \in \sigma(\mathfrak{A}^+)$. To do this it suffices to show that if λ is such that $|\lambda| > q+\varepsilon$ then the operator $\lambda I - \mathfrak{A}^+$ is bijective.

Let $\mathcal{Y} \in E^+$ and $|\lambda| > q+\varepsilon$. Consider the equation

$$\lambda\mathcal{X} - \mathfrak{A}^+\mathcal{X} = \mathcal{Y}.$$

Pick a representative $Y = \{y_n\}$ of the class $\mathcal{Y}$. Then by the construction of the operator $\mathfrak{A}^+$ and the inequality $|\lambda| > q+\varepsilon$ (which guarantees that the operators $\lambda I - A_n(\varepsilon)$ are invertible) there exists a sequence $X = \{x_n\}$ such that $\lambda x_n - A_n(\varepsilon)x_n = y_n$. Since the operators $(\lambda I - A_n(\varepsilon))^{-1}$ converge strongly to $(\lambda I - A_\infty(\varepsilon)^{-1}$, the uniform boundedness principle guarantees that the norms $\|(\lambda I - A_n(\varepsilon))^{-1}\|$ are bounded by a common constant, and consequently so are the norms $\|x_n\|$, i.e., $X \in BE$. Let $\mathcal{X} \in E^+$ be the class with representative X. Then $\lambda\mathcal{X} - \mathfrak{A}^+\mathcal{X} = \mathcal{Y}$. Thus, the map $\lambda I - \mathfrak{A}^+$ is surjective.

Now let us show that zero is the only element that this map takes into zero. Suppose

$$\lambda\mathcal{X} - \mathfrak{A}^+\mathcal{X} = 0 \tag{21}$$

and let $X = \{x_n\}$ be a representative of the class $\mathcal{X}$. Then by relation (21) the sequence $\{y_n\}$, where $y_n = \lambda x_n - A_n(\varepsilon)x_n$, is totally bounded; but then so is the sequence $\{x_n\}$, because $x_n = (\lambda I - A_n(\varepsilon))^{-1}y_n$ and the sequence of operators $\{(\lambda I - A_n(\varepsilon))^{-1}\}$ converges strongly. Hence, $\mathcal{X} = 0$.

Thus, we showed that $R(\mathfrak{A}^+) \le q+\varepsilon$. Since $\varepsilon < \delta$, in E^+ there is a norm $\|\cdot\|_*$ (see **2.6.3**) such that $\|\mathfrak{A}^+\| \le q+\varepsilon$ and $\|\cdot\|_*$ is equivalent to the norm defined by formula (1) of **2.2.1**; in view of formula (2) of **2.2.1**, this means that $\|\cdot\|_*$ is equivalent to the normal MNC generated by the Hausdorff MNC.

Now consider the normal MNC ψ in E defined by the formula $\psi(\{x_n\}) = \|\mathcal{X}\|_*$, where $\mathcal{X}$ is the element of E^+ with representative $X = \{x_n\}$.

Let us estimate $\psi(\{A_n x_n\})$ for an arbitrary bounded sequence $\{x_n\}$. Notice that thanks to the joint compactness of the projectors $P_n(\varepsilon)$, the sequence $\{A_n P_n(\varepsilon)x_n\}$ is totally bounded. Consequently,

$$\psi(\{A_n x_n\}) = \psi(\{A_n(\varepsilon)x_n\}) = \|\mathcal{Y}\|_*, \tag{22}$$

where $\mathcal{Y} \in E^+$ is the element with representative $Y = \{A_n(\varepsilon)x_n\}$. But, by the definition of the operator $\mathfrak{A}^+$, we have $\mathcal{Y} = \mathfrak{A}^+\mathcal{X}$, where $\mathcal{X} \in E^+$ is the element with representative $X = \{x_n\}$. This yields

$$\|\mathcal{Y}\|_* \le \|\mathfrak{A}^+\|_*\|\mathcal{X}\|_* \le (q+\delta)\|\mathfrak{X}\|_* = (q+\delta)\psi(\{x_n\}),$$

which in conjunction with (22) completes the proof of the neccessity part.

Sufficiency. Fix $\delta < \varepsilon$. The joint compactness of the projectors $\{P_n(\varepsilon)\}$ follows from Remark 2.7.12. Now let us show that for any $x \in E$,

$$(\lambda I - A_n(\varepsilon))^{-1}x \to (\lambda I - A_\infty(\varepsilon))^{-1}x$$

when $n \to \infty$. Notice that since the projectors $P_n(\varepsilon)$ are finite-dimensional and jointly compact, the sequence $A_n(\varepsilon) = A_n(I - P_n(\varepsilon))$ satisfies the conditions of **2.7.12** and $\sigma(A_n(\varepsilon)) \subset K(0, q+\varepsilon)$. Now the desired conclusion follows from Lemma 2.7.7. **QED**

2.7.15. Remark. Theorem 2.7.14 remains valid in the case of a real space E, provided the convergence of the resolvents $(\lambda I - A_n(\varepsilon))^{-1}$ is understood as the convergence of the resolvents of the complexifications of the operators $A_n(\varepsilon)$ (see **2.6.1**).

To prove the necessity part it suffices to consider the complexifications E_c and A_{nc} of the space E and the operators A_n, respectively, then construct the requisite normal MNC in E_c with the help of Theorem 2.7.14, and finally consider their restriction to $\hat{E}$ (see **2.6.1**).

In proving the sufficiency part one also has to pass to E_c and A_{nc}, use as the normal MNCs ψ_c in E_c the MNCs constructed from ψ via formula (17), and then remark, as we did in **2.6.12**, that the projectors P_n are real.

2.7.16. Example. By analogy with Theorem 2.6.12, one might conjecture that any sequence $\{A_n\}$ of operators that are jointly condensing with constant k admits for any given $\varepsilon > 0$ a representation

$$A_n = B_n + C_n,$$

where the operators C_n are jointly compact and there is a norm $\|\cdot\|_*$ such that $\|B_n\|_* \leq k+\varepsilon$ for all n. The example given below shows that this is not the case.

In l_2 consider the sequence of operators A_n defined as follows:

$$A_{2k+1}x = \{\xi_{2k+1}\delta_{i,2k+2}\}_{i=1}^{\infty}, \quad k = 0, 1, 2, \dots ,$$

$$A_{2k}x = \{\xi_{2k}\delta_{i,2k-1}\}_{i=1}^{\infty}, \quad k = 1, 2, \dots ,$$

where $x = \{\xi_i\}_{i=1}^{\infty}$ and δ_{ij} is the Kronecker symbol. Clearly, each A_n is nilpotent: indeed, $A_n^2 = 0$. Hence, $\sigma(A_n) = \{0\}$ for all n, and consequently the projector P_n that corresponds to the part of the spectrum of A_n outside an arbitrary disc centered at the origin is equal to zero. Put $A_\infty = 0$. Clearly, A_n and P_n converge strongly to A_∞ and P_∞, respectively. We claim that $(\lambda I - A_n)^{-1}$ converge strongly to $(\lambda I - A_\infty)^{-1}$ for $\lambda \neq 0$. In fact,

$$(\lambda I - A_{2k+1})^{-1}x = \{\lambda^{-1}\xi_i + \lambda^{-2}\xi_{2k+1}\delta_{i,2k+2}\}_{i=1}^{\infty}, \quad k = 0, 1, 2, \dots ,$$

$$(\lambda I - A_{2k})^{-1}x = \{\lambda^{-1}\xi_i + \lambda^{-2}\xi_{2k}\delta_{i,2k-1}\}_{i=1}^{\infty}, \quad k = 1, 2, \dots ,$$

where $x = \{\xi_i\}_{i=1}^{\infty}$. Since $\xi_i \to 0$ when $i \to \infty$, it is obvious that the sequences $\{\lambda^{-1}\xi_i + \lambda^{-2}\xi_{2k+1}\delta_{i,2k+2}\}_{i=1}^{\infty}$ and $\{\lambda^{-1}\xi_i + \lambda^{-2}\xi_{2k}\delta_{i,2k-1}\}_{i=1}^{\infty}$ converge to $\{\lambda^{-1}\xi_i\}$, as needed.

Thus, by Theorem 2.7.14, there is a normal MNC with respect to which the sequence $\{A_n\}$ is jointly condensing, say, with constant 1/2. At the same time, there is no such norm with respect to which both A_{2k} and A_{2k+1} are contractive with a constant smaller than 1.

2.8. SURVEY OF THE LITERATURE

2.8.1. Normal measures of noncompactness. The results described in Sections 2.2 and 2.3 are treated following the paper of B. N. Sadovskiĭ [160] (see also the work of R. R. Akhmerov, M. I. Kamenskiĭ, A. S. Potapov, and B. N. Sadovskiĭ [10]). A somewhat different approach, based on taking the quotient of the algebra of bounded linear operators by the ideal of compact operators, is considered in papers by L. S. Gol′denshteĭn, I. Gohberg, and A. S. Markus [55], L. S. Gol′denshteĭn and A. S. Markus [56], and A. Lebow and M. Schechter [100]. Based on that approach one obtains theorems, analogous to Theorems 2.3.2 and 2.3.3, formulated in terms of elements of the aforementioned quotient algebra. In the case where $\psi_1 = \psi_2 = \chi$ a result close to Theorem 2.3.6 can be found in [100].

2.8.2. (ψ_1, ψ_2)-norms. The content of Section 2.4 is close to the paper of L. S. Gol′denshteĭn and A. S. Markus [56], in which the norms $\|A\|^{(\chi)}, \|A\|^{(\alpha)}, \|A\|^{(\beta)}, \|A\|_T$ were introduced for the first time and the properties of $\|A\|_T$ were investigated in detail. Most of our examples are also borrowed from that paper. V. G. Kurbatov [97] used the MNC χ to obtain estimates of the outer Fredholm radius R_Φ of some operators in the space C, including Example 2.4.8.

2.8.3. Measure of noncompactness of a conjugate operator. Properties of the MNC of a conjugate operator were studied in the aforementioned papers [55], [56], [100], as well as in papers of R. D. Nussbaum [115] and A. A. Sedaev [165]. It is on these sources that our Section 2.5 is based. The equality $\|A^*\|_{(\chi)} = \|A\|_{(\chi)}$ in the Hilbert space case is apparently noticed here for the first time. Alongside with the relations proved here, in the Hilbert space case one also has the relation $\|A\|^{(\chi)} = \|A^*\|^{(\chi)} = (\|AA^*\|^{(\chi)})^{1/2}$ (see the paper by C. A. Stuart [168]).

2.8.4. Fredholm spectrum. Section 2.6., devoted to the study of the Fredholm spectrum of a bounded linear operator, follows a paper by B. N. Sadovskiĭ [160]. Various estimates of and formulas for the computation of $R_\Phi(A)$ in terms of the MNCs χ, α, and also of $\|A\|_T$, are given in papers of L. S. Gol′denshteĭn, I. Gohberg and A. S. Markus [55], L. S. Gol′denshteĭn and A. S. Markus [56], A. Lebow and M. Schechter [100], and R. D. Nussbaum [115]. The theorems on the representation of a condensing operator with constant $k < 1$ as the sum of an operator that is contractive in an equivalent norm, and a finite-rank operator, were obtained A. A. Sedaev [165]. In [55] it was observed that a linear condensing operator does not necessarily admit a decomposition as the sum of a compact operator and an operator that is contractive in the original norm. In connection with the question of the representability of condensing linear operators we give two more results.

Theorem (J. L. R. Webb [180]). *Let $A \in L(E,H)$, where E and H are a Banach and a Hilbert space, respectively. Then $\|A\|^{(\alpha,\alpha)} = k$ if and only if for any $\varepsilon > 0$ there is a representation of A as a sum $K_\varepsilon + L_\varepsilon$ such that K_ε is a compact linear operator and $k \le \|L_\varepsilon\| \le k+\varepsilon$.*

Theorem (I. I. Istrăţescu [63]). *Suppose that $A \in L(H)$, where H is a Hilbert space, has the property that for any bounded set $\Omega \subset H$ there is an n such $\beta(A^n\Omega) \le k\beta(\Omega)$. Suppose also that the norms of the operators A^n $(n = 1, 2, \dots)$ are bounded by the same constant. Then there exists a selfadjoint operator Q such that the operator QAQ^{-1} is nonexpansive.*

To conclude this subsection, we mention one more result (see the papers of R. D. Nussbaum [121] and A. S. Potapov, T. Ya. Potapova, and V. A. Filin [136]) concerning the spectrum of positive condensing operators.

Let K be a normal cone in a Banach space E and suppose that the linear operator $A \in L(E)$ maps K into itself. Set

$$\|A\|_K = \sup\{\|Au\| \colon u \in K,\ \|u\| \le 1\},$$

$$\|A\|_{\alpha K} = \inf\{q \colon A|_K \text{ is } (q,\alpha,\alpha)\text{-bounded }\}.$$

Define the numbers

$$r_K(A) = \lim_{n\to\infty} (\|A^n\|_K)^{1/n},$$

$$\rho_K = \lim_{n\to\infty} (\|A^n\|_{\alpha K})^{1/n}.$$

Theorem. *Suppose $r_K(A) > \rho_K(A)$. Then there is an $x \in K \setminus \{0\}$ such that $Ax = r_K(A)x$.*

2.8.5. Normal measures of noncompactness and perturbation theory. The results of section 2.7 are essentially taken from a paper by M. I. Kamenskiĭ [76]. In connection with Theorem 2.7.14 we mention that other necessary and sufficient conditions for (K) to hold, expressed in terms of the regular convergence of the operators $\lambda I - A_n$ to $\lambda I - A_\infty$, are given in the monograph of G. M. Vaĭnikko [175, p. 72]. Our proof in **2.7.10** also follows [175].

Let us give here one result, again from [175, p. 43], about the connection between normal MNCs and regular convergence. Recall that a sequence of operators B_n that converges to an operator B_∞ is said to *converge regularly* (see [175, p. 32]) if any bounded sequence $\{x_n\}$ such that $\{B_n x_n\}$ is relatively compact is itself relatively compact. From Lemma 2.7.5, for example, it follows that if the operators A_n converge strongly to A_∞ and are regularly jointly condensing with constant q, then the sequence $B_n = \lambda I - A_n$, where $|\lambda| > q$, converges regularly to $B_\infty = \lambda I - A_\infty$.

Theorem [21, p. 43]. *Suppose the operators B_n converge regularly to a Fredholm operator B_∞. Then there is a constant $\gamma > 0$ such that*

$$\psi(\{B_n x_n\}) \geq \gamma \psi(\{x_n\})$$

for any bounded sequence $\{x_n\}$, where ψ is the normal MNC defined as

$$\psi(X) = \inf\{\varepsilon > 0\colon \text{for any } \{n_k\} \text{ there are } \{n_{k_p}\}, x \in E \text{ such that } \|x_{k_p} - x\| \leq \varepsilon\}.$$

We further remark that if the sequence of operators A_n satisfies the conditions of **2.7.1** and is jointly condensing, then by Theorem 2.7.14 the operators $A_n P_n(\varepsilon)$ (see **2.7.13**) compactly approximate $A_\infty P_\infty(\varepsilon)$ in the sense that $A_n P_n(\varepsilon)$ converge strongly to $A_\infty P_\infty(\varepsilon)$ and the set $\{A_n P_n(\varepsilon)x_n - A_\infty P_\infty(\varepsilon)x_n\}$ is relatively compact for any bounded sequence $\{x_n\}$. Concerning questions of compact approximation of operators we refer the reader to the monograph of G. M. Vaĭnikko [174], and also to the paper of Yu. N. Vladimirskiĭ [178]. In the latter one can find proofs of the two theorems given below.

We say that *the operators $T_n, T \in L(E_1, E_2)$ $(n = 1, 2, \dots)$ satisfy condition* (K′) if the sequence $\{T_n x_n - Tx\}$ is relatively compact for any bounded sequence $\{x_n\} \subset E$.

Theorem. *The operators $T_n \in L(E_1, E_2)$ and $T \equiv 0$ satisfy condition* (K′) *if and only if $\chi(\bigcup_{k=n}^{\infty} T_k(B)) \to 0$ when $n \to \infty$, where B is the unit ball in E_1.*

Theorem. *Suppose* $T \in L(E_1, E_2)$ *is a* Φ*-operator* [*resp.* Φ_+*-operator*] *and* T, T_n *satisfy condition* (K$'$). *Then for sufficiently large* n *each* T_n *is also a* Φ*-operator* [*resp.* Φ_+*-operator*] *and* $\operatorname{ind} T_n = \operatorname{ind} T$.

CHAPTER 3

THE FIXED-POINT INDEX OF CONDENSING OPERATORS

In the theory of fixed points and various applied problems an important role is played by the closely related notions of degree of a map, rotation of a vector field, and fixed-point index of an operator. We shall use the terminology connected with the last of these three notions; moreover, whenever there is no danger of confusion, we shall simply speak about the "index" of the operator in question instead of its "fixed-point index".

The index of an operator is a certain integer-valued characteristic that enjoys a number of special properties, listed below. The way in which the index is defined does not play an essential role in its applications. What really matters is the fact that the index is defined and the properties used in those applications.

3.1. DEFINITIONS AND PROPERTIES OF THE INDEX

In this section we introduce the notion of the index of a condensing operator, which is analogous to the classical notion of the index of a compact operator defined on the closure of a bounded domain in a Banach space.

3.1.1. Formulation of the problem. Let E be a Banach space, U a bounded open set in E, and let $\overline{U}$ and ∂U denote the closure and respectively the boundary of U. Suppose the operator $f: \overline{U} \to E$ is χ-condensing and has no fixed points on ∂U (here and in the next sections, up to Section 3.5, we consider only operators that are condensing with respect to the MNC χ; accordingly, we shall simply say "condensing" instead of "χ-condensing"). In this situation we will define the index $\operatorname{ind}(f, U)$, which possesses the properties 1°–5° listed below.

To formulate the first of these properties we need to define the notion of homotopic operators. Two condensing operators $f_0, f_1 : \overline{U} \to E$ are said to be *homotopic* if they can be "connected" by a condensing family of operators $\{f_\lambda : \lambda \in [0,1]\}$, $f_\lambda : \overline{U} \to E$, such that the map $(\lambda, x) \mapsto f_\lambda(x)$ is continuous and f_λ has no fixed points on ∂U for all λ.

1°. *Homotopic condensing operators have equal indices.*

2°. *Let U_i, $i = 1, 2, \dots$ be pairwise disjoint open subsets of U and suppose f has no fixed points on $\overline{U} \setminus \bigcup_{i=1}^{\infty} U_i$. Then the indices* $\operatorname{ind}(f, U_i)$ *are defined for all i, only finitely many of them are different from zero, and*

$$\operatorname{ind}(f, U) = \sum_{i=1}^{\infty} \operatorname{ind}(f, U_i).$$

3°. *If $f(x) \equiv x_0$, where $x_0 \in U$, then* $\operatorname{ind}(f, U) = 1$.

4°. *If $f(x) \equiv x_0$, where $x_0 \notin U$, then* $\operatorname{ind}(f, U) = 0$.

5°. *If* $\operatorname{ind}(f, U) \neq 0$, *then f has at least one fixed point in U.*

The fact that a characteristic with these properties exists in the theory of compact operators is well known: if everywhere above the operators are assumed to be compact, then a characteristic possessing properties 1°–5° exists and is even unique. That characteristics is, for example, the rotation $\gamma(I - f, U)$ of the vector field $I - f$ on ∂U, which in the situation at hand coincides with the Leray-Schauder degree of the map $I - f$ relative to the point zero. For ease of reference we state here the following well-known theorem.

3.1.2. Theorem on the existence and properties of the index of a compact operator. *If in the preceding subsection all operators are assumed to be compact, then a characteristic satisfying conditions* 1°–3° *exists, is unique, and possesses properties* 4°–5°.

3.1.3. Fundamental sets. An important role in the sequel will be played by the notion of a fundamental set, introduced by M. A. Krasnosel′skiĭ, P. P. Zabreĭko, and V. V. Strygin [181].

Definition. A set S in the Banach space E is said to be *fundamental for the operator* $f: M \to E$ if

1) S is nonempty, convex, and compact;

2) $f(M \cap S) \subset S$;

3) if $x_0 \in M \setminus S$, then $x_0 \notin \operatorname{co}[\{f(x_0)\} \cap S]$.

The meaning of the last requirement is that under the action of the operator f the points of the set M are not "repelled" by S: if $x_0 \in M \setminus S$, then $f(x_0)$ does not lie in the shaded cone (see the figure).

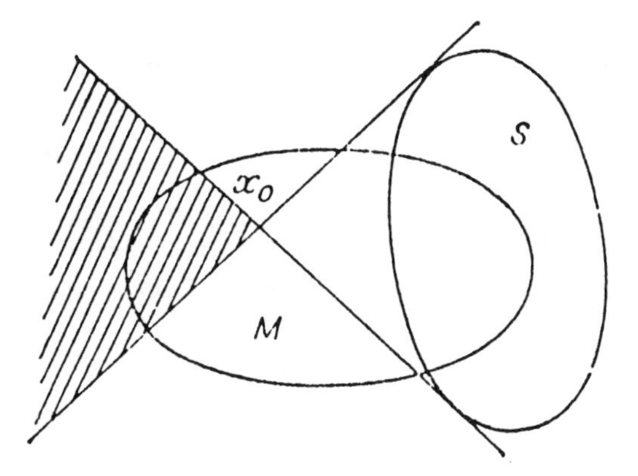

We wish to point out right away that any fundamental set S for the operator f necessarily contains all its fixed points.

The notion of a fundamental set for a family of operators is introduced in analogous manner, as a set that is fundamental for any member of the family. An elementary example of operator possessing a fundamental set is provided by any compact operator $f : M \to E$, where M is a nonempty bounded set. In this case one can take for S the set $\overline{\text{co}}[f(M)]$.

For our purposes it is important that condensing operators also possess fundamental sets.

3.1.4. Theorem on the existence of a fundamental set for condensing families. *Let $f = \{f_\lambda\colon \lambda \in \Lambda\}$ be a condensing family of operators acting from a closed subset M of the space E into E, and let K be an arbitrary compact subset of E. Suppose Λ is a compact topological space and the mapping $(\lambda, x) \mapsto f_\lambda(x)$ is continuous. Then the family f has a fundamental set containing K.*

Proof. Let $\mathfrak{M}$ denote the collection of all compact sets that contain K and satisfy all the requirements defining fundamental sets, except, possibly, for compactness. The family $\mathfrak{M}$ is not empty, since the conditions in question are satisfied, for example, by $T_0 = \overline{\text{co}}[K \cup f(M)]$.

Next, we note that if $T \in \mathfrak{M}$, then $T_1 = \overline{\text{co}}[K \cup f(M \cap T)] \in \mathfrak{M}$, too (with no loss of generality we may assume that $K \neq \emptyset$, so that $T_1 \neq \emptyset$ whenever $T \in \mathfrak{M}$). We need only verify that conditions 2) and 3) in the definition of a fundamental set are satisfied. Clearly, $T_1 \subset T$, because $K \subset T$, $f(M \cap T) \subset T$, and T is closed and convex. Consequently,

$$f(M \cap T_1) \subset f(M \cap T) \subset \overline{\text{co}}[K \cup f(M \cap T)] = T_1,$$

i.e., 2) holds. Further, if $x_0 \notin T$, then 3) follows from the corresponding property of T and the inclusion $T_1 \subset T$. If now $x_0 \in T$, then $f(x_0) \in T_1$ and 3) clearly holds.

Now define S as $S = \bigcap_{T \in \mathfrak{M}} T$. We claim that S is a fundamental set for f. Indeed, it is readily verified that S belongs to $\mathfrak{M}$ and also that S is minimal in $\mathfrak{M}$ (with respect to inclusion). Hence, by the foregoing argument, the set $S_1 = \overline{\text{co}}[K \cup f(M \cap S)]$ coincides with S. But then $\chi(M \cap S) \leq \chi(S) = \chi[f(M \cap S)]$, which in view of the fact that the family f is condensing is possible only when $\overline{M \cap S}$ $(= M \cap S)$ is compact. Since Λ is also compact and the mapping $(\lambda, x) \mapsto f_\lambda(x)$ is continuous, the equality $S = \overline{\text{co}}[K \cup f(M \cap S)]$ implies the compactness of S. **QED**

3.1.5. Remark. In the sequel we shall also need the following similar statement.

Suppose that the operators f_λ of the family f described in Theorem 3.1.4 act from the closure $\overline{M}$ of some subset $M \subset E$ into E. Then under all the remaining assumptions of that theorem the family $f|_M$ has a fundamental set containing K.

The proof is identical to that of Theorem 3.1.4.

3.1.6. Theorem on the extension of continuous operators. In defining the index of a condensing operator we shall also use the well-known theorem of J. Dugundji [33] on the extension of continuous operators.

Theorem. *Any continuous operator that is defined on a nonempty closed subset of a metric space and takes values in a locally convex space admits a continuous extension to the entire space such that the convex hull of the range is preserved.*

3.1.7. Definition of the index of a condensing operator. Let U be a bounded open subset of the Banach space E and let the condensing operator $f: \overline{U} \to E$ have no fixed points on ∂U. Further, let S be some fundamental set for f (the existence of such a set is guaranteed by Theorem 3.1.4). Using Dugundji's theorem, extend f from the closed set $\overline{U} \cap S$ to an operator on the entire space E with values in S (assume for the moment that $\overline{U} \cap S \neq \emptyset$). Let $\tilde{f}$ denote the restriction of the resulting operator to $\overline{U}$. Then $\tilde{f}$ acts from $\overline{U}$ into the compact set S, and hence is compact. Moreover, all its fixed points lie in the set $\overline{U} \cap S$, where $\tilde{f}$ coincides with f, so that $\tilde{f}$, too, has no fixed points on ∂U. Thus, for the compact operator $\tilde{f}$ all conditions of Theorem 3.1.2 are satisfied, and consequently the index $\text{ind}(\tilde{f}, U)$ is defined. Put

$$\text{ind}(f, U) = \text{ind}(\tilde{f}, U). \tag{1}$$

3.1.8. Correctness of the definition of the index. To show that the above definition is correct, we need to check that the index $\text{ind}(f, U)$ does not depend on the choice of the fundamental set S and of the extension $\tilde{f}$.

Suppose the pairs $S_1, \tilde{f}_1$ and $S_2, \tilde{f}_2$ satisfy all the requirements formulated in the definition of the index. Let us show that

$$\operatorname{ind}(\tilde{f}_1, U) = \operatorname{ind}(\tilde{f}_2, U). \tag{2}$$

With no loss of generality we may assume that $S_1 \subset S_2$; indeed, by Theorem 3.1.4, we can find a fundamental set S_3 of f such that S_3 contains the compact set $K = S_1 \cup S_2$, and a corresponding extension $\tilde{f}_3$, and then establish that both sides of (2) are equal to $\operatorname{ind}(\tilde{f}_3, U)$.

Now consider the compact family of operators f_t on U, $f_t(x) = (1-t)\tilde{f}_1(x) + t\tilde{f}_2(x)$, $t \in [0,1]$. In order to show that the family $F = \{f_t : t \in [0,1]\}$ is a homotopy from $\tilde{f}_1$ to $\tilde{f}_2$, it clearly suffices to verify that the operators f_t have no fixed points on ∂U. In fact, if $x \notin S_2$, then the equality $x = (1-t)\tilde{f}_1(x) + t\tilde{f}_2(x)$ is impossible for $t \in [0,1]$, since the right-hand side lies always in S_2. If $x \in S_1$, then the indicated equality is impossible because in this case $\tilde{f}_1(x) = f(x) = \tilde{f}_2(x)$ and f has no fixed points on ∂U by hypothesis. Finally, let $x \in S_2 \setminus S_1$. In this case the relation $x = (1-t)\tilde{f}_1(x) + t\tilde{f}_2(x)$ means that $x \in \operatorname{co}[\{f(x)\} \cap S_1]$, which contradicts the assumption that the set S_1 is fundamental. Thus, the compact operators $\tilde{f}_1$ and $\tilde{f}$ are homotopic on $\overline{U}$, and consequently their indices coincide.

3.1.9. Properties of the index. The index of a condensing operator defined above enjoys all properties 1°–5° formulated in **3.1.1**. All of them are readily derived from the corresponding properties of the index of a compact operator. Let us exemplify this on the first property.

Suppose the operators f_0 and f_1 are homotopic and the homotopy is effected by the condensing family $f = \{f_\lambda : \lambda \in [0,1]\}$, $f_\lambda : \overline{U} \to E$. By Theorem 3.1.4, the family f has a fundamental set S. Extend the operators f_λ from the set $\overline{U} \cap S$ to the entire space E, with preservation of the convex hull of their ranges and of the continuity of the mapping $(\lambda, x) \mapsto f_\lambda(x)$. Let $\tilde{f}_\lambda$ denote the restrictions of these extensions to $\overline{U}$. Then it is readily seen that the family $\tilde{f} = \{\tilde{f}_\lambda : \lambda \in [0,1]\}$ is a compact homotopy from $\tilde{f}_0$ to $\tilde{f}_1$. Hence, by the definition of the index a condensing operator and properties of homotopic compact operators,

$$\operatorname{ind}(f_0, U) = \operatorname{ind}(\tilde{f}_0, U) = \operatorname{ind}(\tilde{f}_1, U) = \operatorname{ind}(f_1, U),$$

as needed.

We omit the verification of the other properties.

3.1.10. Remark. In defining the index of a condensing operator f, when we choosed an extension $\tilde{f}$, we assumed that the intersection of the set $\overline{U}$ with the fundamental set

S is not empty. If the operator f possesses a fundamental set that does not intersect $\overline{U}$, then this means that it has no fixed point in $\overline{U}$. In this case one can put, by definition, $\operatorname{ind}(f, U) = 0$. Incidentally, one can work from the very beginning with fundamental sets S that intersect U (that such sets always exist is guaranteed by Theorem 3.1.4).

3.1.11. Homotopy classes of condensing operators. The homotopy relation on the set of all condensing operators that are defined on a given set $\overline{U}$ and have no fixed points on ∂U is clearly an equivalence relation, i.e., reflexive, symmetric, and transitive. Therefore, it yields a decomposition of the set of all such operators into disjoint classes—homotopy classes of operators. An important fact is that each homotopy class of condensing operators contains a compact operator: for any condensing operator $f: \overline{U} \to E$ with no fixed points on ∂U, the operator $\tilde{f}$ figuring in the definition of the index lies in the same homotopy class as f. The operators f and $\tilde{f}$ are actually linearly homotopic: the condensing family $F = \{f_\lambda: f_\lambda(x) = (1-\lambda)f(x) + \lambda\tilde{f}(x), \lambda \in [0,1]\}$ obviously has no fixed points in ∂U. We give here the following general result.

Theorem. *Let U be a bounded open set in the Banach space E and let $f: \overline{U} \times [0,1] \to E$ be a condensing homotopy, all fixed points of which are contained in some compact set K. Then there exists a condensing homotopy $F: \overline{U} \times [0,1] \times [0,1] \to E$ with the following properties:*

1) *the fixed points of the map $F(\cdot, t, s)$ lie in the same set K for all $t, s \in [0,1]$;*
2) $F(\cdot, \cdot, 0) = f$;
3) $F(\cdot, \cdot, 1)$ *is a compact homotopy.*

Proof. Let S be a fundamental set of the family F such that $K \subset S$ (the existence of such a set is guaranteed by Theorem 3.1.4). Construct an extension of the operator f from the closed subset $(\overline{U} \cap S) \times [0,1]$ of the metric space $E \times [0,1]$ to the entire space and denote by $\tilde{f}$ the restriction of this extension to $\overline{U} \times [0,1]$. Then it is readily verified that $\tilde{f}$ is a compact homotopy, the fixed-point set of which coincides with the fixed-point set of f. The sought-for condensing family can be now obtained as a convex combination of f and $\tilde{f}$:

$$F(x,t,s) = (1-s)f(x,t) + s\tilde{f}(x,t), \quad s \in [0,1]. \quad \textbf{QED}$$

The construction described in the theorem can be used to generalize various homotopy invariants of compact operators to condensing operators.

3.1.12. Generalization of Hopf's homotopy classification theorem. The theorem proved above enables us to extend to condensing operators the classical theorem of H. Hopf about the homotopy of operators with equal indices.

Let U be a bounded domain (i.e., open connected set) in the Banach space E. U is said to be a *Jordan domain* if the set $E \setminus \overline{U}$ is connected. A classical theorem of Hopf asserts that two finite-dimensional or compact operators that are defined on the closure $\overline{U}$ of a Jordan domain U and have equal indices are homotopic. The next result establishes a similar fact for condensing operators.

Theorem. *Let f_0 and f_1 be condensing operators, acting from the closure $\overline{U}$ of a Jordan domain U into E. Suppose f_0 and f_1 have no fixed points on ∂U and* $\operatorname{ind}(f_0, U) = \operatorname{ind}(f_1, U)$. *Then f_0 and f_1 are homotopic.*

Proof. Let $\tilde{f}_0$ and $\tilde{f}_1$ be compact operators in the homotopy classes of the condensing operators f_0 and f_1, respectively. Let Φ and Ψ be condensing homotopies from f_0 to $\tilde{f}_0$ and from f_1 to $\tilde{f}_1$, respectively. For the sake of definiteness we shall assume that $\tilde{f}_0$ and $\tilde{f}_1$ are compact operators figuring in the definition of the index of the operators f_0 and f_1, respectively. Then for Φ and Ψ one can take the linear homotopies

$$\Phi = \{\phi_\lambda : \phi_\lambda(x) = (1-\lambda) f_0(x) + \lambda \tilde{f}_0(x), \quad \lambda \in [0,1]\}$$

and

$$\Psi = \{\psi_\lambda : \psi_\lambda(x) = (1-\lambda) f_1(x) + \lambda \tilde{f}_1(x), \quad \lambda \in [0,1]\},$$

respectively.

Since the compact operators $\tilde{f}_0$ and $\tilde{f}_1$ have equal indices, they are homotopic. Let $G = \{\tilde{f}_\lambda : \lambda \in [0,1]\}$ be a compact homotopy from $\tilde{f}_0$ to $\tilde{f}_1$. Consider the family $F = \{\tilde{f}_\lambda : \lambda \in [0,1]\}$ of condensing operators, where

$$f\lambda(x) = \begin{cases} 3\lambda \tilde{f}_0(x) + (1-3\lambda) f_0(x), & \text{if } \lambda \in [0, 1/3], \\ \tilde{f}_{3\lambda-1}(x), & \text{if } \lambda \in [1/3, 2/3], \\ 3(1-\lambda)\tilde{f}_1(x) + (3\lambda - 2) f_1(x), & \text{if } \lambda \in [2/3, 1]. \end{cases}$$

Clearly, F is a condensing family of operators connecting f_0 and f_1, the mapping $(\lambda, x) \mapsto f_\lambda(x)$ is continuous, and the operators f_λ have no fixed points on ∂U, i.e., F is a condensing homotopy from f_0 to f_1. **QED**

3.2. EXAMPLES OF COMPUTATION OF THE INDEX OF A CONDENSING OPERATOR

In applications it is important to be able to compute the index of a concrete operator or at least to know if it is different from zero, since the latter implies the existence of a fixed point.

The theory of compact operators provides a number of recipes for testing whether the index of an operator is different from zero. A large numbers of those tests carry over to condensing operators.

3.2.1. Theorem. *Suppose the condensing operator $f: \overline{U} \to E$, where U is a convex bounded open set in the Banach space E, is such that $f(\partial U) \subset \overline{U}$ and has no fixed points on ∂U. Then* $\operatorname{ind}(f, U) = 1$.

Proof. Under the assumptions of the theorem f is homotopic to the constant operator $f_0(x) \equiv x_0$, where x_0 is an arbitrary point of U, and a homotopy from f to f_0 is provided, for example, by the condensing family $F = \{f_\lambda : f_\lambda(x) = \lambda f(x) + (1-\lambda)x_0, \lambda \in [0,1]\}$. The needed conclusion follows from properties 1° and 4° of the index. **QED**

3.2.2. Theorem. *Suppose the condensing operator $f: \overline{U} \to E$, where U is a bounded open set in E, has the property that for some interior point $x_0 \in U$ from the equality $f(x) = \lambda x + (1-\lambda)x_0$ with $x \in \partial U$ it follows that $\lambda < 1$. Then* $\operatorname{ind}(f, U) = 1$.

Proof. Clearly, the index of f on U is defined. Let us check, as in the preceding theorem, that f is linearly homotopic to the operator $f_0(x) \equiv x_0$. All we need to show is that the equality $x = \lambda f(x) + (1-\lambda)x_0$ is impossible if $0 < \lambda < 1$ and $x \in U$. Assuming the contrary, we obtain $f(x) = \mu x + (1-\mu)x_0$, where $\mu = \lambda^{-1} > 1$, which contradicts the assumption on f. **QED**

Recall that, by analogy with the linear case, the number λ is called an *eigenvalue* of the nonlinear operator f, and the vector $x \neq 0$ is called an *eigenvector of f corresponding to λ*, if $f(x) = \lambda x$.

Theorem 3.2.2 is often applied in the following simpler formulation.

3.2.3. Theorem. *Let U be a bounded neighborhood of zero in the Banach space E and let f a condensing operator defined on $\overline{U}$ and with no eigenvectors on ∂U corresponding to real eigenvalues ≥ 1. Then* $\operatorname{ind}(f, U) = 1$.

The following result is a straightforward corollary of Theorem 3.2.3.

3.2.4. Theorem. *Let E be Hilbert space, U a bounded neighborhood of zero in E, and $f: \overline{U} \to E$ a condensing operator with the property that*

$$(f(x), x) < \|x\|^2$$

for all $x \in \partial U$. *Then* $\operatorname{ind}(f, U) = 1$.

In fact, under the assumptions of the theorem, the equality $f(x) = \lambda x$ is impossible if $\lambda \geq 1$ and $x \in \partial U$.

Theorems 3.2.1 and 3.2.2 are consequences of a more general assertion about so-called vector fields that do not point in opposite directions.

3.2.5. Theorem. *Suppose the condensing operators* $f_1, f_2: \overline{U} \to E$, *where* U *is a bounded open set in* E, *have no fixed points on* ∂U *and are such that for* $x \in \partial U$ *the equality* $x - f_1(x) = \lambda[x - f_2(x)]$ *is possible only for* $\lambda > 0$. *Then* $\operatorname{ind}(f_1, U) = \operatorname{ind}(f_2, U)$.

The **proof** follows from the obvious fact that f_1 and f_2 are linearly homotopic.

The existence of a linear homotopy also allows us to establish the following test for the equality of the indices of two condensing operators.

3.2.6. Theorem. *Suppose the condensing operators* $f_1, f_2: \overline{U} \to E$, *where* U *is a bounded open set in* E, *have no fixed points on* ∂U *and*

$$||f_1(x) - f_2(x)|| \leq ||x - f_1(x)||$$

for all $x \in \partial U$. *Then* $\operatorname{ind}(f_1, U) = \operatorname{ind}(f_2, U)$.

An important test for the index of a condensing operator to be different from zero is the following theorem on the index of odd operators.

3.2.7. Theorem. *Suppose the condensing operator* $f: \overline{U} \to E$, *where* U *is a convex symmetric bounded neighborhood of zero in* E, *has no fixed points on* ∂U *and is odd on* U. *Then* $\operatorname{ind}(f, U)$ *is odd.*

Proof. Let S be a fundamental set for f that is symmetric with respect to zero (such a set exists because f is odd and U is symmetric), and let $\tilde{f}$ be the extension of f figuring in the definition of the index of a condensing operator. Then the operator g, $g(x) = \frac{1}{2}\tilde{f}(x) - \frac{1}{2}\tilde{f}(-x)$, is also a compact extension of f, is odd on $\overline{U}$, and satisfies all the requirements formulated in the definition of the index. Hence, by properties of the index of an odd compact operator, $\operatorname{ind}(g, U)$, and together with it $\operatorname{ind}(f, U)$, is an odd number. **QED**

3.3. LINEAR AND DIFFERENTIABLE CONDENSING OPERATORS

In this section we obtain formulas for the computation of indices of linear, asymptotically linear, and Fréchet-differentiable χ-condensing operators.

3.3.1. The index of a linear condensing operator. Let A be a linear condensing operator with constant $k < 1$, considered as an operator on the closure $\overline{U}$ of a bounded open set U in a Banach space E, and with no fixed points on ∂U. If A also has no fixed points inside U, then by property 5° its index on U is equal to zero. Therefore, of interest is only the case where A has at least one fixed point x_0 in U. If $x_0 \neq 0$, then A necessarily has at least one fixed point on the boundary ∂U, and so its index, understood as in **3.1.1**, is not defined. Thus, for a linear operator the index theory is nontrivial only in the following situation: the set U contains the zero vector of E, which is the only fixed point of the operator A. The last condition obviously means that 1 is not an eigenvalue of A. To compute the index of a linear condensing operator in this situation one has a formula that is analogous to the formula for compact linear operators.

Theorem. *Let $A: E \to E$ be a linear condensing operator with constant $k < 1$. Suppose that* 1 *is not an eigenvalue of A. Then* $\operatorname{ind}(A, U) = (-1)^\beta$, *where β is the sum of the multiplicities of the real eigenvalues of A that are larger than* 1 *and U is a bounded open neighborhood of zero in E.*

Proof. Decompose E into the direct sum of E_1 and E_2, where the finite-dimensional subspace E_1 is the sum of all root subspaces $E(\lambda)$ of A that correspond to real eigenvalues $\lambda > 1$ and E_2 is an infinite-dimensional complement, which, like E_1, is invariant under A. Let P denote the linear projection of E onto E_1 parallel to E_2, and put $A_1 = AP$, $A_2 = A(I - P)$. Then A_1 has no real eigenvalues larger than 1 other than the respective eigenvalues of A, while A_2 has no real eigenvalues larger than 1 at all. Moreover, A_1 is of finite rank, and hence compact. Let us show that A and A_1 are homotopic, with the homotopy effected by the condensing family $f = \{f_t : t \in [0, 1]\}$, where $f_t(x) = tA_1x + (1 - t)Ax$. It clearly suffices to verify that the equality $(*)$ $x = tA_1x + (1 - t)Ax$ is impossible when $x \in \partial U$ and $t \in [0, 1]$. Assuming the contrary, let x, t be a pair for which $(*)$ holds. Since $x = x_1 + x_2$ with $x_1 \in E_1$ and $x_2 \in E_2$, we can recast $(*)$ in the form

$$x = A_1x_1 + (1 - t)A_2x_2.$$

Next, since $A_1x_1 \in E_1$ and $(1 - t)A_2x_2 \in E_2$, the uniqueness of the direct sum decomposition yields

$$x_1 = A_1x_1 = Ax_1 \text{ and } x_2 = (1 - t)A_2x_2.$$

This implies that $x_1 = x_2 = 0$, and hence $x_2 = 0$. But $0 \notin \partial U$, so the family f is indeed a homotopy from A to A_1. Consequently, $\operatorname{ind}(A, U) = \operatorname{ind}(A_1, U) = (-1)^\beta$, since the sums of the multiplicities of the real eigenvalues larger than 1 of the operators A and A_1 are obviously equal, and for A_1 the claimed index formula holds true. **QED**

3.3.2. Remark. A fixed point x_0 of a condensing operator f is said to be *isolated* if it has a neighborhood that contains no other fixed points of f. In this case the index of the condensing operator f on any ball $B(x_0, \rho)$ centered at x_0 of sufficiently small radius ρ does not depend on ρ. This common value $\text{ind}(f, B(x_0, \rho))$ will be denoted by $\text{ind}(x_0, f)$ and will be referred to as the *index of the fixed point* x_0 of the operator f. In accordance with this terminology, we can say that in the preceding subsection we obtained a formula for the index of the fixed point zero of a linear condesing operator A: $\text{ind}(0, A) = (-1)^{\beta}$.

3.3.3. Theorem on the asymptotic derivative of a condensing operator. We remind the reader that the operator $f : E \to E$ is said to be *asymptotically linear* if there is a linear operator $f'(\infty)$, called the *derivative of f at the point* ∞, or the *asymptotic derivative of* f, such that

$$\lim_{||x||\to\infty} \frac{||f(x) - f'(\infty)x||}{||x||} = 0.$$

The asymptotic derivative of a (k, χ)-bounded operator f inherits the (k, χ)-boundedness property.

Theorem. *The asymptotic derivative of a (k, χ)-bounded operator f is (k, χ)-bounded* (cf. **1.5.9**).

Proof. Denote $A = f'(\infty)$ and $\omega(x) = f(x) - Ax$. Then $Ax = f(x) - \omega(x)$ and $||\omega(x)||/||x|| \to 0$ when $||x|| \to \infty$. Suppose first the bounded set $M \subset E$ is separated from 0: $||x|| \geq \rho > 0$ for all $x \in M$. Then for any $\beta > 0$ one can write

$$A(M) = \frac{1}{\beta}A(\beta M) \subset \frac{1}{\beta}\left[f(\beta M) - \omega(\beta M)\right].$$

Consequently,

$$\chi(A(M)) \leq \frac{1}{\beta}\chi(f(\beta M)) + \frac{1}{\beta}\chi(\omega(\beta M)),$$

or

$$\chi(A(M)) \leq k\chi(M) + \chi\left[\frac{1}{\beta}\omega(\beta M)\right]$$

(here we used the algebraic semi-additivity and the positive homogeneity of the MNC χ). Letting $\beta \to \infty$ in the last inequality and using the fact that M is separated from 0, we conclude that $\chi(A(M)) \leq k\chi(M)$.

Now let M be an arbitrary bounded subset of E. Write $M = M_1 \cup M_2$, where $M_1 = M \cap B(0, \rho)$ and $M_2 = M \setminus M_1$. Given an arbitrary $\varepsilon > 0$, choose ρ small enough to guarantee that $||Ax|| \leq \varepsilon$ for all $x \in M_1$. Then using properties of the MNC χ, we obtain

$$\chi(A(M)) = \chi(A(M_1) \cup A(M_2)) = \max\{\chi(A(M_1)), \chi(A(M_2))\} \leq \max\{\varepsilon, \chi(A(M_2))\}$$

$$\leq \max\{\varepsilon, k\chi(M_2)\} \leq \max\{\varepsilon, k\chi(M)\}.$$

Since ε is arbitrary, we conclude again that $\chi(A(M)) \leq k\chi(M)$, as needed. **QED**

For asymptotically linear χ-condensing operators with constant $k < 1$ the index is determined by the multiplicity of the eigenvalues of the asymptotic derivative.

3.3.4. Theorem on the index of an asymptotically linear condensing operator. *Let $f: E \to E$ be a χ-condensing operator with constant $k < 1$. Suppose f has an asymptotic derivative $f'(\infty)$ and 1 is not an eigenvalue of $f'(\infty)$. Then f has no fixed points on the spheres $S(0, \rho)$ centered at zero of sufficiently large radius and*

$$\operatorname{ind}(f, B(0, \rho)) = (-1)^{\beta},$$

where β denotes the sum of the multiplicities of the real eigenvalues larger than 1 of the operator $f'(\infty)$.

Proof. We use Theorem 3.2.6. Since 1 is not an eigenvalue of $f'(\infty)$ by hypothesis, there is a constant $\alpha > 0$ such that $||x - f'(\infty)|| \geq \alpha||x||$ for all $x \in E$. In fact, otherwise one could find a sequence $x_n \in E$, $||x_n|| = 1$, such that

$$||x_n - f'(\infty)x_n|| < \frac{1}{n}. \tag{1}$$

The last inequality implies that the sets $\{x_n\}$ and $\{f'(\infty)x_n\}$ have the same MNC: $\chi(\{x_n\}) = \chi(\{f'(\infty)x_n\})$, which is possible only for relatively compact sets. Passing, if necessary, to a subsequence and letting $n \to \infty$ in (1), we obtain that $x_0 = f'(\infty)x_0$ for some vector $x_0 \neq 0$, which contradicts the hypothesis. Now let the number ρ be large enough so that

$$||f(x) - f'(\infty)x|| \leq \frac{\alpha}{2}||x||$$

for all $x \in S(0, \rho)$. Then for such x,

$$||x - f'(\infty)x|| > ||f(x) - f'(\infty)x||.$$

Consequently, f has no fixed points on $S(0, \rho)$ and, by Theorem 3.2.6, $\operatorname{ind}(f, B(0, \rho)) = \operatorname{ind}(f'(\infty), B(0, \rho)) = (-1)^{\beta}$. **QED**

3.3.5. The index of a differentiable condensing operator. For an operator that is differentiable at a "finite" point one has the following analogous result.

Theorem. *Let f be a condensing operator with constant $k < 1$ that is defined in a neighborhood of its fixed point x_0 and is Fréchet-differentiable at x_0. Suppose that* 1 *is not an eigenvalue of the linear operator $f'(x_0)$. Then x_0 is an isolated fixed point of f and* $\operatorname{ind}(x_0, f) = (-1)^\beta$, *where β is the sum of the multiplicities of the eigenvalues larger than* 1 *of $f'(x_0)$.*

The **proof** can be carried out following the scheme of the proof of the preceding theorem, with the difference that here one establishes that the operators f and ϕ, $\phi(x) = x_0 + f'(x_0)(x - x_0)$ are homotopic on the closure $\overline{U}$ of some neighborhood U of x_0. The index of ϕ is obviously equal to the index of the condensing linear operator $f'(x_0)$ on the set $W = U - x_0$, and W is already a neighborhood of zero. By Theorem 3.3.1, $\operatorname{ind}(f'(x_0), W)$ equals $(-1)^\beta$. **QED**

3.4. FURTHER PROPERTIES OF THE INDEX

In this section we continue the study of the properties of the index of condensing operators. In the case of compact operators, properties such as the independence of the index on the behavior inside the domain, the restriction principle, the possibility of defining the index of a compact operator that is given only on the boundary, are well known. The question of the extent to which these properties are enjoyed by the condensing operators is addressed here. We also discuss the possibility of defining the index of a condensing operator through the indices of its finite-dimensional approximations, and several other questions.

3.4.1. Theorem on the local constancy of the index. *Let $f = \{f_\lambda : \lambda \in [0,1]\}$ be a condensing family of operators, acting from the closure $\overline{U}$ of a bounded open subset U of a Banach space E into E, and let the mapping $(\lambda, x) \mapsto f_\lambda(x)$ be continuous. Suppose further that the operator f_0 has no fixed points on the boundary ∂U. Then there is a number $\lambda_0 \in (0,1]$ such that for any $\lambda \in [0, \lambda_0]$ the indices of the operators f_λ on U are defined and have the same value.*

Proof. It suffices to show that there exists a $\lambda_0 > 0$ such that for $\lambda \in [0, \lambda_0]$ the operators f_λ have no fixed points on ∂U. This will imply that f_λ is homotopic to f_0 for all $\lambda \in [0, \lambda_0]$.

Suppose the contrary holds. Then there one can find a sequence of numbers λ_n, $\lambda_n \to 0$ as $n \to \infty$, and a sequence of points $x_n \in \partial U$, such that

$$x_n = f_{\lambda_n}(x_n). \tag{1}$$

Since the family f is condensing, (1) means that the sequence $\{x_n\}$ is relatively compact. With no loss of generality we may assume that it converges to some point $x_0 \in \partial U$. Letting $n \to \infty$ in (1) and using the continuity of the mapping $(\lambda, x) \mapsto f_\lambda(x)$ we obtain $x_0 = f_0(x_0)$, which contradicts the hypothesis of the theorem. **QED**

Theorem 3.4.1 can be generalized to the case where λ ranges in an arbitrary compact topological space Λ.

3.4.2. Theorem. *Let Λ be a compact topological space and let $f = \{f_\lambda : \lambda \in \Lambda\}$ be a condensing family of operators, acting from the closure $\overline{U}$ of a bounded open subset of a Banach space E into E, such that the mapping $(\lambda, x) \mapsto f_\lambda(x)$ is continuous. Suppose that for some $\lambda_0 \in \Lambda$ the operator f_{λ_0} has no fixed points on ∂U. Then there is a neighborhood V_{λ_0} of λ_0 such that for any $\lambda \in V_{\lambda_0}$ the index of f_λ on U is defined and coincides with the index of f_{λ_0}.*

Proof. The existence of a neighborhood V'_{λ_0} of the point λ_0 such that for any $\lambda \in V'_{\lambda_0}$ the operator f_λ has no fixed points on ∂U is verified exactly as in the preceding theorem. Let us show that for λ sufficiently close to λ_0 all f_λ have the same index. Let S be a fundamental set for the family f such that $\overline{U} \cap S \neq \emptyset$ (its existence is guaranteed by Theorem 3.1.4). Extend all operators f_λ of the family f from $\overline{U} \cap S$ to the entire space E with preservation of the convex hull of the range of the family and of the continuity of the mapping $(\lambda, x) \mapsto f_\lambda(x)$. Let $\tilde{f} = \{\tilde{f} : \lambda \in \Lambda\}$ be the resulting family of operators. Since $\tilde{f}_{\lambda_0}$ is compact and has no fixed points on ∂U, and the set $\partial U_S = \partial U \cap S$ is compact, there is a number $\alpha > 0$ such that

$$||x - \tilde{f}_{\lambda_0}(x)|| \geq \alpha \tag{2}$$

for all $x \in \partial U_S$. Moreover, since the mapping $(\lambda, x) \mapsto \tilde{f}_\lambda(x)$ is continuous, one can find a neighborhood $V_{\lambda_0} \subset V'_{\lambda_0}$ of λ_0 such that

$$||\tilde{f}_\lambda(x) - \tilde{f}_{\lambda_0}(x)|| < \frac{\alpha}{2} \tag{3}$$

for all $\lambda \in V_{\lambda_0}$ and all $x \in \partial U_S$. Now we claim that for any $\lambda \in V_{\lambda_0}$ the operators f_λ and $\tilde{f}_{\lambda_0}$ are linearly homotopic. It suffices to verify that the equality

$$x = (1 - t)\tilde{f}_{\lambda_0}(x) + t\tilde{f}_\lambda(x) \tag{4}$$

is impossible when $x \in \partial U$, $\lambda \in V_{\lambda_0}$, and $t \in [0,1]$. In fact, if $x \in \partial U \setminus S$ then (4) cannot hold, because its right-hand side always lies in S. If $x \in \partial U_S$, then by (2) and (3) we have

$$||x - (1-t)\tilde{f}_{\lambda_0}(x) - t\tilde{f}_\lambda(x)|| \geq ||x - \tilde{f}_{\lambda_0}(x)|| - t||\tilde{f}_\lambda(x) - \tilde{f}_{\lambda_0}(x)|| > \frac{\alpha}{2}.$$

Thus, for $\lambda \in V_{\lambda_0}$ the compact operators $\tilde{f}_\lambda$ are homotopic. Consequently, their indices coincide, and hence so do the indices of the condensing operators f_λ, $\lambda \in V_{\lambda_0}$. **QED**

3.4.3. Theorem on the independence of the index on the behaviour inside the domain. *Suppose the operators f_1 and f_2, acting from the closure $\overline{U}$ of a bounded open set $U \subset E$ into E, coincide on ∂U and have no fixed points on ∂U. Then* $\operatorname{ind}(f_1, U) = \operatorname{ind}(f_2, U)$.

Proof. Under the assumptions of the theorem, f_1 and f_2 are homotopic on $\overline{U}$, and consequently their indices coincide. **QED**

3.4.4. The restriction principle. Let the condensing operator f be defined on the closure $\overline{U}$ of a bounded open domain U in a Banach space E and have no fixed point on ∂U, and let E_1 be a closed subspace of E. Suppose that f acts from $\overline{U}$ into E_1. Put $U_1 = U \cap E_1$ and denote by $\overline{U}_1 = \overline{U} \cap E_1$ and $\partial U_1 = \partial U \cap E_1$ the closure and the boundary of U_1 in E_1, respectively. We shall assume that $U_1 \neq \emptyset$ (in the opposite case we obviously have $\operatorname{ind}(f, U) = 0$). Let f_1 denote the restriction of f to $\overline{U}_1$. Since f has no fixed points on ∂U, f_1 has no fixed points on ∂U_1. The set U_1 is bounded and open in the Banach space E_1 and f_1 is a condensing operator with no fixed points on its boundary ∂U_1. Therefore, the index $\operatorname{ind}(f_1, U)$ is defined.

Theorem. *The indices of the condensing operators f on U and f_1 on U_1 coincide:* $\operatorname{ind}(f, U) = \operatorname{ind}(f_1, U_1)$.

Proof. Let S be some fundamental set for f such that $S \subset E_1$. Such a set exists: indeed, the intersection of any fundamental set for f with E_1 is again a fundamental set for f and is contained in E_1. Using Dugundji's theorem, extend f (or, which is the same, f_1) from $\overline{U} \cap S$ to the entire space E and denote by $\tilde{f}$ the restriction of this extension to $\overline{U}$ and by $\tilde{f}_1$ — the restriction of $\tilde{f}$ to $\overline{U}_1$. By the definition of the index of a condensing operator,

$$\operatorname{ind}(f, U) = \operatorname{ind}(\tilde{f}, U) \quad \text{and} \quad \operatorname{ind}(f_1, U_1) = \operatorname{ind}(\tilde{f}_1, U_1).$$

To complete the proof it remains to refer to the analogous restriction principle for compact operators (see **3.9.4**). **QED**

3.4.5. Finite-dimensional approximations. Suppose that the Banach space E is equipped with a sequence of finite-dimensional subspaces $\{E_n\colon n = 1, 2, \dots\}$ and a sequence of finite-dimensional operators $\{P_n\colon E \to E_n,\ n = 1, 2, \dots\}$ such that the following conditions are satisfied:

a) $\|P_n x - P_n y\| \leq \|x - y\|$ for all $x, y \in E$ and all $n = 1, 2, \dots$;

b) $\lim_{n\to\infty} P_n x = x$ for all $x \in E$.

Further, let U be a bounded open set in E and let $f\colon \overline{U} \to E$ be a condensing operator with no fixed points on ∂U. The operators P_n and f yield finite-dimensional (and consequently, condensing) operators $P_n f : \overline{U} \to E_n \subset E$. It turns out that for sufficiently large n the operators $P_n f$, too, have no fixed points on ∂U and are homotopic to f. More precisely, we have the following result.

Theorem. *Under the conditions listed above, there is a positive integer N such that for $n \geq N$ the operator $P_n f$ has no fixed points on ∂U and* $\operatorname{ind}(P_n f, U) = \operatorname{ind}(f, U)$.

Proof. We use Theorem 3.4.2. As Λ we take the set of positive integers with the point ∞ adjoined. As neighborhoods of ∞ we take arbitrary sets that contain ∞ and all positive integers, except possibly for finitely many; a neighborhood of n is any subset of Λ that contains n. It is then readily verified that the resulting topological space is compact.

The family of operators f figuring in Theorem 3.4.2 is defined here as follows:

$$f = \{f_\lambda\colon f_\lambda = P_n f \text{ for } \lambda = n \in \{1, 2, \dots\},\ f_\lambda = f \text{ for } \lambda = \infty\}.$$

It is readily verified that the mapping $(\lambda, x) \mapsto f_\lambda(x)$ is continuous. Put $\lambda_0 = \infty$. Then the operator $f_{\lambda_0} = f$ has no fixed points on ∂U.

Let us show that the family f is condensing. If $\Omega \subset \overline{U}$ and $\overline{\Omega}$ is not compact, then $r_2 = \chi(f(\Omega)) < \chi(\Omega) = r_1$, because f is χ-condensing. In particular, this implies that $r_2 < \infty$. We need to show that $\chi\big(\bigcup_{\lambda\in\Lambda} f_\lambda(\Omega)\big) < \chi(\Omega)$. Pick $\varepsilon > 0$ such that $r_2{+}2\varepsilon < r_1$. Let Q be a finite $(r_2{+}\varepsilon)$-net of the set $f(\Omega)$. Put $\Sigma_1 = (\bigcup_{n=1}^{\infty} P_n Q) \cup Q$. By condition b), the set Σ_1 is compact. Moroeover, it is an $(r_2{+}\varepsilon)$-net of $\bigcup_{\lambda\in\Lambda} f_\lambda(\Omega)$. Indeed, if $u = f_\lambda(x)$ and $y \in Q$ are such that $\|f(x) - y\| \leq r_2{+}\varepsilon$, then for $z = P_\lambda$ $(z \in \Sigma_1)$ one has $\|P_\lambda f(x) - P_\lambda y\| \leq \|f(x) - y\| \leq r_2{+}\varepsilon$. Consequently, any finite ε-net Σ of the set Σ_1 is an $(r_2{+}\varepsilon)$-net of $\bigcup_{\lambda\in\Lambda} f_\lambda(\Omega)$. Thus, $\chi\big(\bigcup_{\lambda\in\Lambda} f_\lambda(\Omega)\big) \leq r_2{+}2\varepsilon < r_1 = \chi(\Omega)$, i.e., the family f is indeed condensing. **QED**

3.4.6. Extension of condensing homotopies. In the classical index theory for compact (and finite-dimensional) operators, the operator f under study is assumed to be given on the boundary ∂U of the set U rather than on its entire closure $\overline{U}$. For compact operators in a Banach space this difference is immaterial, since such an operator can always be extended from ∂U to $\overline{U}$ preserving its compactness. For condensing operators an analogous result is not known.

We give here a partial result, namely, a theorem on the extension of a (k, χ)-bounded family f, with preservation of (k, χ)-boundedness, from the unit sphere S of a Banach space E to the entire closed unit ball $\overline{B}$.

3.4.7. Lemma. *Let the family of operators $f = \{f_\lambda: \overline{B} \to E,\ \lambda \in [0,1]\}$ be given by the formula*

$$f_\lambda(x) = \phi(\lambda, x)x,$$

where the function $\phi: [0,1] \times \overline{B} \to \mathbf{R}$ satisfies the condition

$$0 \le \phi(\lambda, x) \le k \quad (\lambda \in [0,1],\ x \in \overline{B}).$$

Then the family f is condensing with constant k.

Proof. The hypotheses imply that

$$f(\Omega) \subset \bigcup_{\mu \in [0,k]} \mu\Omega \subset \operatorname{co}(\{0\} \cup k\Omega)$$

for any $\Omega \subset \overline{B}$. By the properties of the Hausdorff MNC, this implies that

$$\chi(f(\Omega)) \le k\chi(\Omega). \quad \textbf{QED}$$

3.4.8. Theorem. *Suppose the family $f = \{f_\lambda: \lambda \in [0,1]\}$ of operators $f_\lambda : S \to E$ is condensing with constant k and the mapping $(\lambda, x) \mapsto f_\lambda(x)$ is continuous. Then for any $\varepsilon > 0$ there exists a family $g = \{g_\lambda: \lambda \in [0,1]\}$, $g_\lambda: \overline{B} \to E$, such that g is condensing with constant $k+\varepsilon$, the mapping $(\lambda, x) \mapsto g_\lambda(x)$ is continuous, and g coincides with f on S.*

Proof. Choose $\nu \in (0,1)$ such that $k/(1-\nu) \le k+\varepsilon$. Define the family $g = \{g_\lambda: \lambda \in [0,1]\}$ by the formula

$$g_\lambda(x) = \begin{cases} \frac{1}{\nu}(\|x\| - 1 + \nu) f_\lambda\left(\frac{x}{\|x\|}\right), & \text{if } 1 - \nu \le \|x\| \le 1, \\ 0, & \text{if } \|x\| < 1 - v. \end{cases}$$

Clearly, g coincides with f on S and the mapping $(\lambda, x) \mapsto g_\lambda(x)$ is continuous. Let us show that g is condensing with constant $k+\varepsilon$. For an arbitrary set $\Omega \subset \overline{B}$ we have

$$g(\Omega) = \bigcup_{\lambda \in [0,1]} g_\lambda(\Omega) = \{0\} \cup g(\Omega_1),$$

where $\Omega_1 = \{x \in \Omega: ||x|| \geq 1 - \nu\}$. By the properties of the Hausdorff MNC, this implies $\chi(g(\Omega)) = \chi(g(\Omega_1))$. If $x \in \Omega_1$, then

$$g_\lambda(x) = \frac{1}{\nu}(||x|| - 1+\nu)f_\lambda\left(\frac{x}{||x||}\right) = \phi_1(x)f_\lambda(\phi_2(x)x),$$

where

$$0 \leq \phi_1(x) = \frac{1}{\nu}(||x|| - 1+\nu) \leq 1,$$

$$1 \leq \phi_2(x) = \frac{1}{||x||} \leq \frac{1}{1-\nu}.$$

Applying Lemma 3.4.7 twice and using the (k, χ)-boundedness of the family f we obtain

$$\chi(g(\Omega_1)) \leq \frac{k}{1-\nu}\chi(\Omega_1) \leq (k+\varepsilon)\chi(\Omega_1).$$

Finally, since $\chi(\Omega_1) \leq \chi(\Omega)$, we conclude that

$$\chi(g(\Omega_1)) \leq (k+\varepsilon)\chi(\Omega). \quad \textbf{QED}$$

Actually, a more general result holds true. Specifically, it turns out that the extension of a condensing family of operators from the boundary of the unit ball to the entire closed ball, with preservation of the property of being condensing, can be done in such a manner that the resulting family will contain an arbitrarily given condensing operator defined on the entire ball.

3.4.9. Theorem. *Let the family $f = \{f_\lambda: \lambda \in [0,1]\}$ of operators $f_\lambda: S \to E$ be condensing with constant k and such that the mapping $(\lambda, x) \mapsto f_\lambda(x)$ is continuous. Let $g: \overline{B} \to E$ be a condensing operator, with the same constant k, such that $g|_S = f_0$. Then for any $\varepsilon > 0$ there exists family $G = \{g_\lambda: \lambda \in [0,1]\}$ of operators $g_\lambda: \overline{B} \to E$, condensing with constant $k+\varepsilon$, such that the mapping $(\lambda, x) \mapsto g_\lambda(x)$ is continuous, $g_0 = g$, and $g_\lambda(x) = f_\lambda(x)$ for all $\lambda \in [0,1]$ and all $x \in S$.*

Proof. Choose $H > 0$ such that $kH/(H-1) \leq k+\varepsilon$, and set

$$g_\lambda(x) = \begin{cases} g\left(\frac{H}{H-\lambda}x\right), & \text{if } ||x|| \leq \frac{H-\lambda}{H}, \\ f_{H-\frac{H-\lambda}{||x||}}\left(\frac{x}{||x||}\right), & \text{if } ||x|| > \frac{H-\lambda}{H}. \end{cases}$$

It is readily seen that the family $G = \{g_\lambda: \lambda \in [0,1]\}$ so defined is continuous as a mapping $(\lambda, x) \mapsto g_\lambda(x)$ and coincides with f and g on the corresponding sets. It remains to show that G is condensing with constant $k+\varepsilon$.

Let $\Omega \subset \overline{B}$. For $\lambda \in [0,1]$ denote $\Omega(\lambda) = \{x \in \Omega \colon ||x|| \le (H-\lambda)/H\}$. Now write Ω as $\Omega_1 \cup \Omega_2$, where $\Omega_1 = \bigcup_{\lambda\in[0,1]} \Omega(\lambda)$ and $\Omega_2 = \Omega \setminus \Omega_1$. Then for $\lambda \in [0,1]$ and $x \in \Omega_1$,

$$g_\lambda(x) = g\Big(\frac{H}{H-\lambda}x\Big) = g(\phi_1(\lambda,x)x),$$

where $1 \le \phi_1(\lambda,x) = H/(H-\lambda) \le H/(H-1)$. Using the (k,χ)-boundedness of the operator g and Lemma 3.4.7, we get

$$\chi(G(\Omega_1)) \le \frac{kH}{H-1}\chi(\Omega_1) \le (k+\varepsilon)\chi(\Omega). \tag{5}$$

If now $x \in \Omega_2$, then

$$g_\lambda(x) = f_{H-\frac{H-\lambda}{\|x\|}}\Big(\frac{x}{||x||}\Big) = f_{H-\frac{H-\lambda}{\|x\|}}(\phi_2(x)x), \tag{6}$$

where $1 \le \phi_2(x) = 1/||x|| \le H/(H-1)$. Set $f_2(x) = \phi_2(x)x$. It is then easily checked that for $\|x\| > (H-\lambda)/H$ one has the inequalities

$$0 \le H - \frac{H-\lambda}{\|x\|} \le 1.$$

From this and (6) we obtain the inclusion $G(\Omega_2) \subset f(f_2(\Omega_3))$, where $\Omega_3 = \{x \in \Omega \colon (H-1)/H \le ||x|| \le 1\}$. Since the family f is condensing with constant k, applying Lemma 3.4.7 once more we get

$$\chi(G(\Omega_2)) \le \frac{kH}{H-1}\chi(\Omega_1) \le (k+\varepsilon)\chi(\Omega). \tag{7}$$

Combining (5) and (7) and using the semi-additivity of the Hausdorff MNC for the set $G(\Omega) = G(\Omega_1) \cup G(\Omega_2)$, we conclude that

$$\chi(G(\Omega)) \le (k+\varepsilon)\chi(\Omega). \quad \textbf{QED}$$

3.4.10. The index of a condensing operator that is given only on the boundary of a domain. Let us return to the beginning of **3.4.6** and try to define the index of a condensing operator that is given only on the boundary ∂U of a bounded open set U.

In the particular case where the operator f, condensing with constant k, is defined on the unit sphere S of a Banach space E and has no fixed points on S, the index $\operatorname{ind}(f,S)$ on the boundary of the ball B can be defined as $\operatorname{ind}(g,B)$, where g is an arbitrary condensing extension of f from S to B (the existence of such an extension is guaranteed by Theorem

3.4.8). By Theorem 3.4.3, the number $\operatorname{ind}(f, S)$ does not depend on the choice of the extension.

Now let us consider the general case. Suppose the condensing operator f is given on the boundary ∂U of a bounded open subset U of a Banach space E and has no fixed points on ∂U. Let S be some fundamental set for f (with respect to ∂U), such that $\partial U \cap S \neq \emptyset$; its existence is guaranteed by Theorem 3.1.4. Extend f from the closed set $\partial U \cap S$ to the entire space E, preserving the convex hull of its range, and denote the restriction of this extension to $\overline{U}$ by $\tilde{f}$. Define

$$\operatorname{ind}(f, \partial U) = \operatorname{ind}(\tilde{f}, U).$$

The independence of ind $(f, \partial U)$ on the choice of the fundamental set S and of the extension $\tilde{f}$ is verified in exactly the same manner as in **3.1.8**.

If the condensing operator f, given on ∂U, admits a condensing extension g to the entire set U, then its index $\operatorname{ind}(f, U)$ can be defined in one of the two ways indicated above: as $\operatorname{ind}(g, U)$, or as $\operatorname{ind}(\tilde{f}, U)$. However, it is clear that in the present situation the two definitions give the same number, since any set that is fundamental for g (with respect to U) is also fundamental for f with respect to ∂U.

Thus, in particular, if the operator f is given and condensing on $\overline{U}$, then in order to define its index it suffices to use information about f on ∂U. This conclusion is quite natural if one one recalls that, as mentioned in **3.4.3**, the index of a condensing operator does not depend on the behavior of the operator inside the domain.

3.5. GENERALIZATION OF THE NOTION OF INDEX TO VARIOUS CLASSES OF MAPS

In this section we are concerned with the possibility of extending the constructions and results described above to ψ-condensing operators, where ψ is an arbitrary MNC (see **1.2.1**), as well as to other classes of maps.

3.5.1. The index of ψ-condensing maps. In giving the definition of the index of a χ-condensing operator by means of fundamental sets and in establishing its properties a fundamental role was played by the following two facts: first, for any χ-operator (or family of operators) one can construct a fundamental set that contains an arbitrarily given compact set, and, second, linear homotopies of condensing operators are admissible, meaning that if f_0 and f_1 are χ-condensing operators, then the family $f = \{f_\lambda : f_\lambda(x) =$

$(1-\lambda)f_0(x)+\lambda f_1(x),\ \lambda \in [0,1]\}$ is also condensing and the mapping $(\lambda, x) \mapsto f_\lambda(x)$ is continuous. Let us isolate those properties of the Hausdorff MNC thanks to which these two facts hold true.

In Theorem 3.1.4, in the construction of the fundamental set S, we used the montonicity of the Hausdorff MNC χ and its invariance under the adjunction of compact sets (the latter being a consequence of the semi-additivity and regularity of χ). Therefore, if the MNC ψ is monotone and invariant under the adjunction of compact sets, and if f is a ψ-condensing operator, then Theorem 3.1.4 is valid for f.

However, it is readily verified that in the proof of the independence of the index on the choice of a fundamental set, given in **3.1.8**, one could manage without a fundamental set that contains an arbitrarily prescribed compact set. It simply suffices to assume that f admits a fundamental set. The independence of the index on the choice of the fundamental set is then established as follows. If the operator f has no fixed points on the set $\overline{U}$, then the same is true for all extensions $\tilde{f}$ of f from an arbitrary fundamental set S to $\overline{U}$, and consequently all indices $\operatorname{ind}(\tilde{f}, U)$ are equal to zero. Now suppose that f does has at least one fixed point on $\overline{U}$. Then all fundamental sets S intersect and their intersection S_0 is also a nonempty fundamental set. Proceeding exactly as **3.1.8**, one establishes that for any fundamental set S and any extension $\tilde{f}_1$ of f from $\overline{U} \cap S$ one has the equality

$$\operatorname{ind}(\tilde{f}_1, U) = \operatorname{ind}(\tilde{f}_0, U),$$

where $\tilde{f}_0$ is an extension of f from $\overline{U} \cap S_0$.

Fundamental sets also exist for operators that are condensing with respect to a monotone MNC ψ and have fixed points on $\overline{U}$. In verifying properties 1°–5° of the index of a condensing operator f, the other properties of the MNC ψ play no role. Thus, setting, by definition, $\operatorname{ind}(f, U) = 0$ whenever f has no fixed points in $\overline{U}$, one obtains the following result.

Theorem. *Suppose the operator f acts from the closure $\overline{U}$ of a bounded open subset U of a Banach space E into E, is ψ-condensing with respect to a monotone* MNC ψ, *and has no fixed points on the boundary ∂U of U. Then one can define an integer-valued characteristic,* $\operatorname{ind}(f, U)$, *called the index of f on U, which enjoys all properties* 1°–5° *formulated in* **3.1.1**.

3.5.2. Linear condensing homotopies. Let us address now the question whether linear condensing homotopies are admissible. The monotonicity of the MNC ψ alone does not suffice to guarantee that the family of operators

$$f = \{f_\lambda : f_\lambda(x) = (1-\lambda)f_0(x)+\lambda f_1(x),\ \lambda \in [0,1]\} \tag{1}$$

will be ψ-condensing whenever the operators f_0 and f_1 are ψ-condensing (relevant examples are given below). In this respect we recall the following assertion (see **1.5.6**).

Theorem. *Suppose the operators f_0 and f_1 are ψ-condensing, where the* MNC *ψ is semi-additive and its range is linearly ordered. Then the family* (1) *is ψ-condensing.*

Since a semi-additive MNC is also monotone, if one requires that the MNC ψ be semi-additive and have a linearly-ordered range then for ψ-condensing operators the supply of ψ-condensing homotopies is rich enough to allow us the construction of a meaningful index theory. In particular, all theorems of Section 3.2 whose proofs rely only on the transition from one condensing operator to another via a linear homotopy remain valid in this general setting. The requisite constructions are carried out with no difficulty.

Let us give an example of a homotopy for ψ-condensing operators that is different from the linear one. It is known (and also readily verified) that for a family of operators $f = \{f_\lambda : \lambda \in [0,1]\}$, $f_\lambda : M \subset E \to E$ to be compact it suffices that the operator f_λ be compact for any fixed $\lambda \in [0,1]$ and that the family f be continuous in λ uniformly in $x \in M$. An analogous assertion is valid for ψ-condensing operators.

3.5.3. Lemma. *Suppose that in the Banach space E there is given a uniformly continuous real-valued semi-additive* MNC *$\psi: 2^E \to \mathbf{R}$. Let the family of operators $f = \{f_\lambda : \lambda \in [0,1]\}$, $f_\lambda : M \subset E \to E$, be continuous in λ uniformly in $x \in M$. Then for any set $\Omega \subset M$ the function $m(\lambda) = \psi[f_\lambda(\Omega)]$ is continuous on $[0,1]$ and*

$$\psi[f(\Omega)] = \max_{\lambda \in [0,1]} m(\lambda).$$

Proof. Since ψ is uniformly continuous, for any $\varepsilon > 0$ there is a symmetric neighborhood V of zero in E such that the V-closeness of the sets Ω_1 and Ω_2 implies $|\psi(\Omega_1) - \psi(\Omega_2)| \leq \varepsilon$. Next, from the fact that the continuity of the family f in λ is uniform in $x \in M$ it follows that for the indicated neighborhood V there is a $\delta > 0$ such that $|\lambda_1 - \lambda_2| \leq \delta$ implies $f_{\lambda_1}(x) - f_{\lambda_2}(x) \in V$ for all $x \in M$. Let Ω be an arbitrary subset of M. Then the last inclusion means that for $|\lambda_1 - \lambda_2| \leq \delta$ the sets $f_{\lambda_1}(\Omega)$ and $f_{\lambda_2}(\Omega)$ are V-close. But then, by the choice of V,

$$|m(\lambda_1) - m(\lambda_2)| = |\psi[f_{\lambda_1}(\Omega)] - \psi[f_{\lambda_2}(\Omega)]| \leq \varepsilon.$$

Thus, the function m is continuous on $[0,1]$. Let λ^* be a point where m attains its maximum. Pick an arbitrary $\varepsilon > 0$ and find V and δ as above. Now choose points

$0 = \lambda_1 < \lambda_2 < \ldots < \lambda_k = 1$ on $[0, 1]$ such that, first, the distance between neighbors does not exceed 2δ and, second, $\lambda_i = \lambda^*$ for some i. Then it is readily seen that

$$\bigcup_{i=1}^{k} f_{\lambda_i}(\Omega) \subset f(\Omega) \subset \bigcup_{i=1}^{k} f_{\lambda_i}(\Omega) + V.$$

Therefore, the sets $f(\Omega)$ and $\bigcup_{i=1}^{k} f_{\lambda_i}(\Omega)$ are V-close, and consequently

$$\left|\psi[f(\Omega)] - \psi\Big[\bigcup_{i=1}^{k} f_{\lambda_i}(\Omega)\Big]\right| \le \varepsilon. \tag{2}$$

It remains to observe that

$$\psi\Big[\bigcup_{i=1}^{k} f_{\lambda_i}(\Omega)\Big] = \max_i \psi\big[f_{\lambda_i}(\Omega)\big] = m(\lambda^*) = \max_{\lambda \in [0,1]} m(\lambda),$$

thanks to the semi-additivity of ψ and the choice of λ^*. Since ε is arbitrary, the needed equality now follows from (2). **QED**

3.5.4. Theorem (example of a nonlinear condensing homotopy). *Let ψ be a uniformly continuous real-valued semi-additive* MNC *in a Banach space E. Suppose the family of operators $f = \{f_\lambda : \lambda \in [0, 1]\}$, $f_\lambda : M \subset E \to E$, is continuous in λ uniformly with respect to $x \in M$ and f_λ is ψ-condensing for any fixed $\lambda \in [0, 1]$. Then the family f is ψ-condensing.*

Proof. We need to show that $\psi(f(\Omega)) < \psi(\Omega)$ for any set $\Omega \subset M$ whose closure is not compact. By the preceding lemma, there is a point $\lambda^* \in [0, 1]$ such that $\psi(f(\Omega)) = \psi(f_{\lambda^*}(\Omega))$. Since the operator f_{λ^*} is ψ-condensing, we get $\psi(f_{\lambda^*}(\Omega)) < \psi(\Omega)$. **QED**

3.5.5. The index of a ψ-condensing operator that is given only on the boundary. Let us now examine the case where the ψ-condensing operator f, where ψ is some MNC, is given only on the boundary ∂U of a bounded open subset U of a Banach space E. To the authors' knowledge, no theorems about the extension of such an operator from ∂U to U with the preservation of the ψ-condensing property are available, even in the case where U is the unit ball in E. Hence, if one follows the scheme of **3.4.10**, the first of the recipes indicated therein of defining the index of an arbitrary ψ-condensing operator cannot be implemented in the present setting.

Let us determine the extent to which the method of fundamental sets can be employed for ψ-condensing operators that are defined only on the boundary. As it turns out, here neither the existence of fundamental sets for an operator, nor the monotonicity of the MNC ψ in question, is by itself sufficient for establishing the independence of the index on the choice of a fundamental set.

When the operator f is given on the whole set $\overline{U}$ and possesses a fundamental set S that does not intersect $\overline{U}$, f has no fixed points on $\overline{U}$ and therefore it is correct to set its index equal to zero. In the case of an operator f given only on the boundary, the situation is more complicated, the reason being that f can have two types of fundamental sets that do not intersect ∂U, namely sets contained in U and sets contained in the complement of $\overline{U}$. If the ψ-condensing operator f possesses a fundamental set $S \subset U$ [resp. $S \subset E \setminus \overline{U}$] then it is natural to set its index equal to one [resp. zero]. Hence, if f simultaneously possesses fundamental sets of the two types, then its index is not correctly defined.

Let us give an example of a ψ-condensing operator that possesses fundamental sets of the two aforementioned types, where the MNC ψ is monotone.

Define the operator F on the unit sphere T_1 of the Banach space c_0 by the formula (see **1.6.8**)

$$F(x) = F(x_1, x_2, \dots) = (1, x_1, x_2, \dots). \tag{3}$$

Then F has no fixed points. In fact, if $f(x) = x$, then (3) implies that $x_n = 1$ for all n, but this sequence does not converge to zero. It is readily verified that F is condensing with respect to the MNC ψ considered in **1.2.7**:

$$\psi(\Omega) = \frac{1}{n(\Omega)+1},$$

where $n(\Omega) = \min_{x \in \Omega} n(x)$ and $n(x)$ is the number of coordinates of the vector x that are not smaller than 1. The MNC ψ is semi-additive, and hence monotone.

Since $F(T_1) \subset T_1$, the one-element set $S = \{0\}$ is fundamental for F with respect to T_1 and is contained inside the unit ball (the latter is regarded here as the domain on the boundary of which the operator F is given). But the set $S_2 = \{x^*\}$, where $x^* = (2, 0, 0, \dots)$ is also fundamental. To see this we need only verify that condition 3) in the definition of a fundamental set is satisfied (see **3.1.3**). Suppose that for some point $x = (x_1, x_2, \dots)$ one has $x = \operatorname{co}[\{F(x)\} \cup S_2]$; this means that x is an interior point of the segment connecting $F(x)$ and x^*: $x = \lambda F(x) + (1-\lambda)x^*$, $0 < \lambda < 1$. But from this last equality it follows, in particular, that $x_1 = 2 - \lambda > 1$, which is impossible because $\|x\| = \max\{|x_i| : i = 1, 2, \dots\} = 1$.

One of the requirements on the MNC ψ allowing for a correct definition of the index of a ψ-condensing operator that is given only on the boundary of a domain is the invariance of ψ under the adjunction of one-element sets (and hence of arbitrary finite sets, too). The same arguments as those used in **3.1.4** yield a proof of the following result.

Theorem. *Let $f = \{f_\lambda : \lambda \in \Lambda\}$ be a ψ-condensing family of operators that act from a closed subset M of a Banach space E into E, where the* MNC ψ *is assumed to be monotone and invariant under the adjunction of one-element sets. Suppose Λ is a compact topological space and the mapping $(\lambda, x) \mapsto f_\lambda(x)$ is continuous. Then f has a fundamental set that contains an arbitrarily prescribed finite subset $N \subset E$.*

In the case where the MNC ψ has the properties indicated in the theorem, the index of a ψ-condensing operator $f: \partial U \to E$ with no fixed points on ∂U is defined in the usual manner, as follows. Let S be a fundamental set for f such that $S \cap \partial U \neq \emptyset$ (the existence of such sets is guaranteed by the preceding theorem). Use Dugundji's theorem to extend f from the set $S \cap \partial U$ to the entire space with preservation of the convex hull of its range and denote by $\tilde{f}$ the restriction of this extension to $\overline{U}$ (and also to ∂U). Clearly, the compact operator $\tilde{f}$ has no fixed points on ∂U and so its index $\mathrm{ind}(\tilde{f}, U)$ is defined. Set, by definition,

$$\mathrm{ind}(f, \partial U) = \mathrm{ind}(\tilde{f}, U).$$

The correctness of this definition is established in almost the same manner as in **3.1.8**: for two arbitrary fundamental sets S_1, S_2 that intersect ∂U and two corresponding extensions $\tilde{f}_1, \tilde{f}_2$ one shows that $\mathrm{ind}(\tilde{f}_1, U) = \mathrm{ind}(\tilde{f}_2, U)$. The case where $S_1 \subset S_2$ is dealt with exactly as in **3.1.8**. The case where S_1 and S_2 are in arbitrary relative position can be reduced to the preceding case as follows. Construct a fundamental set S_3 that contains points $x_1 \in \partial U \cap S_1$ and $x_2 \in \partial U \cap S_2$. Since the intersection of any two fundamental sets is again fundamental, to pass from S_1 to S_2 one can consider successively the following pairs of fundamental sets contained in one another: S_1 and $S_1 \cap S_3$, $S_1 \cap S_3$ and S_3, S_3 and $S_3 \cap S_2$, $S_3 \cap S_2$ and S_2, together with corresponding extensions of the operator f, and apply to each pair the already known arguments.

A ψ-condensing operator f, where the MNC ψ is monotone and semi-additively nonsingular (i.e., invariant under the adjunction of one-element sets), given only on the boundary ∂U of an open set U, may of course possess fundamental sets that do not intersect ∂U. However, it is readily verified that it cannot have simultaneously fundamental sets lying inside and outside U.

The existence of a fundamental set that does not intersect ∂U simplifies considerably the computation of the index of a ψ-condensing operator f on ∂U. Indeed, if S_0 is such a set, then for any fundamental set S satisfying $S \cap \partial U \neq \emptyset$ and a corresponding extension $\tilde{f}$, the operators $\tilde{f}$ and $f_0(x) \equiv x_0$, where x_0 is an arbitrary point of S_0, are linearly homotopic on ∂U. Consequently, if $S_0 \subset U$, then $\mathrm{ind}(f, \partial U) = 1$, whereas if $S_0 \subset E \setminus \overline{U}$, then $\mathrm{ind}(f, \partial U) = 0$.

3.5.6. The index of an ultimately compact operator. There is yet another class of operators for which the notion of index can be defined by using fundamental sets, namely, the class of ultimately compact operators (see **1.6.3**).

Let f be an ultimately compact operator given on the closure $\overline{U}$ of a bounded open subset U of a Banach space E. If the ultimate range $f^\infty(\overline{U})$ of f is empty then, in particular, f has no fixed points on $\overline{U}$. Hence, in attempting to define the index of the ultimately compact operator f in this case it is natural to set $\operatorname{ind}(f, U) = 0$.

Now suppose that the set $f^\infty(\overline{U})$ is not empty. Then it is fundamental for f with respect to $\overline{U}$. Indeed, conditions 1) and 2) of the definition of a fundamental set are obviously satisfied for $f^\infty(\overline{U})$. It remains to verify condition 3). To this end it suffices to show that if for some point $x_0 \in \overline{U}$ one has the inclusion

$$x_0 \in \operatorname{co}[\{f(x_0)\} \cup f^\infty(\overline{U})], \tag{4}$$

then necessarily $x_0 \in f^\infty(\overline{U})$. Let us check by transfinite induction that if the point x_0 satisfies (4), then it belongs to all sets T_α of the transfinite sequence figuring in the construction of the ultimate range (see **1.6.1**). For $\alpha = 0$ the inclusion $x_0 \in T_0 = \overline{\operatorname{co}}\, f(\overline{U})$ is obvious. Suppose $x_0 \in T_\alpha$ for all $\alpha < \alpha_0$ and let us show that $x_0 \in T_{\alpha_0}$. Two cases are possible:

a) The ordinal number $\alpha_0 - 1$ exists. Then we have successively

$$x_0 \in \operatorname{co}[\{f(x_0)\} \cup f^\infty(\overline{U})] \subset \operatorname{co}[f(\overline{U} \cap T_{\alpha_0 - 1}) \cup f^\infty(\overline{U})] \subset \overline{\operatorname{co}}[f(\overline{U} \cap T_{\alpha_0-1})] = T_{\alpha_0}.$$

b) The ordinal $\alpha_0 - 1$ does not exist. Then $T_{\alpha_0} = \bigcap_{\alpha < \alpha_0} T_\alpha$ and the inclusion $x_0 \in T_{\alpha_0}$ is obvious.

Therefore, $x_0 \in T_\alpha$ for any α, and consequently the point x_0 belongs to the ultimate range of the operator f, since $f^\infty(\overline{U})$ is one of the sets T_α.

Thus, the (nonempty) set $f^\infty(\overline{U})$ is fundamental for the ultimately compact operator f, and so one can follow the scheme used earlier and define the index of f, which will enjoy all properties 1°–5°. If f also has fundamental sets different from its ultimate range, then the independence of the index on the choice of the fundamental set is established exactly as in **3.5.1**.

Incidentally, since any ultimately compact operator in a Banach space is condensing with respect to a semi-additive (and hence monotone) MNC (Theorem 1.7.4), the possibility of defining an index for ultimately compact operators follows already from the results of **3.5.1**.

The supply of ultimately compact homotopies is not so rich. As we remarked in **1.6.8**, even the simple linear homotopy from an ultimately compact operator to another such operator is not always ultimately compact. This fact, established in **1.6.8**, can be proved in a considerably simpler manner using the index theory for ultimately compact operators. In fact, let

$$f_0(x) = (0, 0, \dots),$$

$$f_1(x) = f_1(x_1, x_2, \dots) = (1, x_1, x_2, \dots),$$

$$x = (x_1, x_2, \dots) \in \overline{B} \subset c_0.$$

Clearly, $\operatorname{ind}(f_1, B) = 0$, since the operator f_1 has no fixed points (see **3.5.5**), whereas $\operatorname{ind}(f_0, B) = 1$ (property 4° of the index). From this it immediately follows that the family of operators $f = \{f_\lambda : \lambda \in [0,1]\}$, $f_\lambda(x) = (1-\lambda)f_0(x) + \lambda f_1(x)$, is not ultimately compact, since otherwise it would effect a homotopy from f_0 to f_1 and consequently the indices $\operatorname{ind}(f_0, B)$ and $\operatorname{ind}(f_1, B)$ would coincide.

3.5.7. Example of the computation of the index of an ultimately compact operator. As an example of such a computation we give the following result.

Theorem. *Let R be a nonempty bounded open subset of a Banach space E and let the continuous operator $f : \overline{R} \to \overline{R}$ be ultimately compact and have no fixed points on ∂R. Suppose that one of the following conditions is satisfied:*

a) *$f^\infty(\overline{R}) \neq \emptyset$;*

b) *in $\overline{R}$ there is a nonempty subset Ω such that $\Omega \subset \overline{\text{co}}\, f(\Omega)$;*

c) *in $\overline{R}$ there is a nonempty compact subset K such that $f(K) \subset K$;*

d) *for some point $x_0 \in \overline{R}$ the set $\{f^n(x_0) : n = 0, 1, 2, \dots\}$ is relatively compact;*

e) *the space E is semi-reflexive.*

Then $\operatorname{ind}(f, R) = 1$.

Proof. We first show that the conclusion of the theorem holds true under condition a), and then verify that any of the conditions b) through e) implies a).

Suppose $S = f^\infty(\overline{R})$ is a fundamental set for the operator f and let $\tilde{f} : \overline{R} \to S \subset \overline{R}$ be a compact extension of f from S. By the properties of the index of a compact operator, $\operatorname{ind}(\tilde{f}, R) = 1$, and consequently $\operatorname{ind}(f, R) = 1$, too.

If condition b) is satisfied, then obviously the nonempty set Ω is contained in $f^\infty(\overline{R})$, and hence a) holds.

Suppose condition c) is satisfied. We construct a transfinite sequence of sets $\{\tau_\alpha\}$ as follows: $\tau_0 = f(K)$, $\tau_\alpha = f(\tau_{\alpha-1})$ if $\alpha - 1$ exists, and $\tau_\alpha = \bigcap_{\beta<\alpha} \tau_\beta$ if $\alpha - 1$ does not exist. It is readily verified that $\tau_\alpha \neq \emptyset$ for all α. Further, it is clear that $\tau_\alpha \subset T_\alpha$, where $\{T_\alpha\}$

is the sequence of sets constructed for the operator f on $\overline{R}$ when definining its ultimate range $f^\infty(\overline{R})$. Consequently, $f^\infty(\overline{R}) \neq \emptyset$, i.e. a) holds.

d) implies c). Indeed, set $K = \overline{K_1}$, where $K_1 = \{f^n(x_0)\colon n = 0, 1, 2, \dots\}$. Then the set K is compact and contained in $\overline{R}$. Clearly, $f(K_1) \subset K_1$. Hence, $f(K) = f(\overline{K_1}) \subset \overline{f(\overline{K_1})} \subset \overline{f(K_1)} \subset \overline{K_1} = K$.

Finally, e) implies a). Indeed, in a semi-reflexive Banach space E any centered family of convex, closed, and bounded (and hence weakly compact) sets has a nonempty intersection. From this it readily follows that $T_\alpha \neq \emptyset$ for any α; in particular, $f^\infty(\overline{R}) \neq \emptyset$. **QED**

3.5.8. The ultimate range of a ψ-condensing operator. For a continuous ultimately compact operator f that maps a convex, closed, and bounded set $\overline{R}$ into itself the set $f^\infty(\overline{R})$ may be empty, as we saw on the example of the operator $F(x) = F(x_1, x_2, \dots) = (1, x_1, x_2, \dots)$ from the closed unit ball $\overline{B}$ of the space c_0 into itself. Therefore, even the requirement that the MNC ψ be semi-additive does not suffice to guarantee that a ψ-condensing operator f will have a nonempty ultimate range. Two conditions on the MNC ψ guaranteeing that ψ-condensing operators have a nonempty ultimate range are formulated in the following theorem.

Theorem. *Let R be a nonempty, convex, bounded, and open subset of a Banach space E and let $f\colon \overline{R} \to \overline{R}$ be a ψ-condensing operator with no fixed points on ∂R. Suppose that one of the following conditions is satisfied:*

a) *the MNC ψ is invariant under the adjunction of one-element sets;*

b) *ψ is semi-additive and invariant under translations.*

Then $\operatorname{ind}(f, R) = 1$.

Proof. Under condition a) or b) the operator f is ultimately compact, and so the preceding theorem applies. Suppose a) holds. Consider the set $K_1 = \{f_n(x_0)\colon n = 0, 1, 2, \dots\}$, where x_0 is an arbitrary point in $\overline{R}$. Clearly, $f(K_1) \neq \emptyset$ and $f(K_1) \cup \{x_0\} = K_1$. Consequently, $\psi(f(K_1)) = \psi(K_1)$. But then from the definition of condensing operators it follows that the set $K = K_1$ is compact, i.e., condition d) of the preceding theorem is satisfied.

Now let us show that b) implies a). Let $\Omega \subset \overline{R}$, $x_1 \in \Omega$, $x_0 \in \overline{R}$. Since ψ is monotone (thanks to its semi-additivity) and invariant under translations, $\psi(\Omega) \geq \psi(\{x_1\}) = \psi(\{x_0\})$. Using the semi-additivity of ψ, we conclude that

$$\psi(\Omega \cup \{x_0\}) = \max\{\psi(\Omega), \psi(\{x_0\})\} = \psi(\Omega),$$

as needed. **QED**

3.5.9. The index of a K_2-operator. Let us consider again the case where the operator f is given only on the boundary of a bounded open set U. In **3.5.5** we gave sufficient conditions on the MNC ψ under which an index can be defined for a ψ-condensing operator f given on ∂U, namely, monotonicity and invariance under the adjunction of one-element sets.

A notion that generalizes the index of a condensing operator with such an MNC ψ is that of the index of a K_2-operator (see **1.7.1**).

Theorem. *Suppose the K_2-operator f is given on the boundary ∂U of a bounded open subset U of a Banach space E and has no fixed points. Then one can define an integer-valued characteristic* $\operatorname{ind}(f, \partial U)$, *called the index of f on ∂U, which enjoys the natural properties of an index.*

The index of a K_2-operator on ∂U can also be introduced by means of fundamental sets. In verifying that this index is correctly defined, the crucial role in establishing that the definition is independent of the choice of the fundamental set is played by the fact that for a K_2-operator one can always find a fundamental set S containing two arbitrarily prescribed points.

3.5.10. Theorem on the existence of a fundamental set for a K_2-operator. *Let $f = \{f_\lambda : \lambda \in \Lambda\}$ be a family of K_2-operators, $f_\lambda : M \to E$, where M is a subset of a Banach space E and Λ is some set. Then f admits a fundamental set S that contains an arbitrarily prescribed pair of points $x_1, x_2 \in E$.*

Proof. Here one can argue almost word-for-word as in Theorem 3.1.4. In E consider the family $\mathfrak{M}$ of all sets that satisfy the conditions of the definition of a fundamental set, except possibly for compactness, and contain the points x_1 and x_2. This family includes, for example, the set $T_0 = \overline{\text{co}}\,[\{x_1, x_2\} \cup f(M)]$; it also includes, together with any $T \in \mathfrak{M}$, the set $T_1 = \overline{\text{co}}\,[\{x_1, x_2\} \cup f(M \cap T)]$ (and $T_1 \subset T$).

Obviously, the set $S = \bigcup_{T \in \mathfrak{M}} T$ also belongs to $\mathfrak{M}$, and, by minimality, it satisfies

$$S = \overline{\text{co}}\,[\{x_1, x_2\} \cup f(M \cap S)].$$

By the definition of the K_2-operators, this implies that S is compact, and hence is a fundamental set for f. **QED**

3.5.11. K_2- and K_c-homotopies. The theorem proved above permits us to employ the same arguments as in **3.5.5** for operators that are condensing with respect to a monotone semi-additive and nonsingular MNC ψ, in order to establish that the index of a K_2-operator given only on the boundary of a domain, does not depend on the choice of the fundamental set.

It goes without saying that one can also define the index for a K_2-operator that is given not only on the boundary ∂U, but also on the closure of the domain U.

The supply of K_2-homotopies is also rather poor. For example, the linear homotopy between two arbitrary K_2-operators is not necessarily a K_2-homotopy. If one considers the whole chain (see **1.7.2**) of classes of K_2-, K_3-,...,K_∞-, K_c-operators, then quite naturally the index of K_c-operators will enjoy the best properties. The most important fact is that any K_c-operator f can be linearly connected to a compact operator, and the implementing homotopy is a K_c-family of operators. For the classes of K_n-operators with $n = 2, 3, \ldots$ even this assertion is not true.

The last circumstance concerning K_c-operators permits us to extend to the class of such operators many of the properties of the index of χ-condensing operators. For instance, for K_c-operators one can prove that every homotopy class contains a compact operator, an analogue of Hopf's classification theorem, as well as other results.

Definition. Two K_c-operators $f_0, f_1: \overline{U} \to E$ are said to be *homotopic* if there exists a K_c-family of operators $f = \{f_\lambda: \lambda \in [0,1]\}$, $f_\lambda: \overline{U} \to E$, such that $f_\lambda(x) \neq x$ for all $\lambda \in [0,1]$ and all $x \in \partial U$, and the mapping $(\lambda, x) \mapsto f_\lambda(x)$ is continuous. The operators f_0, f_1 are said to be *weakly homotopic* if there is a finite collection of K_c-operators $g_1, \ldots, g_n$ such that f_0 is homotopic to g_1, g_i is homotopic to g_{i+1} for $i = 1, \ldots, n-1$, and g_n is homotopic to f_1.

Theorem. *Let f_0 and f_1 be K_c-operators that act from the closure $\overline{U}$ of a Jordan domain of a Banach space E into E. Suppose that f_0 and f_1 have no fixed points on ∂U and* $\operatorname{ind}(f_0, U) = \operatorname{ind}(f_1, U)$. *Then f_0 and f_1 are weakly homotopic.*

3.6. THE INDEX OF OPERATORS IN LOCALLY CONVEX SPACES

The main difference here from the Banach space case is that for maps that act in locally convex spaces (LCSs) no results are available of the type of Dugundji's theorem, which plays an important role in the contruction of the index in the Banach space setting by the method of fundamental sets. However, for such maps one can prove assertions about "quasi-extension" with preservation of desired properties, where the new operator does not coincide exactly with the given one on its domain, being instead only sufficiently close to it. As it turns out, this already allows us to introduce a correctly-defined index.

Thus, let E be a Hausdorff LCS, $\{p\}$ a sufficient family of continuous seminorms that define the locally convex topology of E, U an open subset of E, $\overline{U}$ and ∂U its closure and boundary, respectively. For a compact operator f given on $\overline{U}$ (or even only on ∂U) and with no fixed points on ∂U the notion of index $\operatorname{ind}(f, U)$ is meaningful and completely analogous to that of the index for operators acting in Banach spaces. For example, in the present setting the index of a compact operator f in a LCS E enjoys all properties 1° through 5° listed in **3.1.1.**

The notion of a fundamental set, and the theorems asserting the existence of such sets for χ-condensing operators, operators that are ψ-condensing with respect to a monotone MNC ψ, and K_2-operators can also be extended to operators acting in LCSs, with no modifications compared to **3.1.3.**

3.6.1. The "quasi-extension" theorem. *Let f be a continuous map from a compact subset M of an* LCS *E into E. Then for any continuous seminorm p on E there exists a continuous map $g: E \to E$ such that $g(E) \subset \overline{\mathrm{co}}\,[f(M)]$ and $p(f(x) - g(x)) \leq 1$ for all $x \in M$.*

Proof. The map f, being continuous on the compact set M, is uniformly continuous there. Hence, there exists a continuous seminorm p_0 on E such that $p(f(x) - f(y)) \leq 1/2$ whenever $x, y \in M$ and $p_0(x - y) \leq 1$. In M find a finite $(1/2)$-net Q for M with respect to the seminorm p_0. Next, for each $y \in Q$ define a continuous positive function $\mu_y : E \to \mathbf{R}$ by the rule

$$\mu_y(x) = \begin{cases} 1 + \varepsilon - p_0(x - y), & \text{if } p_0(x - y) \leq 1, \\ \varepsilon, & \text{if } p_0(x - y) > 1. \end{cases}$$

The positive number ε will be chosen below.

Now define the sought-for map g as follows:

$$g(x) = \Big[\sum_{y \in Q} \mu_y(x)\Big]^{-1} \sum_{y \in Q} \mu_y(x) f(y).$$

Clearly, g is continuous, finite-dimensional, and maps E into $\overline{\mathrm{co}}\,[f(M)]$ (more precisely, into $\overline{\mathrm{co}}\,[f(Q)]$).

Let $x \in M$. Then

$$p(f(x) - g(x)) \leq \Big[\sum_{y \in Q} \mu_y(x)\Big]^{-1} \sum_{y \in Q} \mu_y(x) p(f(x) - f(y))$$

$$= \Big[\sum_{y \in Q} \mu_y(x)\Big]^{-1} \sum_{y \in Q_1} \mu_y(x) p(f(x) - f(y)) + \Big[\sum_{y \in Q} \mu_y(x)\Big]^{-1} \sum_{y \in Q_2} \mu_y(x) p(f(x) - f(y)), \quad (1)$$

where $Q_1 = \{y \in Q\colon p_0(x-y) \le 1\}$ and $Q_2 = Q \setminus Q_1$.

Since Q is a $(1/2)$-net for M with respect to p_0, there is an $y \in Q$ such that $p_0(x-y) \le 1/2$, i.e., $\mu_y(x) \ge 1/2+\varepsilon$. Consequently, $\left[\sum_{y\in Q} \mu_y(x)\right]^{-1} \le 2$. Further, $\mu_y(x) = \varepsilon$ for all $y \in Q_2$. Denoting the number of elements in Q by n and the p-diameter of the (compact) set $f(M)$ by d, we obtain

$$\left[\sum_{y\in Q} \mu_y(x)\right]^{-1} \sum_{y\in Q_2} \mu_y(x)p(f(x)-f(y)) \le 2dn\varepsilon.$$

Hence, if $\varepsilon \le 1/(4dn)$ (for $d=0$ there is no restriction on ε), then the second term in the right-hand side of (1) is smaller than or equal to $1/2$. The first term is also smaller than or equal to $1/2$:

$$\left[\sum_{y\in Q} \mu_y(x)\right]^{-1} \sum_{y\in Q_1} \mu_y(x)p(f(x)-f(y))$$

$$\le \frac{1}{2}\left[\sum_{y\in Q} \mu_y(x)\right]^{-1} \sum_{y\in Q_1} \mu_y(x) \le \frac{1}{2}\left[\sum_{y\in Q} \mu_y(x)\right]^{-1} \sum_{y\in Q} \mu_y(x) = \frac{1}{2}. \quad \textbf{QED}$$

3.6.2. Definition. An operator g with the properties indicated in Theorem 3.6.1 will be referred to as a *quasi-extension of the operator f, with respect to the seminorm p, from the set M to the entire space.*

3.6.3. Definition of the index of a K_2-operator. Now relying on the notion of the index of compact (or finite-dimensional) operators in LCSs, the existence of fundamental sets with the requisite properties, and the quasi-extension theorem, one can proceed according to a unified scheme and introduce an index for ultimately compact operators (in particular, for operators that are condensing with respect to monotone MNCs in quasi-complete LCSs) defined on the closure $\overline{U}$ of an open subset of a LCS E, as well as an index for K_2-operators (in particular, operators in quasi-complete LCSs that are condensing with respect to "sufficiently nice" MNCs, for instance, MNCs that are invariant under adjunction of one-element sets and monotone or invariant under translations, and semi-additive MNCs), given only on the boundary ∂U of an open set U. Let us describe this scheme for, say, K_2-operators.

Let $f\colon \partial U \to E$ be a continuous K_2-operator, defined on the boundary of an open subset U of an LCS E, and with no fixed points.

Consider some fundamental set S for f relative to ∂U, such that $\partial U \cap S \ne \emptyset$. Since the set $\partial U \cap S$ is compact and f has no fixed points, there is a continuous seminorm p such that $p(x - f(x)) > 1$ for all $x \in \partial U \cap S$.

Let g be a quasi-extension of f, with respect to p, from $\partial U \cap S$ to the entire space E: $g: E \to S$, $p(f(x) - g(x)) \leq 1$ for all $x \in \partial U \cap S$.

The operator g also has no fixed points on ∂U. Indeed, if $x \in \partial U \setminus S$, then $g(x) \neq x$, since $g(x) \in S$. If now $x \in \partial U \cap S$, then

$$p(x - g(x)) \geq p(x - f(x)) - p(f(x) - g(x)) > 0.$$

Hence, for the finite-dimensional operator g, considered on $\overline{U}$, there is defined an index $\operatorname{ind}(g, U)$, which enjoys, as we already remarked, the natural properties.

Now put, by definition,

$$\operatorname{ind}(f, \partial U) = \operatorname{ind}(g, U).$$

To check the correctness of this definition, we need to show that the number $\operatorname{ind}(f, \partial U)$ does not depend on the choice of the seminorm p, the quasi-extension g, and the fundamental set S. We break the proof into three separate lemmas.

3.6.4. Lemma on the independence of the index on the quasi-extension. *Let S be a fundamental set for f with respect to ∂U and let p be a continuous seminorm such that $p(x - f(x)) > 1$ for all $x \in \partial U \cap S$. Then for any two quasi-extensions g_0 and g_1 of f, with respect to p, from the set $\partial U \cap S$ to the space E,*

$$\operatorname{ind}(g_0, U) = \operatorname{ind}(g_1, U).$$

Proof. In the present case the operators g_0 and g_1 are homotopic on $\overline{U}$, and a homotopy is provided, for example, by the family $g = \{g_\lambda\colon\ g_\lambda(x) = (1 - \lambda)g_0(x) + \lambda g_1(x),\ \lambda \in [0, 1]\}$. It suffices to verify that

$$(1 - \lambda)g_0(x) + \lambda g_1(x) \neq x$$

whenever $x \in \partial U$ and $\lambda \in [0, 1]$. If $x \in \partial U \setminus S$, then this is obvious. If now $x \in \partial U \cap S$, we have

$$p(x - (1 - \lambda)g_0(x) - \lambda g_1(x))$$
$$\geq p(x - f(x)) - (1 - \lambda)p(f(x) - g_0(x)) - \lambda p(f(x) - g_1(x)) > 0. \quad \textbf{QED}$$

3.6.5. Lemma on the independence of the index on the seminorm. *Let S be a fundamental set for f with respect to ∂U, and let p_1 and p_2 be continuous seminorms* *such that* $p_1(x - f(x)) > 1$ *and* $p_2(x - f(x)) > 1$ *for all* $x \in \partial U \cap S$. *Then for any quasi-extensions* g_1 *and* g_2 *of* f, *with respect to the seminorms* p_1 *and* p_2, *respectively, from the set* $\partial U \cap S$ *to* E,

$$\operatorname{ind}(g_1, U) = \operatorname{ind}(g_2, U).$$

Proof. The function $p(x) = \max\{p_1(x), p_2(x)\}$ is again a continuous seminorm and clearly $p(x - f(x)) > 1$ for all $x \in \partial U \cap S$. Let g be a quasi-extension of f, with respect to p, from $\partial U \cap S$: $g(E) \subset S$ and $p(g(x) - f(x)) \leq 1$ for all $x \in \partial U \cap S$. But then we also have $p_1(g(x) - f(x)) \leq 1$ and $p_2(g(x) - f(x)) \leq 1$ for all $x \in \partial U \cap S$, i.e., g is also a quasi-extension of f with respect to the seminorms p_1 and p_2. We are therefore in the conditions of the previous lemma, and so

$$\operatorname{ind}(g_1, U) = \operatorname{ind}(g, U) = \operatorname{ind}(g_2, U). \quad \textbf{QED}$$

3.6.6. Lemma on the independence of the index on the fundamental set. *The index of the* K_2*-operator* f *does not depend on the choice of the fundamental set.*

Proof. Let S_1 and S_2 be two fundamental sets for f with respect to ∂U, such that $\partial U \cap S_1 \neq \emptyset$ and $\partial U \cap S_2 \neq \emptyset$. Suppose $S_1 \subset S_2$.

Let $S_1(x)$ denote the union of all rays emanating from x, the extensions of which pass through points of S_1. Clearly, the set $S_1(x)$ is closed. Since S_1 is fundamental, $\partial U \cap S_2$ is compact, and $S_1(x)$ is closed, there exists a continuous seminorm p such that $p(x - f(x)) > 1$ for $x \in \partial U \cap S$ and $\inf\{p(f(x) - y) : y \in S_1(x)\} > 1$ for $x \in S_2 \setminus S_1$. With respect to this seminorm construct two quasi-extensions of f, g_1 and g_2, from the sets $\partial U \cap S_1$ and $\partial U \cap S_2$, respectively:

$$g_1(E) \subset S_1, \quad p(g_1(x) - f(x)) \leq 1 \quad \text{for } x \in \partial U \cap S_1,$$

and

$$g_2(E) \subset S_2, \quad p(g_2(x) - f(x)) \leq 1 \quad \text{for } x \in \partial U \cap S_2.$$

We claim that the operators g_1 and g_2 are linearly homotopic on $\overline{U}$. In fact, it suffices to check that

$$(1 - \lambda)g_1(x) + \lambda g_2(x) \neq x$$

whenever $x \in \partial U$ and $\lambda \in [0, 1]$. If $x \in \partial U \setminus S_2$ this relation holds because its left-hand side belongs to S_2.

For $x \in \partial U \cap S_1$ one can verify, as in Lemma 3.6.4, that $p(x-(1-\lambda)g_1(x)-\lambda g_2(x)) > 0$, i.e., the needed relation holds again.

It remains to examine the case $x \in S_2 \setminus S_1$. Suppose that $x = (1-\lambda)g_1(x) + \lambda g_2(x)$ for some point $x \in S_2 \setminus S_1$. This means that the point $g_2(x)$ belongs to $S_1(x)$, i.e.,

$$\inf\{p(g_2(x) - y)\colon\ y \in S_1(x)\} = 0.$$

On the other hand,

$$\inf\{p(g_2(x) - y)\colon\ y \in S_1(x)\} \geq$$
$$\inf\{p(f(x) - y)\colon\ y \in S_1(x)\} - p(f(x) - p(f(x) - g_2(x)) > 0,$$

and so we reached a contradiction. Thus, the operators g_1 and g_2 are indeed homotopic on $\overline{U}$, and consequently their indices coincide.

The case where the fundamental sets S_1 and S_2 are disjoint can be reduced to the case considered above because for a K_2-operator one can always construct a third fundamental set, S_3 that intersects both S_1 and S_2. The arguments needed in the proof are not different from those used in the Banach space case (see, for example, **3.5.5**). **QED**

The properties of the integer-valued characteristic introduced above are completely similar to those of the index of operators in Banach spaces.

3.7. THE RELATIVE INDEX

In the theory of compact operators there is a well-known and widely used notion of a relative index, which generalizes the notion of the index of a compact operator. An analogous generalization is possible for condensing operators (as well as for all the other classes of operators considered in the preceding section). The present section is devoted to this generalization.

3.7.1. Formulation of the problem. Consider the following setting. Let E be a Banach space, R a closed convex subset of E, and V a subset of R that is open in the induced topology on R. We let $\overline{V}_R$ and ∂V_R denote the closure and respectively the boundary of V in R. Suppose the operator $f\colon \overline{V}_R \to R$ is χ-condensing and the fixed-point set $\mathrm{Fix}(f|_V)$ of its restriction to V is compact. We note right away that a sufficient (but not necessary) condition for the latter to hold is that the operator f have no fixed points on ∂V_R. For such an operator we define below an integer $\mathrm{ind}_R(f, V)$, called the *fixed-point index of the operator f on V relative to R*, or simply the *relative index of f*.

3.7.2. Definition of the relative index. We use the notion of a fundamental set and the index theory for compact operators.

First, we note that if in the setting described above one assumes that the operator f is compact and acts from an open (not necessarily bounded) subset U of the Banach space E into E, then one can define a characteristic $\operatorname{ind}(f, U)$ that will enjoy properties analogous to properties 1°–5° given in **3.1.1**. In fact, it suffices to put $\operatorname{ind}(f, U) = \operatorname{ind}(f, W)$, where W is a bounded open domain in E such that $\operatorname{Fix}(f) \subset W$, $\overline{W} \subset U$, and then refer to Theorem 3.1.2.

Let S be a fundamental set of the condensing operator $f|_V$ (which exists by Remark 3.1.5) and let U be an open subset of E such that $U \cap R = V$. Using Dugundji's theorem, extend f from the closed subset $U \cap S$ of the space U to the entire U, keeping the range in S. The case of interest is that where $U \cap S \neq \emptyset$: the opposite case is trivial, since f has no fixed points and $\operatorname{ind}_R(f, V)$ is set, by definition, equal to zero. The resulting extension $\tilde{f}$ acts from U into S, and hence is compact; moreover, $\operatorname{Fix}(\tilde{f}) = \operatorname{Fix}(f|_V) \subset S$. Put

$$\operatorname{ind}_R(f, V) = \operatorname{ind}(\tilde{f}, U).$$

Our immediate goal is to show that the index $\operatorname{ind}_R(f, V)$ defined in this manner does not depend on the choice of U, S, and $\tilde{f}$. Thus, suppose both triples $U_1, S_1, \tilde{f}_1$ and $U_2, S_2, \tilde{f}_2$ satisfy the requirements stated in the definition of the index. Let us show that

$$\operatorname{ind}(\tilde{f}_1, U_1) = \operatorname{ind}(\tilde{f}_2, U_2). \tag{1}$$

With no loss of generality we may assume that $S_1 \subset S_2$, since one can always construct a fundamental set S_3 that contains the compact set $K = S_1 \cup S_2$, together with a corresponding pair $U_3, \tilde{f}_3$, and then establish that both sides of (1) are equal to $\operatorname{ind}(\tilde{f}_3, U_3)$. Next, put $U = U_1 \cup U_2$. Clearly,

$$\operatorname{ind}(\tilde{f}_1, U_1) = \operatorname{ind}(\tilde{f}_1, U)$$

and

$$\operatorname{ind}(\tilde{f}_2, U_2) = \operatorname{ind}(\tilde{f}_2, U).$$

This follows from property 2° of the index of a compact operator. Therefore, to prove (1) it suffices to show that

$$\operatorname{ind}(\tilde{f}_1, U) = \operatorname{ind}(\tilde{f}_2, U).$$

On U consider the linear homotopy $\tilde{f}_t(x) = (1-t)\tilde{f}_1(x) + t\tilde{f}_2(x)$ from $\tilde{f}_1$ to $\tilde{f}_2$. The ranges of the operators $\tilde{f}_t$ lie in S_2, and so the family $F = \{f_t : t \in [0, 1]\}$ is compact. Let us show that $\text{Fix}(F) = \text{Fix}(f|_V)$; this will imply that this set is compact. If $x_0 \in \text{Fix}(F)$, i.e., $x_0 = (1-t_0)\tilde{f}_1(x_0) + t_0\tilde{f}_2(x_0)$, then $x_0 \in S_2$. It follows that $\tilde{f}_2(x_0) = f(x_0)$ and $x_0 \in \text{co}\,[\{f(x_0)\} \cup S_1]$, which implies that $x_0 \in S_1$, and hence $\tilde{f}_1(x_0) = f(x_0)$. We conclude that $x_0 = f(x_0)$, i.e., $x_0 \in \text{Fix}(f|_V)$. Thus, $\text{Fix}(F) \subset \text{Fix}(f|_V)$; the opposite inclusion is plain. This establishes the correctness of the definition of the relative index.

3.7.3. Properties of the relative index. The relative index defined above for condensing operators enjoys the usual properties 1°–5° formulated in **3.1.1**. We recall them, stressing that in the present setting the definition of homotopic operators takes a slightly different form.

1°. *Homotopic condensing operators have equal relative indices.*

Here two condensing [resp. compact] operators $f_0, f_1 : \overline{V}_R \to R$ are said to be *homotopic* if there exists a condensing [resp. compact] family $f = \{f_\lambda : \lambda \in [0,1]\}$, $f_\lambda : \overline{V}_R \to R$, such that the map $(\lambda, x) \mapsto f_\lambda(x)$ is continuous and the set $\text{Fix}(f|_V) = \bigcup_{\lambda\in[0,1]} \text{Fix}(f_\lambda|_V)$ is compact.

2°. *Let V_i, $i = 1, 2, \dots$ be pairwise disjoint subsets of V that are open in R. Suppose the operator f has no fixed points on $V \setminus \bigcup_{i=1}^{\infty} V_i$. Then the indices $\text{ind}_R(f, V_i)$ are defined for all i, only finitely many of them are different from zero, and*

$$\text{ind}_R(f, V) = \sum_{i=1}^{\infty} \text{ind}_R(f, V_i).$$

3°. *If $f(x) \equiv x_0 \in V$, then $\text{ind}_R(f, V) = 1$.*

4°. *If $f(x) \equiv x_0 \notin V$, then $\text{ind}_R(f, V) = 0$.*

5°. *If $\text{ind}_R(f, V) \neq 0$, then f has at least one fixed point in V.*

All these properties are easy consequences of the corresponding properties of the index for compact operators. For illustration, let us prove property 1°.

Suppose the condensing operators f_0 and f_1 are homotopic and the homotopy is effected by the condensing family $f = \{f_\lambda : \lambda \in [0,1]\}$. By Remark 3.1.5, the family $f|_V$ possesses a fundamental set S. Moreover, by the definition of a homotopy, the set $\text{Fix}(f|_V)$ is compact. Let U be an open subset of E such that $U \cap R = V$. Extend the operators f_λ of the family f from $U \cap S$ to U, preserving the convex hull of the range of the family and the continuity of the mapping $(\lambda, x) \mapsto f_\lambda(x)$. This yields a compact homotopy $\tilde{f} = \{\tilde{f}_\lambda : \lambda \in [0,1]\}$, for which $\text{Fix}(\tilde{f}) = \text{Fix}(f|_V)$ is compact. We thus see that

$$\text{ind}_R(f_0, V) = \text{ind}(\tilde{f}_0, U) = \text{ind}(\tilde{f}_1, U) = \text{ind}_R(f_1, V).$$

To conclude this section we note that once the notion of the relative index $\text{ind}_R(f, V)$ of a condensing operator f is introduced in the setting described above, one can define an analogous notion under more general conditions; specifically, it suffices to assume that f is defined and χ-condensing on V, rather than on the closure $\overline{V}$. In fact, one can put, by definition, $\mathrm{ind}_R(f, V) = \mathrm{ind}_R(f, W)$, where the relatively open set W is such that $\mathrm{Fix}(f) \subset W$ and $\overline{W} \subset V$. The fact that this definition does not depend on W follows from property 2°; properties 1°–5° are obviously inherited, where in 1° it suffices to consider that f_0, f_1 and f are defined on V.

3.7.4. Remark. A sufficient condition for the compactness of the fixed-point set $\mathrm{Fix}(f|_V)$ of the restriction of a χ-condensing operator f to V is that f have no fixed points on the boundary ∂V_R of V. Indeed, in this case the set $G = \mathrm{Fix}(f|_V) = \mathrm{Fix}(f)$ is obviously closed, and hence relatively compact, since f is χ-condensing and $f(G) = G$. Therefore, if the χ-condensing operator $f: \overline{V}_R \to R$ has no fixed points on ∂V_R, then its (relative) index is defined. Moreover, two condensing operators $f_0, f_1: \overline{V}_R \to R$, both with no fixed points on ∂V_R, are homotopic if there exists a condensing family $f = \{f_\lambda: \lambda \in [0, 1]\}$, $f_\lambda: \overline{V}_R \to R$, such that the mapping $(\lambda, x) \mapsto f_\lambda(x)$ is continuous and all operators f_λ have no fixed points on ∂V_R.

3.7.5. Generalizations. The notion of a relative index, like that of an index, can be generalized to various wider classes of maps acting in Banach as well as locally convex topological vector spaces. Let us describe several possible directions of generalization.

First, one can consider maps f that are condensing with respect to MNCs other than χ, for instance, MNCs ψ in the sense of Definition 1.2.1, with a specific set of properties. In this respect an analogue of Theorem 3.5.1 holds true for the relative index: in order to be able to define the index of a ψ-condensing operator given on the relative closure of a relatively open set, it already suffices that the MNC ψ be monotone. In particular, such a definition is possible for ultimately compact operators.

Second, one can consider the case where the maps (condensing or of class $\mathcal{K}_c, \mathcal{K}_n$) are given not on the closure of a relatively open set, but only on its relative boundary. In this direction one can obtain a generalization of Theorem 3.5.9 for a K_2-operator f, given on the relative boundary ∂V_R of a relatively open set V and with no fixed points there, and hence define for f an index with the usual properties. In particular, such a definition is possible for condensing operators with respect to an MNC that is monotone and invariant under adjunction of one-element sets.

Third, a notion of relative index can be introduced for operators that act in LCSs. Here, too, one can investigate the case when the operator is given on the closure of a relatively open set (and is condensing with respect to a monotone MNC), as well as the case when it is given on the relative boundary of such a set (and is of class $\mathcal{K}_2$). In constructing a theory of the relative index in LCSs, instead of Dugundji's theorem one can use the quasi-extension theorem proved in **3.6.1**.

Concerning such generalizations, the reader is referred to **3.9.1, 3.9.7** and the references given therein.

3.8. THE INDEX OF POSITIVE OPERATORS

In this section the theory of the relative index of condensing operators is applied to study positive χ-condensing operators.

3.8.1. Preliminary remarks. Recall (see, for example, [85]) that a *cone* K in a Banach space E is a closed and convex set that is invariant under multiplication by nonnegative scalars and does not contain opposite elements. An operator f is said to be *positive* on a set M if $f(M) \subset K$.

Let E be a Banach space, K a cone in E, f a positive χ-condensing operator, defined on the (relative) closure $\overline{\Omega}_K$ of a set $\Omega \subset K$ that is open in the induced topology on K, and with no fixed on the boundary $\partial\Omega_K$. In what follows, whenever there is no danger of confusion, the relative index of the operator f on Ω relative to K, $\mathrm{ind}_K(f,\Omega)$, will be denoted simply by $\mathrm{ind}(f,\Omega)$. By the discussion in the preceding section, the number $\mathrm{ind}(f,\Omega)$ is defined and enjoys all properties 1°–5° of the relative index.

Suppose now that Ω contains the vertex of the cone K (the point 0), for example, $\Omega = K \cap B_r$, where $B_r = B(0,r) \subset E$ is the ball of radius r centered at zero. Consider the operators f_1 and f_2, defined as follows:

$$f_1(x) \equiv x, \quad f_2(x) \equiv h_0 \quad (h_0 \in K, \ \|h_0\| > r).$$

3.8.2. Lemma. *The indices of the operators f_1 and f_2 on Ω relative to K are defined and*

$$\mathrm{ind}(f_1,\Omega) = 1, \quad \mathrm{ind}(f_2,\Omega) = 0.$$

Proof. Clearly, f_1 and f_2 are χ-condensing on Ω and have no fixed points on $\partial\Omega_K$, so that their indices are defined. The needed equalities follow from properties 3° and 4° of the index. **QED**

3.8.3. Theorem. *Suppose the positive condensing operator f is defined on $\overline{\Omega}_K$ and for any $x \in \partial\Omega_K$ one has that $f(x) \neq \alpha x$ if $\alpha \geq 1$. Then* $\operatorname{ind}(f, \Omega) = 1$.

Proof. Under the assumptions of the theorem f is homotopic to the operator f_1 introduced above, and the homotopy can be effected by the condensing family $F = \{\lambda f : \lambda \in [0,1]\}$. Indeed, in the present situation, to prove that F is a homotopy from f to f_1 it suffices to check that F is condensing (which is obvious), and that the operators of the family F have no fixed points on $\partial\Omega_K$, i.e., $\lambda f(x) \neq x$ whenever $x \in \partial\Omega_K$ and $0 \leq \lambda \leq 1$. But the last requirement is satisfied by the hypothesis. **QED**

3.8.4. Theorem. *Suppose the positive condensing operator f is defined on $\overline{\Omega}_K$ ($\Omega = K \cap B_r$) and for any $x \in \partial\Omega_K$ one has that $x \neq f(x) + \alpha h_0$ for $\alpha \geq 0$, $h_0 \in K$, $\|h_0\|$ sufficiently large. Then* $\operatorname{ind}(f, \Omega) = 0$.

Proof. We claim that f is homotopic to the operator f_2 introduced above provided $\|h_0\| > 2\max\{r, \sup_{x\in\partial\Omega_K} \|f(x)\|\}$, and the homotopy is effected by the family $F = \{F_\lambda : \lambda \in [0,1]\}$, where

$$F_\lambda(x) = \begin{cases} f(x) + 2\lambda h_0, \text{ if } 0 \leq \lambda \leq 1/2, \\ 2(1-\lambda)f(x) + h_0, \text{ if } 1/2 \leq \lambda \leq 1. \end{cases}$$

Indeed, the family F is condensing since, obviously,

$$F(A) \subset \operatorname{co}[f(A) \cup (f(A) + h_0)] \cup \operatorname{co}[f(A + h_0) \cup \{h_0\}]$$

for any set A. Consequently, $\chi[F(A)] \leq \chi[f(A)] < \chi(A)$ provided A is not relatively compact.

Let us check that the operators F_λ have no fixed points on $\partial\Omega_K$. In fact, the equality $x = f(x) + 2\lambda h_0$ is forbidden by hypothesis, while the equality $x = 2(1-\lambda)f(x) + h_0$ is forbidden thanks to the assumption $\|h_0\| > 2\max\{r, \sup_{x\in\partial\Omega_K} \|f(x)\|\}$. **QED**

Theorems 3.5.7 and 3.5.8 can be used to generalize the well-known theorems of M. A. Krasnosel′skiĭ on the existence of fixed points for operators that are contractions or expansions of a cone.

We remind the reader that a positive operator f ($f(0) = 0$) is called a *contraction* (or *compression*) *of the cone* K if there are numbers $R > r > 0$ such that $f(x) \nleq x$ for $x \in K$, $\|x\| \leq r$, $x \neq 0$ and $f(x) \ngeq (1+\varepsilon)x$ for $x \in K$, $\|x\| \geq R$, $\varepsilon > 0$; f is called an *expansion of the cone* K if there are numbers $r, R > 0$ such that $f(x) \ngeq (1+\varepsilon)x$ for all $\varepsilon > 0$, $x \in K$, $\|x\| \leq r$, $x \neq 0$ and $f(x) \nleq x$ for $x \in K$, $\|x\| \geq R$.

3.8.5. Theorem. *Let the condensing operator f be a contraction of the cone K. Then f has at least one nonzero fixed point in K.*

Proof. If f is a contraction of K then, in particular, there is a nonzero element $h_0 \in K$ (the norm of which may be assumed sufficiently large) such that $x \neq f(x) + \alpha h_0$ for all $\alpha > 0$ and all $x \in K$, $\|x\| \leq r$ and such that $f(x) \neq (1+\varepsilon)x$ for all $\varepsilon > 0$ and all $x \in K$, $\|x\| \geq R$. Hence, $\operatorname{ind}(f, \Omega_1) = 0$ and $\operatorname{ind}(f, \Omega_2) = 1$, where $\Omega_1 = K \cap B_r$ and $\Omega_2 = K \cap B_R$. Using property 2° of the index, we obtain $\operatorname{ind}(f, \Omega) = 1$, where $\Omega = \Omega_1 \setminus \Omega_2$. Therefore, the operator f has a fixed point in K satisfying $r < \|x\| < R$. **QED**

The following fixed-point theorem for an operator that is an expansion of a cone is established in analogous manner.

3.8.6. Theorem. *Let the operator f be an expansion of the cone K. Then f has at least one nonzero fixed point in K.*

3.8.7. The derivative of a positive condensing operator with respect to a cone. A positive operator f is said to be *differentiable with respect to* (or *on*) *the cone K at the point $x_0 \in K$* if there exists a bounded linear operator $f'(x_0)$ such that

$$\lim_{h \in K,\ h \to 0} \frac{\|f(x_0+h) - f(x_0) - f'(x_0)h\|}{\|h\|} = 0.$$

If this is the case, then $f'(x_0)$ is called the *derivative of the operator f with respect to* (or *on*) *the cone K at the point x_0*. In what follows we will be interested in the derivative of f with respect to a cone at the point 0, $f'(0)$. The operator $f'(0)$ is also positive if $f(0) = 0$.

An operator f is said to be *differentiable with respect to* (or *on*) *the cone K at infinity* or to be *strongly asymptotically linear with respect to* (or *on*) *the cone K* if there exists a bounded linear operator $f'(\infty)$ such that

$$\lim_{x \in K,\ \|x\| \to \infty} \frac{\|f(x) - f'(\infty)x\|}{\|x\|} = 0.$$

If this is the case, then $f'(\infty)$ is called the *strong asymptotic derivative of f with respect to K*; $f'(\infty)$ is positive together with f.

The derivative with respect to a cone of an operator f that is condensing with constant $q < 1$ is also condensing with constant q. The proof of this assertion is an almost word-for-word repetition of the proof of Theorem 1.5.9 on the Fréchet derivative of a (q, χ)-bounded operator.

An analogous result holds for the asymptotic derivative of a condensing operator f: if the positive operator $f\colon K \to K$ is condensing with constant q, then its asymptotic derivative $f'(\infty)$ with respect to K is also condensing with constant q.

Let E be a Banach space with a cone K and $f: K \to K$ a positive operator that is condensing with constant $q < 1$. We shall assume that f is differentiable with respect to K at infinity. We are interested in conditions ensuring the existence of fixed points of f (in particular, of nonzero fixed points), formulated in terms of the asymptotic derivative of f.

3.8.8. Theorem. *Suppose the operator $A = f'(\infty)$ has no nonzero eigenvectors in K corresponding to eigenvalues $\lambda \geq 1$. Then f has a fixed point in K.*

Proof. Let $\Omega = K \cap B_\rho$. By Theorem 3.8.3, $\operatorname{ind}(A, \Omega) = 1$ for all $\rho > 0$. Next, since $Ax \neq x$ for all $x \in K$, there exists a $\gamma > 0$ such that $\|x - Ax\| \geq \gamma x$ for all $x \in K$. Indeed, assuming the contrary and setting $\gamma_n = 1/n$, one could find a sequence of elements $\{x_n\}$ such that $\|x_n - Ax_n\| \leq \|x_n\|/n$. This implies that the bounded sets $\{u_n\} = \{x_n/\|x_n\|\}$ and $\{Au_n\}$ have the same MNC. Since A is condensing, it follows that $\{u_n\}$ is a relatively compact set. With no loss of generality one can assume that $u_n \to u_0 \in K$ when $n \to \infty$, where $\|u_0\| = 1$. Letting $n \to \infty$ in the inequality $\|u_n - Au_n\| < 1/n$, we conclude that u_0 is a nonzero eigenvector of A with eigenvalue 1, which contradicts the hypothesis.

Now let ρ_0 be such that $\|f(x) - Ax\| \leq \frac{\gamma}{2}\|x\|$ whenever $\|x\| \geq \rho_0$. Let us show that the operators f and A are homotopic on $\Omega = K \cap B_{\rho_0}$ and a homotopy is provided by the family $f_\lambda(x) = \lambda f(x) + (1 - \lambda)Ax$. To this end it suffices to verify that $x \in K$, $\|x\| = \rho_0$, $0 \leq \lambda \leq 1$ imply $f_\lambda(x) \neq x$. We have

$$\|x - f_\lambda(x)\| = \|(x - Ax) - \lambda[f(x) - Ax]\|$$
$$\geq \|x - Ax\| - \lambda\|f(x) - Ax\| \geq \gamma\|x\| - \lambda\frac{\gamma}{2}\|x\| \geq \frac{\gamma}{2}\rho_0 > 0.$$

Thus, f and A are indeed homotopic, and consequently $\operatorname{ind}(f, \Omega) = \operatorname{ind}(A, \Omega) = 1$, which implies that f has a fixed point in K. **QED**

A sufficient condition for the operator $f'(\infty)$ to have no positive eigenvectors corresponding to eigenvalues larger than or equal to 1 is that $r[f'(\infty)] < 1$.

If the positive operator f has no fixed points on the intersection of the cone K and the set $\{\|x\| \geq \rho_0\}$, we say that *infinity is an isolated singular point of the vector field $I - f$*. Accordingly, the theorem proved above can be regarded as a condition for the singular point at infinity of the vector field $I - f$ to be isolated and a rule for computing its index.

The following result provides another example of computation of the index of the singular point at infinity of a condensing vector field $I - f$ with a positive operator f.

3.8.9. Theorem. *Suppose the operator $A = f'(\infty)$ has no eigenvectors with eigenvalue 1 in the cone K and has an eigenvector with an eigenvalue $\lambda_0 > 1$ in K. Then*

$\operatorname{ind}(f, \Omega) = 1$ *for all sets* $\Omega = K \cap B_\rho$ *with* $\rho \geq \rho_0$.

Proof. As in the preceding theorem, one establishes that the operators A and f are homotopic on Ω. Next, if $Ah_0 = \lambda_0 h_0$, $h_0 \in K$, $||h_0|| > \rho_0$, then $\alpha \geq 0$ and $\|x\| = \rho$ ($\rho \geq \rho_0$) obviously imply $x \neq Ax + \alpha h_0$: otherwise, $y = x - \alpha(1 - \lambda_0)^{-1} h_0 \in K$ would be an eigenvector of A with eigenvalue 1. Now it remains to refer to Theorem 3.8.4. **QED**

If the operator f satisfies $f(0) = 0$, the analogous results can be formulated in terms of $f'(0)$.

Let, as above, $f: K \to K$ be a positive operator that is condensing with constant $q < 1$ and satisfies $f(0) = 0$. Then, as we already remarked, $f'(0)$ and $f'(\infty)$ are also positive and condensing with constant q.

3.8.10. Theorem. *Suppose the operator* $C = f'(0)$ *has no eigenvectors with eigenvalues* $\lambda \geq 1$ *in the cone* K. *Then* $\operatorname{ind}(f, \Omega) = 1$ *for all sets* $\Omega = K \cap B_\rho$ *with sufficiently small* ρ.

3.8.11. Theorem. *Suppose the operator* $C = f'(0)$ *has no eigenvectors with eigenvalue* 1 *in* K *and has an eigenvector with an eigenvalue* $\lambda_0 > 1$ *in* K. *Then* $\operatorname{ind}(f, \Omega) = 0$ *for all sets* $\Omega = K \cap B_\rho$ *with sufficiently small* ρ.

Theorems 3.8.9 and 3.8.10 may serve as tests for verifying whether the fixed point 0 of the operator f (or the singular point 0 of the vector field $I - f$) is isolated in K and as rules for the computation of its index.

Notice that neither Theorem 3.8.10, nor Theorem 3.8.8 in the case $f(0) = 0$, guarantee that f has a nonzero fixed point in K. Results in that direction can be obtained by combining the hypotheses of Theorems 3.8.8–3.8.11 as follows.

3.8.12. Theorem. *Let the operator* $f: K \to K$ $(f(0) = 0)$ *be positive, condensing with constant* $q < 1$, *and differentiable with respect to the cone* K *at zero and at infinity. Suppose* $f'(\infty)$ *has no positive eigenvectors with eigenvalues* ≥ 1, *while* $f'(0)$ *has an eigenvector with an eigenvalue* $\lambda_0 > 1$ *in* K, *and has no eigenvectors with eigenvalue* 1 *in* K. *Then* f *has at least one nonzero fixed point in* K.

3.8.13. Theorem. *Let the operator* $f: K \to K$ $(f(0) = 0)$ *be positive, condensing with constant* $q < 1$, *and differentiable with respect to the cone* K *at zero and at infinity. Suppose* $f'(0)$ *has no positive eigenvectors with eigenvalues* ≥ 1, *while* $f'(\infty)$ *has an eigenvector with an eigenvalue* $\lambda_0 > 1$ *in* K , *and has no positive eigenvectors with eigenvalue* 1. *Then* f *has at least one nonzero fixed point in* K.

3.9. SURVEY OF THE LITERATURE

3.9.1. Fundamental sets and the construction of the index theory for condensing maps. As we already mentioned, the notion of a fundamental set was introduced by P. P. Zabreĭko, M. A. Krasnosel′skiĭ, and V. V. Strygin in [181] in order to obtain invariance principles for the rotation of compact vector fields. Subsequently, V. V. Obukhovskiĭ [122], Yu. I. Sapronov [163], M. A. Krasnosel′skiĭ and P. P. Zabreĭko [92], A. S. Potapov [132, 133], V. V. Obukhovskiĭ and E. V. Gorokhov [124], R. R. Akhmerov, M. I. Kamenskiĭ, A. S. Potapov, and B. N. Sadovskiĭ [10], Yu. G. Borisovich and V. V. Obukhovskiĭ [20], relying on the idea of fundamental set, proposed various schemes of defining the rotation of condensing (including also multi-valued), as well as compactly- and fundamentally-supported vector fields. The compactly- and fundamentally-supported operators are operators for which there exist fundamental sets, possibly with certain additional properties (see **3.5.12**). Methods close in spirit for defining the index by means of compact restriction to convex invariant sets or restriction to compact invariant sets were proposed even earlier by B. N. Sadovskiĭ [155] and by Yu. G. Borisovich and Yu. I. Sapronov [21, 22].

Theorem 3.1.2 on the existence and properties of the index of a compact operator, based on Brouwer's degree theory for maps in finite-dimensional spaces, was proved in various versions by J. Leray and J. P. Schauder, M. A. Krasnosel′skiĭ, E. H. Rothe, M. Nagumo, and F. E. Browder. Its proof can be obtained, for example, using Theorems 20.1 through 20.4 in the monograph of M. A. Krasnosel′skiĭ and P. P. Zabreĭko [92]: it suffices to set $\operatorname{ind}(f, U) = \gamma(I - f, W)$, where $\gamma(I - f, W)$ is the rotation of the vector field $I - f$ on the boundary of a bounded domain W that satisfies the conditions $\operatorname{Fix}(f) \subset W$, $\overline{W} \subset U$.

The exposition of the results of **3.1.4** and **3.1.5** on the existence of fundamental sets for condensing families of operators follows the paper [10]. The existence of fundamental sets, possessing various properties, for condensing and related maps was noted by many authors (see, e.g., [20–22, 92, 122, 163, 181]).

Theorem 3.1.11 on homotopy classes of condensing operators was proved by Yu. I. Sapronov in [163]. A generalization of this theorem is provided by the following result [135].

Theorem. *Let Λ be a compact topological space, M a subset of a locally convex space E, and $f: \Lambda \times M \to E$ a continuous K_c-operator whose fixed points are contained in a compact set T. Then there exists a continuous K_c-operator $F: [0,1] \times \Lambda \times M \to E$ such* *that*

1) $F(0,\cdot,\cdot) = f$;

2) $F(1,\cdot,\cdot)$ *is a compact operator;*

3) *all fixed points of F are contained in an arbitrarily prescribed neighborhood of the set T.*

An analogous theorem for fundamentally-supported multi-valued maps in metrizable locally convex spaces E with the property that the closure of any compact subset of E is compact was proved by V. V. Obukhovskiĭ and E. V. Gorokhov in [124].

Theorem 3.1.12 (the analogue of Hopf's theorem) on the homotopy of condensing operators with equal indices is a simple consequence of a theorem of Yu. I. Sapronov asserting the existence of a compact operator in any homotopy class of condensing operators and of the generalization of Hopf's theorem to compact operators [163]. There are many extensions of this theorem to various classes of operators, both single- and multi-valued, in Banach as well as in locally convex spaces. In these extensions, alongside with indices of operators, relative indices were considered [10, 20, 124, 135].

3.9.2. Computation of the index, and fixed-point theorems. Theorems 3.2.1–3.2.7 on the computation of the index of a condensing operator in various situations were given here in their simplest formulations. Variants and generalizations of these theorem, as well as other results concerning the computation of the index of condensing operators can be found in works by Yu. G. Borisovich, J. Daneš, P. M. Fitzpatrick, A. I. Istrăţescu, V. I. Istrăţescu, M. A. Krasnosel′skiĭ, M. Martelli, R. D. Nussbaum, V. V. Obukhovskiĭ, W. V. Petryshyn, S. Reich, J. Reinermann, A. Vignoli, J. R. L. Webb, P. P. Zabreĭko, and others (see the list of references).

Let us give some results concerning the computation of indices of condensing maps. The next theorem represents an extension of the well-known principle of H. Schaefer [164] for compact compact operators to condensing operators.

Theorem (Schaefer's principle). *Let f be a condensing operator in a Banach space E. Suppose that all solutions of the equation $x = \lambda f(x)$ $(0 < \lambda < 1)$ are uniformly bounded. Then this equation admits a solution in E for any $\lambda \in [0,1]$.*

In the framework of index theory this theorem admits a very transparent proof: if B_ρ is a ball of sufficiently large radius centered at the origin, then the condensing family $\{\lambda f\}$ effects a homotopy on B_ρ from the operator f to the operator $f_0 \equiv 0$, the index of which on B_ρ is equal to 1.

Various variants and generalizations of Schaefer's principle to condensing operators can be found in works of J. A. Gatica [50], O. Hadžić [59], G. B. Lyal′kina [105], W. V.

Petryshyn [128].

In [166] H. Steinlein obtained a generalization of the well-known result of M. A. Krasnosel'skiĭ and P. P. Zabreĭko (see, e.g., [92]) on the connection between the index of an operator f and the index of its iterates, to the case of condensing maps with constant $q < 1$.

Theorem. *Let the operator f, defined on a domain V, be condensing with constant $q < 1$. Let p be a prime number and suppose that the domain of the operator f^p contains V together with its boundary. Let* $\mathrm{Fix}(f^p)$ *denote the set of fixed points of f^p in $\overline{V}$. Suppose f^p has no fixed points on ∂V and $f[\mathrm{Fix}(f^p)] \subset V$. Then*

$$\mathrm{ind}(f, V) \equiv \mathrm{ind}(f^p, V) \pmod p.$$

With the help of this theorem one readily obtains a generalization of F. E. Browder's principle (see, e.g., [92]) to condensing maps.

Theorem (Browder's principle). *Suppose that the χ-condensing operator f with constant $q < 1$ and all its iterates f^n ($n = 1, 2, \dots$) are defined on $\overline{U}$, where U is a bounded convex domain in E, and that $f^n(\overline{U}) \subset \overline{U}$ and $f^n(x) \neq x$ for all $x \in \partial U$ and all $n \geq n_0$, where n_0 is some natural number. Then* $\mathrm{ind}(f, U) = 1$, *and consequently f has at least one fixed point in U.*

To conclude this subsection we give a variant of the theorem on the "product of rotations" for condensing operators, obtained by H. Mönch and G.-F. von Harten [110] (see also J. W. Thomas' paper [173]), which generalizes Theorem 22.2 of the monograph of M. A. Krasnosel'skiĭ and P. P. Zabreĭko [92].

Theorem. *Let V_1 and V_2 be bounded domains in a Banach space E and let $f: \overline{V}_1 \to E$ and $g: \overline{V}_2 \to E$ be operators such that the operator $h = f + g(I - f)$ is condensing and $(I - f)V_1 \subset V_2$. Let $\{U_\lambda : \lambda \in \Lambda\}$ be the collection of all connected components of the set $V_2 \setminus (I - f)(\partial V_1)$. Let* $\mathrm{ind}(f, V_1, \lambda)$ *denote the index of the (condensing) operator $x \mapsto f(x) + z$ on V_1, where z is an arbitrary point of U_λ. Then*

$$\mathrm{ind}(h, V_1) = \sum_{\lambda \in \Lambda} \mathrm{ind}(g, U_\lambda)\, \mathrm{ind}(f, V_1, \lambda).$$

3.9.3. The index of linear and of linearizable condensing operators. The exposition of the results in **3.4.1–3.4.5** on the computation of the index of the fixed point 0 of a linear operator f that is χ-condensing with constant $q < 1$, of the index of an asymptotically linear condensing operator, and of the index of an isolated fixed point x_0 at which the operator f has a Fréchet derivative $f'(x_0)$ follows the works of J. Daneš [28], M. A. Krasnosel′skiĭ and P. P. Zabreĭko [92], C. A. Stuart [167], and C. A. Stuart and J. F. Toland [169].

3.9.4. Further properties of the index. Theorems 3.4.1 and 3.4.2 on the local constancy of the index are treated following the papers of R. R. Akhmerov, M. I. Kamenskiĭ, A. S. Potapov, and B. N. Sadovskiĭ [10] and of B. N. Sadovskiĭ [160]. We remark that in [160] Theorem 3.4.2 is proved for ultimately compact families of operators.

Theorem 3.4.3 on the independence of the index on the behavior of the operator inside its domain of definition is taken from [160], where it is proved for ψ-condensing operators that admit linear homotopies, where ψ is an arbitrary MNC. As we already remarked, for the latter to hold is suffices that ψ be real-valued and semi-additive.

Theorem 3.4.4—the restriction principle for condensing operators—is a simple consequence of the analogous principle for compact operators and was observed by many authors dealing with the definition of an index and the development of index theory.

Our treatment of the question of approximation of the index of a condensing operator by indices of finite-dimensional operators follows papers of G. M. Vaĭnikko and B. N. Sadovskiĭ [177] and of B. N. Sadovskiĭ [160]. In [177] this approach is used to define the index of a condensing operator, acting in a Banach space, for which there exists a sequence of finite-dimensional operators P_n as in **3.4.5**. To establish that the index introduced in this manner is well-defined it suffices to remark that the value of the stabilizing quantity $\operatorname{ind}(P_n f, U)$ is the same for all sequences $\{P_n\}$ with the properties indicated in **3.4.5**. In various settings this method of defining the index of a condensing operator by means of indices of "approximating maps" (in some specific sense) was discussed by many authors (see, e.g., [48, 114, 120, 176]).

The questions discussed in **3.4.6–3.4.8** concerning the extension of condensing operators with preservation of the condensing property have been considered in [161] in connection with the definition of the index of a condensing operator that is given only on the boundary of a domain.

3.9.5. Definitions of a notion of index for various classes of maps. As we already remarked in **3.9.1**, the notion of a fundamental set was used by many authors to define an index for various classes of maps. In the references listed in **3.9.1** one can find the definition of the index, studies of its properties, and various examples of computation and application of this notion. We wish to emphasize once more that practically all assertions of Sections 3.1–3.4, stated and proved by us for χ-condensing operators acting in Banach spaces, can be restated for operators that are condensing with respect to other MNCs, as well as for K-, compactly-, and fundamentally-supported operators. In each concrete case, the reader will find no difficulty in isolating those properties of the Hausdorff MNC and those requirements on the operators or families of operators that play an essential role in the proofs.

Our exposition of the results of **3.5.3** and **3.5.4** follows the paper of B. N. Sadovskiĭ [160].

Questions regarding the possibility of defining an index for operators that are condensing with respect to an arbitrary MNC, in particular, for K_n-operators (with $n = 0, 1, 2$) were considered by A. S. Potapov [133], and in the case of multi-valued maps—by Yu. G. Borisovich and V. V. Obukhovskiĭ [20], V. V. Obukhovskiĭ [123] (see also [22]), and A. S. Potapov [132].

The index of an ultimately compact operator f (the rotation, more precisely, the relative rotation of an ultimately compact vector field $I - f$) was introduced and studied by B. S. Sadovskiĭ in [155, 160]. Theorems 3.5.7 and 3.5.8 are taken from [160].

As an example of result that can be proved with the aid of the index theory for ultimately compact operators one can give the generalization of the Krasnosel′skiĭ-Perov conectedness principle (see **4.3.5**).

A systematic exposition of the index theory for ultimately compact operators is given, in addition to the aforementioned papers of B. N.Sadovskiĭ [155, 160], in the interesting survey of J. Daneš [28].

A notion of index for operators that ressemble in their properties the ultimately compact operators (including multi-valued maps) were considered in [114, 129, 130].

3.9.6. Index theory in locally convex spaces. The material in Section 3.5 follows the paper of A. S. Potapov [133]. In connection with Theorem 3.6.1 on quasi-extension we remark that it is based on the well-known construction of a "quasi-projector", proposed by Schauder and apparently used for the first time for LCSs by M. Nagumo [112]. Closely related types of quasi-extensions were constructed by V. V. Obukhovskiĭ and A. G. Skaletskiĭ [125]. These constructions found application in the definition of the index of compactly- and fundamentally-supported operators in LCSs (see [19]). A definition of the index for multi-valued K_2-compact operators in LCSs that is based on Theorem 3.5.1 was also given in [132].

3.9.7. The relative index of condensing operators. In our exposition of the theory of the relative index of condensing operators (**3.7.1–3.7.3**) we followed [10]. The notion of a relative index for various classes of maps was considered by many authors.

Thus, almost all works listed at the beginning of **3.6.1** introduced and studied precisely the relative index of operators [10, 19–22, 28, 122, 124, 133, 155, 160]. Two approaches can be distinguished in the aforementioned works . The first is to reduce the operator f in question to a compact (or even to a finite-dimensional) operator $\tilde{f}$ that acts in a closed convex set R, and then use the theory of the relative index of compact maps (see, e.g., the paper of Yu. G. Borisovich [18]). The second is the approach implemented in the present book in the definition the index, as well in that of the relative index—one first extends (or quasi-extends) the operator $\tilde{f}$ to some domain in space with preservation of its compactness property, and then use the theory of the classical index (rotation, degree) of compact operators.

We notice that most of the theorems concerning the computation and properties of the index, given in the preceding sections of this chapter, can be carried over with practically no modifications to the case of the relative index of condensing (generally speaking, with respect to an arbitrary MNC ψ with some specified set of properties) operators (see also the surveys [10, 28, 160]).

The theory of the relative index is used effectively in proving fixed-point theorems for operators given on nonsolid sets: one constructs a homotopy from the given operator to a constant operator $f(x) \equiv x_0$. For example, in this manner one obtains the following result.

Theorem. *Suppose the K_c-operator f maps a nonempty convex closed and bounded subset T of a Banach space E into itself. Then f has at least one fixed point in T.*

We remark that the attempt of using the same method to generalize this theorem to wider classes of operators (K_∞- or K_n-operators, say) for which the notion of a relative index is defined fails. Incidentally, the theorem is valid (see **1.7.3**) for K_1-operators (and hence for all K_n-operators with $1 \leq n \leq \infty$) (see [133]). It is not valid, however, for K_0-(ultimately compact) operators, as shown by the example given in **1.7.4**.

3.9.8. The index of positive operators. Subsections **3.8.3–3.8.6** follow the paper of A. S. Potapov, T. Ya. Potapova, and V. A. Filin [136]. The theorems on indices of derivatives with respect a cone are taken in part from H. Amann's paper [12]. In addition to the aforementioned theorems, analogous generalizations can be given of the well-known results on the existence of positive eiegenvectors for positive compact operators.

Theorem. *Suppose the positive condensing operator f has no fixed points on $\partial\Omega_K$, where Ω is a bounded domain in E such that $0 \in \Omega$. Suppose* $\operatorname{ind}(f, K \cap \Omega) = 1$. *Then f has at least one positive eigenvector in $\partial\Omega_K$ that corresponds to a (positive) eigenvalue $\lambda > 1$.*

In addition to the sources mentioned above, a number of results concerning positive condensing operators are given in papers by Yu. G. Borisovich and Yu. I. Sapronov [22], A. A. Kalmykov [69], G. B. Lyal'kina [106], I. V. Misyurkeev [109], and D. E. Edmunds, A. J. B. Potter, and C. A. Stuart [39].

In the following subsections we describe some other classes of maps for which one can define a notion that is close in its properties to the notion of the index of a condensing operator.

3.9.9. Nussbaum's generalized index. The generalized index introduced and studied mainly by R. D. Nussbaum (see [113, 114, 116, 120]) will be described here in a setting and notation that are convenient to us.

Let A be a compact subset of a Banach space E that can be written as the union of finitely many convex sets D_i: $A = \bigcup_{i=1}^{m} D_i$, let G be a subset of A that is open in the relative topology of A, and denote by $\overline{G}$ and ∂G the closure and the boundary of G in A, respectively. Further, let $f: \overline{G} \to A$ be a continuous operator with no fixed points on ∂A. Then (see [24]) there exists an integer-valued quantity $\operatorname{ind}_A(f, G)$ that enjoys the main properties of the index, namely, is additive with respect to the domain, invariant under homotopies, and equal to zero if f has no fixed points in G.

Now let P be a closed subset of a Banach space E that can be written as a locally finite union of closed convex sets C_i, i.e., $P = \bigcup_{i=1}^{\infty} C_i$, and for any point $x \in P$ there is a neighborhood $U(x)$ of x which intersects only finitely many of the C_i's. Let W be a set open in P, let $\overline{W}$ and ∂W denote its (relative) closure and boundary, respectively, and let $f: \overline{W} \to P$ be an ultimately compact operator on W. The ultimate range $f^{\infty}(W)$ is a closed convex set, which for the moment we shall assume nonempty. Denote $A = P \cap f^{\infty}(W)$. It is then readily verified that A is compact and can be written as a finite union of closed convex sets of the form $C_i \cap f^{\infty}(W)$. Renumbering these sets, if necessary, and denoting them by D_i, we have $A = \bigcup_{i=1}^{m} D_i$. The set $G = W \cap A \subset A$ is open in A. Let $\overline{G}$ and ∂G denote its closure and boundary in A, respectively. Clearly, f acts from $\overline{G}$ into A and has no fixed points on ∂G. Thus, we are in the situation described above, and the index $\operatorname{ind}_A(f, G)$ is defined. We put, by definition,

$$\operatorname{ind}_P(f, W) = \operatorname{ind}_A(f, G)$$

and call $\operatorname{ind}_P(f, W)$ *the generalized index of the ultimately compact operator* f. If the ultimate range of f is empty, then, in particular, f has no fixed points and one can put, by definition, $\operatorname{ind}_P(f, W) = 0$. The characteristic thus defined inherits all properties of $\operatorname{ind}_A(f, G)$ in the case where A is compact. Finally, $\operatorname{ind}_P(f, W)$ can also be defined in a somewhat more general setting, namely, in that considered when we defined the relative index of a condensing operator. Suppose f is defined not on the closure of W, but only on the set W itself, and its fixed-point set $\mathrm{Fix}(f)$ is compact. Let V be a set open in P such that $\mathrm{Fix}(f) \subset V$ and $\overline{V} \subset W$. Then $\mathrm{ind}_P(f, V)$ is defined and has the same value for all sets V with the indicated properties. Put, by definition,

$$\mathrm{ind}_P(f, W) = \mathrm{ind}(f, V).$$

Notice that if the transfinite sequence $\{T_\alpha\}$ (see **1.6.1**), used in the construction of the ultimate range $f^\infty(W)$, becomes compact at the first transfinite step (i.e., $T_\infty = \bigcap_{n=0}^{\infty} T_n$ is compact), then to define $\mathrm{ind}_P(f, G)$ one can use instead of $f^\infty(W)$ the set T_∞, as well as any compact set K that contains $f^\infty(W)$ and is invariant under f: $f(K \cap W) \subset K$. In R. D. Nussbaum's works [114, 120] this recipe for defining a generalized index was implemented precisely for operators for which the set $T_\infty = \bigcap_{n=0}^{\infty} T_n$ is compact. In particular, he considered operators that are α-condensing with a constant $q < 1$, and also strictly locally condensing maps. Let us describe the recipe used in [114] to define a generalized index for strictly locally condensing maps. As we remarked in **1.8.12**, a strictly locally condensing map f with compact fixed-point set $\mathrm{Fix}(f)$ is ultimately compact in some neighborhood V of $\mathrm{Fix}(f)$, and in the sequence of **1.6.1** constructed for f on V, already the set T_∞ is compact. Hence, if $f: W \to P$ is a strictly locally condensing operator such that the set $\mathrm{Fix}(f) = \{x \in W: x = f(x)\}$ is compact, and if V is a neighborhood of $\mathrm{Fix}(f)$ with the property that $f^\infty(V)$ is compact, then $\mathrm{ind}_P(f, W)$ can be set equal to $\mathrm{ind}_P(f, V)$, by definition. The independence of this definition on the choice of the neighborhood V is an obvious consequence of the properties of the generalized index.

3.9.10. The index of $(1, \alpha)$-bounded operators. Imposition of some additional conditions on $(1, \alpha)$- (or $(1, \chi)$-) bounded maps allows one to introduce a characteristic enjoying some properties of an index even for such maps. Let us describe a construction proposed by W. V. Petryshin [127]. Let V be bounded open subset of a Banach space E and let $f: \overline{V} \to E$ be a $(1, \alpha)$-bounded operator with no fixed points on ∂V and with the property that any sequence $\{x_n\} \subset \overline{V}$ such that $\{x_n - f(x_n)\}$ converges contains a convergent subsequence $\{x_{n_k}\}$ (such maps are termed *demicompact*). Since f is demicompact and has no fixed points on ∂V, $\|x - f(x)\| \geq d > 0$ for all $x \in \partial V$, where d is some fixed number. Let g be an arbitrary operator given on ∂V which is condensing with constant $q < 1$ and has the property that $\|f(x) - g(x)\| < d$ for all $x \in \partial V$. It is readily seen that the condensing operator g has no fixed points on ∂V, and so its index $\mathrm{ind}(g, V)$ is defined. Now put

$$\mathrm{ind}(f, V) = \mathrm{ind}(g, V).$$

Properties and applications of the notion of an index thus introduced are described in [127]. An analogous idea of approximating an $(1, \alpha)$-bounded map f by a condensing map was also used, under different assumptions on f, by R. D. Nussbaum [114] to define a generalized index for f.

CHAPTER 4

APPLICATIONS

The aim of this chapter is to describe the most typical examples of application of the notions and facts connected with MNCs and condensing operators to the theory of differential and integral equations. We do not aim at giving the results in their maximal generality; rather, we regard each of the problems discussed here as an illustration of the methodology based on the use of condensing operators.

The chapter is divided into nine sections. In Section 1 we consider the Cauchy problem for an ordinary differential equation in an infinite-dimensional Banach space. In Section 2 we examine a problem that can be handled by similar ideas, namely, the solvability of a stochastic equation with deviating argument. Sections 3–8 are devoted to various problems of the theory of functional-differential equations of neutral type. We discuss the Cauchy problem, study periodic solutions, examine the applicability of the averaging principle, consider questions of stability theory, construct a generalization of Floquet's theory. Section 9, the last one, contains a study of the Hammerstein integral operator in L_p-spaces from the standpoint of the theory of condensing operators, and it is shown that in those limiting situations where this operator looses the compactness property, it is, as a rule, locally condensing.

In contrast to the preceding chapters, here the comments on the relevant literature are made separately in each section.

4.1. DIFFERENTIAL EQUATIONS IN BANACH SPACE

For the equation

$$x' = f(x,t) \tag{1}$$

in an infinite-dimensional Banach space E a sufficient condition for the existence of a solution to the Cauchy problem is that f be Lipschitz in x and continuous in t; the Lipschitz condition can be replaced by absolute continuity. These facts can be proved, for example, using the contraction mapping principle and the Schauder principle, applied to the related integral operator. Utilization of the theory of condensing operators leads to more general situations, which are described in the present section.

4.1.1. Preliminary remarks. A well-known example of Bourbaki shows that in the case of infinite-dimensional Banach spaces equation (1) with a continuous right-hand side may have no solutions.

Let $f: c_0 \to c_0$ be given by the formula $f(x) = \{|x_n|^2 + 1/n\}$, where $x = \{x_n\}$. Then one can show that the map f is continuous on c_0, but the problem

$$x' = f(x), \quad x(0) = 0$$

has no solution.

A. N. Godunov showed that a similar example can be exhibited in any infinite-dimensional Banach space, including Hilbert space (see [53]). The construction of such examples is possible due to the fact that in infinite-dimensional spaces balls are not compact.

Among the results on the solvability of the Cauchy problem we mention a theorem of M. A. Krasnosel′skiĭ and S. G. Kreĭn [89] (dealing with the case in which $f(t,x) = f_1(t,x) + f_2(t,x)$, where f_1 is compact and f_2 satisfies the Lipschitz condition with respect to x) and a theorem of M. A. Krasnosel′skiĭ, A. V. Kibenko, and Ya. D. Mamedov [87] (in this last paper it is established that the one-sided estimates on f that are sufficient for the uniqueness of the solution also guarantee its existence).

The main result of this section can be formulated as follows: if for any t and any set Ω the MNC of the set $f(t,\Omega)$ does not exceed "too much" the MNC of Ω, then the integral operator

$$(Jx)(t) = x_0 + \int_0^t f(s, x(s))ds \tag{2}$$

is condensing and equation (1) with the initial condition

$$x(0) = x_0 \tag{3}$$

has a solution.

In what follows $C(P,E)$ denotes, as usual, the space of continuous functions on the interval $P \subset \mathbf{R}$ with values in the Banach space E, equipped with the norm $\|x\|_{C(P,E)} = \sup_{t\in P} \|x(t)\|$, and $C^1(P,E)$ denotes the space of continuously differentiable functions $x \in$

$C(P,E)$, equipped with the norm $\|x\|_{C^1(P,E)} = \|x\|_{C(P,E)} + \|x'\|_{C(P,E)}$. Also, $B(x_0,r)$ is the ball in E of radius r and center x_0, and R is the cylinder $\{(t,x): t \in [0,b],\ x \in B(x_0,r)\}$ $(b, r > 0,\ x_0 \in E)$.

4.1.2. Theorem on the solvability of the Cauchy problem. *Let the operator f be uniformly continuous in the cylinder R. Suppose that for any set $M \subset B(x_0,r)$ and any $t \in [0,b]$ one has the inequality*

$$\psi[f(t,M)] \leq k\psi(M) \tag{4}$$

where k does not depend on t and M and the MNC ψ in E is semi-additive, invariant under translations, continuous with respect to the Hausdorff metric, semi-homogeneous, and takes positive values on noncompact sets. Then for some $b_1 \in (0,b]$ the problem (1), (3) *has a solution in $C^1([0,b_1],E)$.*

The essential ingredient in the proof of this theorem is the following lemma.

4.1.3. Lemma. *Suppose the conditions of Theorem 4.1.2 are satisfied. Let $\Omega \subset C([0,b],E)$ be an equicontinuous set of functions, the values of which lie in $B(x_0,r)$. Then for any $t \in [0,b]$ one has the inequality*

$$\psi\left[x_0 + \left\{\int_0^t f(s,x(s))ds : x \in \Omega\right\}\right] \leq bk \max_{s\in[0,b]} \psi[\Omega(s)].$$

Proof. The family of functions $f(\cdot,y(\cdot))$ $(y \in \Omega)$ is equicontinuous, and so the integrals of these functions can be uniformly approximated by integral sums

$$\frac{t}{n}\sum_{i=1}^{n} f(s_i,y(s_i)),\quad s_i = i\frac{t}{n},\ y \in \Omega.$$

In view of the homogeneity, continuity, and translation-invariance of the MNC ψ, to prove the lemma is suffices to verify that

$$\psi(\Gamma_n) \leq k \max_{s\in[0,b]} \psi[\Omega(s)],$$

where

$$\Gamma_n = \left\{z : z = \frac{1}{n}\sum_{i=1}^{n} f(s_i,y(s_i)), y \in \Omega\right\}.$$

But $\Gamma_n \subset \operatorname{co} Q_n$, where $Q_n = \bigcup_{i=1}^n f(s_i, y(s_i))$, and so in view of the semi-additivity, monotonicity, and invariance of ψ under the passage to the convex hull we obtain

$$\psi(\Gamma_n) \leq \psi(Q_n) = \max_{1 \leq i \leq n} \psi[f(s_i, \Omega(s_i))] \leq k \max_{s \in [0,b]} \psi[\Omega(s)]. \quad \textbf{QED}$$

4.1.4. Proof of Theorem 4.1.2. With no loss of generality one can assume that f is bounded in R. Pick $b_1 \in (0, b]$ such that $b_1 k < 1$ and

$$\|f(t,x)\| \leq r/b_1 \quad \text{for } (t,x) \in R.$$

Let $T = \{x \in C([0,b_1], E) : x(0) = x_0,\ x$ satisfies the Lipschitz condition with constant $r/b_1\}$. It is not hard to verify that T is convex, closed, bounded, and nonempty, and that $TJ \subset T$. Using Lemma 4.1.3, it is easy to show that the operator J is condensing on T with respect to the MNC ψ_C, defined by the rule

$$\psi_C(\Omega) = \psi(\Omega[0, b_1]) = \max_{t \in [0,b_1]} \psi[\Omega(t)]$$

(see **1.2.4**). In view of **1.5.11** and **1.5.12**, J has at least one fixed point. **QED**

4.1.5. Example. In the space c_0 consider the Cauchy problem

$$x_k' = x_{k+1}\sqrt{|x_1|} + 1/k, \ x_k(0) = x_k^0 \quad (k = 1, 2, \ldots). \tag{5}$$

Let $f(x) = \Phi(x,x)$, $\Phi(x,y) = \{x_{k+1}\sqrt{|y_1|} + 1/k\}_{k=1}^\infty$, where $x = \{x_k\}$, $y = \{y_k\} \in c_0$ and $x_0 = \{x_k^0\}$. It is readily checked that for any $r > 0$ the operator $\Phi: c_0 \times B(x_0, r) \to c_0$ is uniformly continuous in (x,y), compact in y for any fixed x, and satisfies the Lipschitz condition in x for any fixed y. It follows from Theorem 1.5.7 that the operator $f: B(x_0, r) \to c_0$ is $(\sqrt{r}, \chi)$-bounded. Hence, by Theorem 4.1.2, problem (5) is locally solvable.

Theorem 4.1.2 uses as an estimate (see (4)) a linear function ku; the latter can be replaced by an Osgood function $L(u)$ or even by a Kamke function $L(t,u)$. However, as Bourbaki's example (see **4.1.1**) shows, one cannot use as an estimate the Hölder function Hu^ρ, $\rho \in (0,1)$. If instead of the ordinary differential equation (1) one considers the equation with deviating argument

$$x'(t) = f(t, x(h(t))), \tag{6}$$

then the situation changes. Here the connection between the delay $h(t)$ and the way in which f depends on the second argument already plays a role.

4.1.6. Existence theorem for equations with deviating argument. *Let the operator f be uniformly continuous in the cylinder R. Suppose that for any $M \subset B(x_0, r)$ and any $t \in [0, b]$ the following inequality holds*

$$\psi[f(t, M)] \leq H[\psi(M)]^\rho, \tag{7}$$

where $\rho \in (0, 1]$, $H \geq 0$, and the MNC *ψ in E is algebraically semi-additive, invariant under translations, continuous, semi-homogeneous, and takes positive values on noncompact sets. Finally, suppose the function h is continuous and $0 \leq h(t) \leq t^{1/\rho}$ for all $t \in [0, b]$. Then there is a $b_1 \in (0, b]$ such that problem* (6), (3) *has a solution in $C^1([0, b_1], E)$.*

Proof. First of all we remark that if the set $\Omega \subset C([0, b], E)$ is equicontinuous, then using the semi-homogeneity, algebraic semi-additivity, and continuity of the MNC ψ in conjunction with the fact that $\int_0^t x(s)ds$ (for $x \in \Omega$) can be uniformly approximated by integral sums one can easily show that

$$\psi\Big(\Big\{\int_0^t x(s)ds : x \in \Omega\Big\}\Big) \leq \int_0^t \psi[\Omega(s)]ds.$$

Now choose $b_1 \leq \min\{b, 1\}$ such that $\|f(t, x)\| \leq r/b_1$ for all $(t, x) \in R$. As in **4.1.4**, denote $T = \{x \in C([0, b_1], E) : x(0) = x_0$, x satisfies the Lipschitz condition with constant $r/b_1\}$. To complete the proof it remains to convince ourselves that the integral operator

$$(Jx)(t) = x_0 + \int_0^t f(s, x(h(s)))ds, \quad 0 \leq t \leq b_1,$$

is condensing on T with respect to the MNC ψ_C^1 defined by the rule $[\psi_C^1(\Omega)](t) = \psi[\Omega(t)]$ (see **1.2.4**). Notice that ψ_C^1 is an MNC that takes values in the set of nonnegative continuous functions on $[0, b_1]$.

Thus, suppose that for some $\Omega \subset T$ one has

$$\psi_C^1(\Omega) \leq \psi_C^1(J\Omega). \tag{8}$$

We claim that Ω is relatively compact. Indeed,

$$[\psi_C^1(J\Omega)](t) = \psi[(J\Omega)(t)] = \psi\Big[\int_0^t f(s, \Omega(h(s)))ds\Big]$$

$$\leq \int_0^t \psi[f(s, \Omega(h(s)))]ds \leq \int_0^t H[\psi(\Omega(h(s)))]^\rho ds. \tag{9}$$

From inequalities (8) and (9) it follows that the function $m(t) = \psi[\Omega(t)]$, which is nonnegative and bounded on $[0, b_1]$, satisfies

$$m(t) \leq \int_0^t H[m(h(s))]^\rho ds.$$

Let us show that $m(t) \equiv 0$. Iterating n times and estimating the integral in the right-hand side we obtain

$$m(t) \leq \int_0^t H\Big(\int_0^{h(s_1)} H[m(h(s_2))]^\rho ds_2\Big)^\rho ds_1 \leq \dots$$

$$\dots \leq \int_0^t H\Big(\int_0^{h(s_1)} H\Big(\int_0^{h(s_2)} H \dots \Big(\int_0^{h(s_{n-1})} H[m(h(s_n))]^\rho ds_n\Big)^\rho ds_{n-1} \dots\Big)^\rho ds_2\Big)^\rho ds_1$$

$$\leq \int_0^t H\Big(\int_0^{s_1^{1/\rho}} H \dots \Big(\int_0^{s_{n-1}^{1/\rho}} H[m(h(s_n))]^\rho ds_n\Big)^\rho ds_{n-1} \dots\Big)^\rho ds_1$$

$$\leq H^{\sum_{i=0}^n \rho^i}\Big(\sup_{0\leq t\leq b_1} m(t)\Big)^{\rho^n} \frac{b_1^n}{2^{\rho^{n-2}} \cdot \dots \cdot (n-2)^{\rho^2}(n-1)^\rho n} \to 0 \quad \text{when } n \to \infty. \quad \textbf{QED}$$

4.1.7. Equations with a uniformly continuous operator. Here we consider equation (1) under the assumption that the operator $f: \mathbf{R} \times E \to E$ is bounded and uniformly continuous on every bounded subset of $\mathbf{R} \times E$. This condition is not sufficient for local solvability. It turns out, however, that (1) can be in a certain sense extended to an equation

$$\mathcal{X}' = F(t, \mathcal{X}) \tag{10}$$

in a specially constructed Banach space E_+, such that equation (10) is locally solvable for any initial condition. Then from the behavior of the solutions of (10) one can draw conclusions about the solvability of equation (1). In particular, the uniqueness of the solutions of (10) already guarantees the local solvability of (1). This approach is only outlined here; for details the reader is referred to [157].

To get started, let BE be the space of bounded sequences X in E, $\|X\| = \sup_n \|x_n\|$, and let CE be the subspace of the sequences converging to zero. Set $E_+ = BE/CE$ (compare with the space E^+ introduced in Section 2.2). For $\mathcal{X} \in E_+$ set $\|\mathcal{X}\| = \lim_{n\to\infty} \|x_n\|$, where $\{x_n\} = X$ is any representative of the class $\mathcal{X}$. Let $\tilde{E}$ denote the subspace of E_+ consisting of the classes $\mathcal{X}$ generated by stationary sequences $(x, x, \dots)$. Define an isometric isomorphism $T : \tilde{E} \to E$ by the formula

$$T\mathcal{X} = x, \text{ where } (x, x, \dots) \in \mathcal{X}.$$

Now define the operator $F: \mathbf{R} \times E_+ \to E_+$ as follows: if $\mathcal{X} \ni \{x_n\}$, then $F(t, \mathcal{X}) \ni \{f(t, x_n)\}$. It is a straightforward matter to check that F is well defined and $f(t, T\mathcal{X}) = TF(t, \mathcal{X})$.

It is readily seen that $\mathcal{X}(t) \in \tilde{E}$ is a solution of equation (10) if and only if $x(t) = T\mathcal{X}(t)$ is a solution of equation (1). Using functions of the type of Tonelli approximants, one can verify that in the space E_+ equation (10) admits a solution for any initial condition. This yields a solvability test for equation (1): if every solution of equation (10) with initial condition

$$\mathcal{X}(t_0) = \mathcal{X}_0, \tag{11}$$

where $\mathcal{X}_0 \in \tilde{E}$, does not leave the space $\tilde{E}$, then equation (1) has a solution with initial condition

$$x(t_0) = T\mathcal{X}_0. \tag{12}$$

Clearly, in order for problem (1), (12) to have a solution it suffices that problem (10), (11) admit at least one solution lying in $\tilde{E}$. Another sufficient condition for the solvability of problem (1), (12) is that all solutions of problem (10), (11) remain in E^c, the subspace of E_+ consisting of the classes of relatively compact sequences. To guarantee that the latter holds, one can require that $\chi[f(t, X)] \leq L(t, \chi(X))$ for all $X \in BE$ and all $t \in \mathbf{R}$, where the function $L(t, u)$ has the following property:

$$\textit{the problem } z' = L(t, z),\ z(0) = 0 \textit{ has a unique nonzero solution.} \tag{13}$$

Here the Hausdorff MNC χ can be replaced by any normal MNC ψ in E, provided only that the seminorm ψ_1 on E_+ associated with ψ (according to the rule $\psi_1(\mathcal{X}) = \psi(X)$, $X \in \mathcal{X}$) is continuous on E_+.

4.1.8. Notes on the references. Theorem 4.1.2 is quoted from the paper of B. N. Sadovskiĭ [156], from which we also took Example 4.1.5. Close results can be found in the papers of A. Ambrosetti [13] and S. Szufla [170]. In the paper of B. N. Sadovskiĭ [160] instead of inequality (4) it is required that

$$\psi[f(t, M)] \leq L(t, \psi(M)), \tag{14}$$

for any bounded set $M \subset E$ and any $t \in [0, T]$, where the continuous function L satisfies condition (13) and the MNC ψ is sufficiently nice (monotone, semi-additive, and so on). Analogous results can be found in the papers of G. Pianigiani [131], S. Szufla [171], and A. Cellina [25]. In particular, in the last paper it is required that the function

$$L(\varepsilon) = \sup_{M \subset B(x_0, r),\ \alpha(M) \geq \varepsilon} \{\alpha[f(t, M)]/\alpha(M)\}$$

satisfy

$$\int_{0+} (\varepsilon L(\varepsilon))^{-1} d\varepsilon = \infty.$$

The function L in (14) is allowed to satisfy, instead of Osgood's condition (13), the less restrictive Kamke condition (see, e.g., [171]). In that paper, an in a number of others, it is not required that the operator f be uniformly continuous.

Theorem 4.1.6 generalizes a result of Ya. I. Gol′tser and A. M. Zverkin [57]. A similar result was obtained by Yu. A. Dyadchenko in [35], where instead of (7) it was required that

$$\|f(t,u) - f(t,v)\| \leq L_1(t, \|u - v\|),$$

$$\|f(t,u)\| \leq L_2(t, \|u\|),$$

where the functions L_i are such that the inequality

$$z(t) \leq L_i\Big(t, \int_0^{h(t)} z(s)ds\Big)$$

has no nonzero solutions for $i = 1$, and has only bounded solutions for $i = 2$.

The exposition in **4.1.7** follows the paper by B. N. Sadovskiĭ [157] (see also [161]).

To conclude let us mention several other works. In [182] P. P. Zabreĭko and I. B. Ledovskaya prove a theorem on the continuous dependence on a parameter λ of the solution of the Cauchy problem

$$x' = f(t, x, \lambda), \quad x(0) = x_0.$$

It is assumed that f satisfies the Carathéodory condition for each fixed λ, is integral-continuous in λ, and satisfies

$$\|f(t, x_1, \lambda) - f(t, x_2, \lambda)\| \leq n(t)\phi(\|x_1 - x_2\|),$$

$$\chi\Big[\bigcup_{\lambda \in \Lambda} R_\lambda(t_1, t_2)(\Omega)\Big] \leq \Big[\int_{t_1}^{t_2} \rho(s)ds\Big]\chi(\Omega),$$

where

$$R_\lambda(t_1, t_2)(x) = \int_{t_1}^{t_2} f(s, x, \lambda)ds,$$

the functions n and ρ are integrable on $[0,T]$, $\phi(u)$ is nondecreasing, and $\phi(u) \to 0$ as $u \to 0$.

In the paper of V. M. Gershteĭn [51] it is shown that if in equation (1) f is T-periodic in t, and if equation (1) is dissipative and the operator of translation along its trajectories is condensing, then (1) admits at least one λ-center that is bounded in $C(E)$ and compact in the sense of uniform convergence on any time interval, and among the λ-centers of (1) there is minimal one.

Various generalizations of Peano's theorem in Banach spaces can be found in the papers of P. P. Zabreĭko and A. I. Smirnov [183] and M. I. Kamenskiĭ [72].

4.2. ITÔ STOCHASTIC EQUATIONS WITH DEVIATING ARGUMENT

In this section we prove a theorem on the existence and uniqueness of the solution of the Itô stochastic equation with deviating argument

$$x(t) = x_0 + \int_0^t a(s, x(h(s)))ds + \int_0^t b(s, x(h(s)))dw(s), \tag{1}$$

where a and b are Borel-measurable functions, $0 \leq h(t) \leq t$. In the case where a and b satisfy the Lipschitz condition in the second argument, analogous theorems for $h(s) \equiv s$ were proved for the first time by I. I. Gikhman and K. Itô (see, e.g., [52]). For that purpose an integral operator was associated with (1), which turned out to be contractive, and the solution was obtained by the method of successive approximations. In the theorem given below the Lipschitz condition is replaced by a less stringent requirement—a condition of type (14), Section 4.1—and it is proved that the integral operator associated with equation (1) acts in a certain Banach space where it is condensing.

4.2.1. Definitions. We recall a number of definitions and facts from the theory of stochastic processes (see, e.g., [52], [103]).

A *probability space* $(\Omega, \mathcal{U}, \mathbf{P})$ is a set Ω together with a σ-algebra of subsets $\mathcal{U}$ and a probability measure $\mathbf{P}$ defined on $\mathcal{U}$ (i.e., a nonnegative countably-additive function of the elements of $\mathcal{U}$ such that $\mathbf{P}(\Omega) = 1$).

A *random variable* ξ is a $\mathbf{P}$-measurable function on the space Ω with values in $\mathbf{R}^n$. A family of random variables $\xi(t)$ $(t \geq 0)$ is termed a *stochastic process.* For each fixed $\omega \in \Omega$ there arises the function of time $t \mapsto [\xi(t)](\omega)$, called a *path* (*trajectory*, or *sample function*) *of the stochastic process* $\xi(t)$). The *expectation* or *mean value* $\mathbf{M}\xi$ *of the random variable* ξ is defined to be the integral

$$\mathbf{M}\xi = \int_\Omega \xi(\omega)d\mathbf{P}(\omega),$$

provided it exists.

Two random variables ξ and η are said to be *independent* if, for any $a, b \in \mathbf{R}^n$,

$$\mathbf{P}\{\xi < a, \eta < b\} = \mathbf{P}\{\xi < a\}\mathbf{P}\{\eta < b\}.$$

The random variable $\xi = (\xi_1, \ldots, \xi_n)$ is said to have a *Gaussian (or normal) distribution* if its characteristic function $\psi(\tau_1, \ldots, \tau_n) = \mathbf{M}e^{i(\tau,\xi)}$ admits the representation $\psi(\tau_1, \ldots, \tau_n) = e^{i(m,\tau)-(\Lambda\tau,\tau)/2}$, where $m = (m_1, \ldots, m_n)$, $\tau = (\tau_1, \ldots, \tau_n)$ are vectors and Λ is a real, symmetric, nonnegative definite matrix.

A *Wiener process* $w(t)$ $(t \geq 0)$ is a stochastic process with the following properties:

a) $w(0) = 0$, and for $0 < t_1 < \ldots < t_n$ the random variables $w(t_2) - w(t_1), w(t_3) - w(t_2), \ldots, w(t_n) - w(t_{n-1})$ are independent;

b) the random variable $w(t+s) - w(t)$ has a Gaussian distribution with parameters $m = (0, \ldots, 0)$ for all t, s and $\Lambda = I$ (where I is the identity matrix).

If $w(t)$ is a Wiener process, then for almost every ω the paths of $w(t)$ are continuous in t, satisfy Hölder's condition with an exponent smaller than $1/2$, are nowhere differentiable, and have infinite variation on any bounded interval.

From the enumerated properties of a Wiener process it follows that for almost any path $w(t)$ the Stieltjes integral $\int f(t)dw(t)$ has no meaning even for continuous functions f. Nevertheless, an integral with respect to a Wiener process, known as the Itô integral, can indeed be defined.

4.2.2. The Itô integral (see, e.g., [52], [103]). With a Wiener process one usually associates a nondecreasing family of σ-algebras $\mathcal{F}_t \subset \mathcal{U}$ ($\mathcal{F}_{t_1} \subset \mathcal{F}_{t_2}$ for $t_1 \leq t_2$), where $\mathcal{F}_t$ is the minimal σ-algebra with respect to which $w(s)$ is measurable for $s \leq t$. Let $\mathfrak{M}_2[a, b]$ $(0 \leq a \leq b)$ denote the set of all functions $f(t, \omega)$ that are jointly measurable in the variables t, ω, are defined for $t \in [a, b]$ and $\omega \in \Omega$, are measurable with respect to $\mathcal{F}_t$ for all $t \in [a, b]$, and such that the integral $\int_a^b |f(t, \cdot)|^2 dt$ is finite with probability one. We first define the integral $\int_a^b f(t)dw(t)$ for step functions $f(t)$. Thus, suppose $f(t) = f(t_k)$ for $t \in [t_k, t_{k+1})$, where $a = t_0 < t_1 < \ldots < t_{r-1} < t_r = b$. Set

$$\int_a^b f(t)dw(t) = \sum_{k=0}^{r-1} f(t_k)[w(t_{k+1}) - w(t_k)].$$

If now $f(t)$ is an arbitrary function in $\mathfrak{M}_2[a, b]$ then, as is shown in [52], there exists a sequence of step functions $f_n(t)$ such that $\int_a^b [f(t) - f_n(t)]^2 dt \to 0$ in probability and the sequence $\int_a^b f_n(t)dw(t)$ converges in probability to some limit, called the *Itô stochastic integral of the function* $f(t)$ and denoted $\int_a^b f(t)dw(t)$.

One can show that the Itô stochastic integral has the linearity property, that $\int_0^t f(s)ds$ almost surely has continuous paths, and that if $\int_a^b \mathbf{M}|f(t)|^2 dt < \infty$, then

$$\mathbf{M}\int_a^b f(t)dw(t) = 0,$$

$$\mathbf{M}\sup_{0\le s\le \tau}\left|\int_0^s f(t)dw(t)\right|^2 \le 4\int_0^\tau \mathbf{M}|f(t)|^2 dt, \quad a \le 0 \le \tau \le b. \tag{2}$$

Now let us turn to equation (1). Suppose that $a, b\colon [0,T]\times \mathbf{R}^n \to \mathbf{R}^n$ are Borel-measurable functions, $0 \le h(t) \le t$, and the random variable x_0 is $\mathcal{F}_0$-measurable.

The process $x(t)$ is called a *solution* of equation (1) (a *strong solution*—see [103]) if $x(t)$ is $\mathcal{F}_t$-measurable, the integrals in (1) exist, and (1) holds with probability one for all $t \in [0,T]$.

Let B_T denote the space of measurable random functions $\zeta(t,\omega)$ with almost surely continuous paths, which are measurable with respect to the σ-algebra $\mathcal{F}_t$ for any $t\in[0,T]$, and put $\|\zeta\|_{B_T} = (\mathbf{M}\sup_{0\le s\le T}|\zeta(s,\omega)|^2)^{1/2}$.

4.2.3. Lemma (see [52]). *B_T is a Banach space.*

Proof. Clearly, it suffices to show that B_T is complete. Let $\{\zeta_n\}$ be a Cauchy sequence in B_T. Then from $\{\zeta_n\}$ one can extract a subsequence that converges for almost every (t,ω). Now pick a sequence of positive integers $\{m_r\}$ such that

$$\mathbf{M}\sup_{0\le s\le t}|\zeta_n(s)-\zeta_{n'}(s)|^2 < 2^{-r} \text{ for } t\in[0,T],\ n,n' \ge m_r.$$

Then by the Borel-Cantelli theorem (see [52]), the series $\zeta_{m_1}(t) + \sum_{r=0}^{\infty}|\zeta_{m_{r+1}}(t)-\zeta_{m_r}(t)|$ almost surely converges uniformly in $t\in[0,T]$. Hence, the sequence $\{\zeta_{m_r}(t)\}$ almost surely converges uniformly to a continuous process $\zeta(t)$ which, as one can readily verify, belongs to B_T. A standard argument then easily yields the conclusion that the whole sequence $\{\zeta_n\}$ converges in B_T to ζ. **QED**

Now consider the conditions

$$|a(t,u)|^2 \le K(|u|^2+1), \quad |b(t,u)|^2 \le K(|u|^2+1). \tag{3}$$

4.2.4. Lemma. *Suppose conditions* (3) *are satisfied. Then the operator G, defined by the formula,*

$$(Gx)(t) = \int_0^t a(s, x(h(s)))ds + \int_0^t b(s, x(h(s)))dw(s), \tag{4}$$

maps B_τ into itself for any $\tau \in [0,T]$.

Proof. Let $x \in B_\tau$, $t \in (0,T]$. Using Hölder's inequality and the estimates (2) and (3), we obtain

$$\mathbf{M} \sup_{0 \le s \le \tau} |(Gx)(s)|^2 \le 2\mathbf{M}\tau \int_0^\tau |a(\theta, x(h(\theta)))|^2 d\theta + 2\mathbf{M} \sup_{0 \le s \le \tau} \Big(\int_0^s b(\theta, x(h(\theta)))dw(\theta)\Big)^2$$

$$\le 2\tau \int_0^\tau \mathbf{M}|a(\theta, x(h(\theta)))|^2 d\theta + 8 \int_0^\tau \mathbf{M}|b(\theta, x(h(\theta)))|^2 d\theta$$

$$\le 2K\Big[\tau \int_0^\tau \mathbf{M}(|x(h(\theta))|^2 + 1)d\theta + 4\int_0^\tau \mathbf{M}(|x(h(\theta))|^2 + 1)d\theta\Big]$$

$$\le 2K\Big[\tau \int_0^\tau \|x\|^2_{B_{h(\theta)}} d\theta + \tau^2 + 4\int_0^\tau \|x\|^2_{B_{h(\theta)}} d\theta + 4\tau\Big]. \tag{5}$$

Inequality (5) yields the desired estimate:

$$\|Gx\|^2_{B_\tau} \le 2K(\tau^2 + 4\tau)(\|x\|^2_{B_\tau} + 1). \quad \textbf{QED}$$

4.2.5. Lemma. *Suppose conditions* (3) *are satisfied. Then there is an $H > 0$ such that $\|x\|^2_{B_{T'}} < H$ for any solution $x(t)$ of problem* (1) *on an arbitrary interval $[0,T']$ with $T' \le T$.*

Proof. Set $\mu(t) = \|x\|^2_{B_t}$. Obviously, inequalities (5) yield the estimate

$$\mu(t) \le K_1 + K_2 \int_0^t \mu(\tau)d\tau,$$

where K_1, K_2 are positive constants. Hence, the function $\mu(t)$ is bounded on $[0,T]$. **QED**

Now let us introduce the following conditions:

$$|a(t,u) - a(t,v)|^2 \le L(t, |u-v|^2), \qquad |b(t,u) - b(t,v)|^2 \le L(t, |u-v|^2), \tag{7}$$

where $L(t,z)$ is jointly continuous and is nondecreasing and convex in z;

$$\textit{for any } A \ge 0 \textit{ the inequality } z(t) \le A\int_0^t L(s, z(h(s)))ds, \quad 0 \le t \le T \tag{8}$$

has no nonzero nonnegative solutions;

$\operatorname{mes} h^{-1}(e) \to 0$ *when* $\operatorname{mes} e \to 0$, *where* mes *stands for Lebesgue measure and $e \subset [0,1]$.* (9)

Define an MNC $\psi: B_T \to \mathfrak{M}[0,T]$ (see **1.2.4**) by the formula

$$[\psi(\Omega)](t) = \chi_t(\Omega_t),$$

where χ_t is the Hausdorff MNC in the space B_t and $\Omega_t = \{x_t = x|_{[0,t]} : x \in \Omega\} \subset B_t$.

4.2.6. Lemma. *Suppose conditions* (3) *and* (6)–(8) *are satisfied. Then the operator G defined by formula* (4) *is condensing with respect to the* MNC ψ *on any bounded subset of the space B_T.*

Proof. Suppose $\psi(\Omega) \leq \psi(G\Omega)$ for some bounded set $\Omega \subset B_T$. Let us show that in this case $\psi(\Omega) = 0$. To this end we notice that the function $t \mapsto [\psi(\Omega)](t)$ is nondecreasing and bounded. Consequently, for any $\varepsilon > 0$ it has only a finite number of jumps of magnitude $\geq \varepsilon$. Remove the points corresponding to these jumps together with their disjoint δ_1-neighborhoods from the segment $[0, T]$, and using points β_j, $j = 1, \ldots, m$, divide the remaining part into intervals on which the oscillation of the function $\psi(\Omega)$ is smaller than ε. Now surround the points β_j by disjoint δ_2-neighborhoods and consider the family of all functions $Z = \{z_k : k = 1, \ldots, l\}$, continuous with probability one, constructed as follows: z_k coincides with an arbitrary element of a $[(\psi(\Omega))(\beta_j) + 1]$-net of the set Ω_{β_j} on the segment $\sigma_j = [\beta_{j-1} + \delta_2, \beta_j - \delta_2]$, $j = 1, \ldots, m$, and is linear on the complementary segments.

Let $u \in (G\Omega)_t$; then $u = (Gz)_t$ for some $z \in \Omega$ and

$$[(\psi(\Omega))(\beta_j) + \varepsilon] \geq \|z - z_p^{\beta_j}\|^2_{B_{\beta_j}},$$

where $z_p^{\beta_j}$ is some element of a $[(\psi(\Omega))(\beta_j) + \varepsilon]$-net of Ω_{β_j}. Since $z_p^{\beta_j}|_{\sigma_j} = z_k|_{\sigma_j}$ (where z_k is some element of the set Z), it follows that for $s \in \sigma_j$ we have

$$\mathbf{M}|z(s) - z_k(s)|^2 \leq \mathbf{M} \sup_{\beta_{j-1}+\delta_2 \leq s \leq \beta_j - \delta_2} |z(s) - z_k(s)|^2$$

$$\leq \|z - z^{\beta_j}\|^2_{B_{\beta_j}} \leq [(\psi(\Omega))(\beta_j) + \varepsilon]^2 \leq [(\psi(\Omega))(s) + 2\varepsilon]^2.$$

Set $\sigma_j^t = \sigma_j \cap [0, \overline{h}(t)]$, where $\overline{h}(t) = \sup_{0 \leq s \leq t} h(s)$. Then

$$\mathbf{M} \sup_{0 \leq s \leq t} |(Gz)(s) - (Gz_k)(s)|^2 \leq 2t \int_0^t L(s, \mathbf{M}|z(h(s)) - z_k(h(s))|^2) ds$$

$$+ 8 \int_0^t L(s, \mathbf{M}|z(h(s)) - z_k(h(s))|^2) ds$$

$$\leq A \sum_{j=1}^m \int_{h^{-1}(\sigma_j^t)} L(s, \mathbf{M}|z(h(s)) - z_k(h(s))|^2) ds$$

$$+ A \int_{[0,t] \setminus \bigcup_{j=1}^m h^{-1}(\sigma_j^t)} L(s, \mathbf{M}|z(h(s)) - z_k(h(s))|^2) ds.$$

Choosing $\delta_1 > 0$ and $\delta_2 > 0$ sufficiently small, we can ensure that

$$[(\psi(\Omega))(t)]^2 \le [(\psi(G\Omega))(t)]^2 \le \varepsilon + A\int_0^t L(s,(\psi(\Omega))(h(s)) + 2\varepsilon)ds.$$

In conjunction with (7), this implies that $\psi(\Omega) \equiv 0$. The continuity of the operator G is established in a similar manner. **QED**

4.2.7. Theorem. *Suppose conditions* (3) *and* (6)–(8) *are satisfied. Then equation* (1) *is uniquely solvable in* B_T.

Proof. The solvability of problem (1) in B_T follows from Theorem 1.5.11 (see also **1.5.12**). Indeed, it is readily verified that the MNC ψ possesses the requisite properties and the operator G is condensing. Then estimate (5) and Lemma 4.2.5 permit us first to find a solution on some small segement $[0, T_1]$, and then extend it to $[0, T]$. The uniqueness of the solution is established using condition (1). **QED**

4.2.8. Notes on the references. As we mentioned above, the existence and uniqueness theorem for equation (1) in the case $h(t) \equiv t$ can be found in the book of I. I. Gikhman and A. V. Skorokhod [52]. Corresponding theorems for Itô stochastic equations with distributed delay (see **4.3.1** below) of the form:

$$x(t) - f(t, x_t) = x(0) - f(0, x_0) + \int_0^t a(s, x_s) + \int_0^t b(s, x_s)dw(s), \tag{9}$$
$$t \ge 0;\ x(t) = \phi(t),\ t \le 0,$$

where the functions a and b satisfy the Lipschitz condition in the second variable, can be found in the books of R. Sh. Liptser and A. N. Shiryaev [103] (for the case $f(t,u) \equiv 0$) and of B. V. Kolmanovskiĭ and V. R. Nosov [81] (for the case where f satisfies the Lipschitz condition with constant $k < 1$ in the second variable). Theorem 4.2.7 is taken from the paper of A. E. Rodkina [149]. Therein she considered also the case $f(t,u) \not\equiv 0$. In [148] A. E. Rodkina proved the existence and uniqueness of the solution of equation (9) under requirements similar to (3) and (6)–(9) on the functions a and b and the assumption that the operator F, $(Fx)(t) = x(t) - f(t, x_t)$, is invertible.

4.3. THE CAUCHY PROBLEM FOR EQUATIONS OF NEUTRAL TYPE

This section is devoted to the study of the structure of the set of solutions of the Cauchy problem for a functional-differential equation of neutral type.

4.3.1. The basic types of equations. We shall consider a Cauchy problem of the type

$$x'(t) = f(t, x_t, x'_t),\ t \geq 0, \tag{1}$$

$$x(t) = \phi(t),\ t \leq 0. \tag{2}$$

The right-hand side in (1) depends on the prehistory x_t [resp. x'_t] of the unknown function x [resp. of its derivative x'], defined as

$$x_t(s) = x(t+s),\ x'_t(s) = x'(t+s).$$

Here $s \in [-h, 0]$ in the case of a finite delay, and $s \in (-\infty, 0]$ in the case of infinite delay.

Among the equations that can be cast in the form (1) we mention

$$x'(t) = f_1(t, x(g(t,x)), x'(h(t))) \tag{3}$$

(in which $g(t,x), h(t) \in (-\infty, t]$ for all t and x), and

$$x'(t) = f_2(t, x(g_1(t)), \dots, x(g_r(t)), x'(h_1(t)), \dots, x'(h_r(t))) \tag{4}$$

(in which $g_i(t), h_i(t) \in (-\infty, t]$). Equation (1) also includes various equations with delay distributed over a finite or infinite interval.

As one can readily verify, for the existence of a continuously differentiable solution of problem (1), (2) it is necessary that the following "sewing" or "matching" condition be satisfied:

$$\phi'_-(0) = f(0, \phi, \phi'). \tag{5}$$

However, condition (5) often turns to be too restrictive. For that reason, instead of or alongside with (1) one often considers the equation

$$x'(t) = f(t, x_t, x'_t) + \nu_\eta(t)[x'_-(0) - f(0, x_0, x'_0)], \tag{6}$$

where $\eta \in (0, \infty)$ is a parameter and $\nu_\eta(t) = 1 - \eta^{-1}t$ for $t \in [0, \eta]$, $\nu_\eta(t) = 0$ for $t \in (\eta, \infty)$. For equation (6) the sewing condition is obviously satisfied, and already for any initial function ϕ.

J. Hale and collaborators studied various problems for equations of neutral type of the form

$$\frac{d}{dt} D(t, x_t) = F(t, x_t).$$

This way of writing the equation allows one to interpret a solution of the Cauchy problem as a continuous function x on $[-h, T]$ $(T > 0)$ such that $x_0 = \phi$ and

$$D(t, x_t) = D(0, \phi) + \int_0^t F(s, x_s) ds$$

for $t \in [0, T]$.

We will not describe here the multitude of results concerning this equation, refering instead the reader to J. Hale's monograph [60]. We only mention that its study makes effective use of the theory of MNCs and condensing operators.

4.3.2. Main assumptions. Below we shall use the following notation: $C_T = C((-\infty, T], \mathbf{R}^m)$, $C_T^1 = C^1((-\infty, T], \mathbf{R}^m)$, in particular, $C_0 = C((-\infty, 0], \mathbf{R}^m)$, $C_0^1 = C^1((-\infty, 0], \mathbf{R}^m)$; the norm in C_T is denoted by $\|\cdot\|_T$. We shall assume that the functions f and ϕ satisfy the following conditions:

$$f \colon [0, \infty) \times C_0 \times C_0 \to \mathbf{R}^m, \text{ and } \phi \in C_0^1; \tag{7}$$

the operator $\tilde{f} \colon [0, T] \times C_T \times C_T \to \mathbf{R}^m$, *given by the formula* $\tilde{f}(t, u, v) = f(t, u_t, v_t)$, *is jointly continuous for any* $T > 0$; (8)

f *satisfies the Lipschitz condition with constant* $k < 1$ *in the third argument*; (9)

$$\phi'_-(0) = f(0, \phi, \phi'). \tag{10}$$

4.3.3. Lemma: an estimate of a solution. *Suppose conditions* (7)–(10) *are satisfied. Then there exist numbers* $M, T > 0$ *such that any solution* x *of the problem* (1), (2) *satisfies* $|x'(t)| < M$ *for all* $t \in [0, T]$.

Proof. Set $M = \max\{\|\phi'\|, (1 + H)/(1 - k) + 1\}$, $T = \min\{T_1, \delta/M\}$, where the numbers H, T_1, and δ are such that $|f(t, u, 0) - f(t, \hat{\phi}_t, 0)| \le 1$, $|f(t, \hat{\phi}_t, 0)| \le H$ for all $t \in [0, T_1]$, $\|u - \hat{\phi}_t\|_t \le \delta$, with $\hat{\phi}$ defined as

$$\hat{\phi}_t(s) = \begin{cases} \phi(s), & \text{if } s \le 0, \\ \phi(0), & \text{if } 0 \le s \le t \end{cases}$$

(the existence of such H, T_1, δ follows from conditions (8), (9)). Now suppose that for the numbers M and T thus defined the assertion of the lemma is false, i.e., there exist a solution x of problem (1), (2) and a point $t_0 \in [0, T)$ such that $|x'(t)| < M$ for $t \in [0, t_0)$ and $|x'(t_0)| = M$. Then

$$M = |x'(t_0)| = |f(t_0, x_{t_0}, x'_{t_0})|$$

$$\le |f(t_0, x_{t_0}, x'_{t_0}) - f(t_0, x_{t_0}, 0)| + |f(t_0, x_{t_0}, 0) - f(t_0, \hat{\phi}_{t_0}, 0)| + |f(t_0, \hat{\phi}_{t_0}, 0)|$$

$$\le k\|x'\|_{t_0} + 1 + H \le kM + 1 + H < M. \tag{11}$$

This contradiction proves the lemma. **QED**

Further, set

$$R = \{y \in C_T : y_0 = \phi', \ ||y||_T \leq M\}, \tag{12}$$

$$(Jy)(t) = \begin{cases} \phi(0) + \int_0^t y(s)ds, & \text{if } t \geq 0, \\ \phi(t), & \text{if } t < 0, \end{cases}$$

$$[G(x,y)](t) = \begin{cases} \phi'(t) + f(0,(Jx)_0, y_0) - \phi'(0), & \text{if } t \leq 0, \\ f(t,(Jx)_t, y_t), & \text{if } 0 \leq t \leq T, \end{cases} \tag{13}$$

$$Fx = G(x,x). \tag{14}$$

where $x, y \in R$.

The next lemma is an easy consequence of Theorem 1.5.7.

4.3.4. Lemma on the integral operator F. *Suppose conditions* (7)–(10) *are satisfied. Then the operator F defined by formula* (14) *is* (k,χ)*-bounded.*

To continue our exposition we need a version of the Krasnosel′skiĭ-Perov conectedness principle (see [90]), which is based on the local constancy property of the index (see **3.4.1**).

Let U and R be an open and respectively a closed convex subset of a Banach space E. Let $\overline{U}_R$ and ∂U_R denote the closure and the boundary of $U \cap R$ in the relative topology of R. Let $F: \overline{U}_R \to R$, $F_n: \overline{U}_R \to R$, $F_n x \to Fx$ as $n \to \infty$, for all $x \in \overline{U}_R$, $x^* = Fx^*$ and $a \in A = \{N, N+1, \ldots, \infty\}$. Put

$$\tilde{F}(a,x) = \begin{cases} F_n x + x* - F_n x*, & \text{if } a = n, \\ Fx, & \text{if } a = \infty. \end{cases}$$

4.3.5. Theorem (connectedness principle). *Suppose the following conditions are satisfied:*

a) *the operator F is continuous;*

b) *$\tilde{F}$ is jointly condensing;*

c) *$Fx \neq x$ for all $x \in \partial U_R$ and* $\operatorname{ind}(F,U) \neq 0$*;*

d) *for any n and any fixed point x^* of F the only fixed point of the operator $\tilde{F}(n,\cdot)$ is x^*;*

Then the fixed-point set K of F is nonempty, compact, and connected.

Proof. One proceeds according to the same scheme as in [90]. Specifically, suppose that K is not connected (condition c) implies that K is nonempty and compact). Then there exist open sets V and W such that $V \cup W \supset K$, $V \cap W = \emptyset$, $V \cap K \neq \emptyset$, $W \cap K \neq \emptyset$;

there exist open sets V and W such that $V \cup W \supset K$, $V \cap W = \emptyset$, $V \cap K \neq \emptyset$, $W \cap K \neq \emptyset$; consequently, $\mathrm{ind}_R(F, U) = \mathrm{ind}_R(F, V) + \mathrm{ind}_R(F, W)$. Now if one picks $x^* \in W \cap K$, then by condition d) the operators $\tilde{F}(n, \cdot)$ will have no fixed points in V, and so $\mathrm{ind}_R(\tilde{F}(n, \cdot), V) = 0$ for $n \geq N(x^*)$. Applying Theorem 3.4.1 to the operator $\tilde{F}$ on the set $A \times \overline{V}_R$ one concludes that $\mathrm{ind}_R(F, V) = 0$. In a similar manner one shows that $\mathrm{ind}_R(F, W) = 0$. The contradiction one reaches proves the theorem. **QED**

4.3.6. Theorem on the solution set being nonempty and connected. *Suppose conditions* (7)–(10) *are satisfied. Then there exists a* $T > 0$ *such that the solution set of problem* (1), (2) *in the space* C_T^1 *is nonempty, compact, and connected.*

Proof. Let us verify that the conditions of Theorem 4.3.5 are satisfied for $U = C_T$ and the set R defined by formula (12). Clearly, $U_R = R$. By Lemma 4.3.3, the operator F has no fixed points on $\partial U_R = R$. By the same lemma, $\mathrm{ind}_R(F, U) = 1$. To prove this is suffices to consider the linear homotopy from F to the operator F_1,

$$(F_1 x)(t) = \begin{cases} \phi'(t), & \text{if } t \leq 0, \\ \phi'(0), & \text{if } 0 < t < T. \end{cases}$$

Condition c) is thus verified.

Now for any $y \in C_T$ put $(S_n y)(t) = y(t - 1/n)$, $F_n = F S_n$. Condition a) is checked directly, while b) follows from Lemma 4.3.4. Finally, condition d) is satisfied because for any $z \in C_T$ the equation $y = F_n y + z$ has exactly one solution, which can be found by the step method. **QED**

4.3.7. Remark. It is not difficult to see that in order to ensure the local solvability of problem (1), (2) it suffices to impose a Lipschitz-type condition only on the dependence of f on the recent prehistory of x_t'. For example, conditions (7)–(9) are satisfied for equation (4) provided each of the functions g_i, h_i and $f_2 \colon \mathbf{R} \times \mathbf{R}^r \times \mathbf{R}^r \to \mathbf{R}$ is jointly continuous and $\sum k_i < 1$, where the sum contains only the Lipschitz constants k_i for f_2 with respect to those of the last n arguments for which $h_i(0) = 0$, $0 \leq h_i(t) \leq t$, $0 \leq g_i(t) \leq t$.

G. A. Kamenskiĭ (see [71]) observed that under some restrictions on the delay of the argument of the derivative of the unknown function in the right-hand side of equation (4) one can relax the condition on the Lipschitz constant with respect with that derivative. More precisely, it suffices to require that for $t \in [0, T]$:

$$|f_2(t_1, u_1^1, \ldots, u_r^1, v_1^1, \ldots, v_r^1) - f_2(t_2, u_1^2, \ldots, u_r^2, v_1^2, \ldots, v_r^2)|$$

$$\leq p|t_1 - t_2| + \sum_{i=1}^{r} a_i |u_i^1 - u_i^2| + \sum_{i=1}^{r} b_i |v_i^1 - v_i^2|; \tag{15}$$

$$0 \le h_i(t) \le q_i t, \ i = 1, \dots, r; \quad \sum_{i=1}^{r} q_i b_i < 1; \quad 0 \le g_i(t) \le t; \tag{16}$$

$$f_2(0, 0, \dots, 0, 0, \dots, 0) = 0. \tag{17}$$

In the next two subsections we consider a problem that is not directly connected with MNCs and condensing operators, but is closely related to the existence theorems given here and in Section 4.1.

4.3.8. Kamenskiĭ's existence and uniqueness theorem. *Suppose h_i, g_i are continuous, $h_i(0) = g_i(0) = 0$ $(i = 1, \dots, r)$, and conditions (15)–(17) are satisfied. Then the Cauchy problem for the equation (4) with initial condition $x(0) = 0$ has a unique solution on $[0, T]$ whose derivative satisfies the Lipschitz condition.*

For the **proof** it suffices to introduce the metric space $E_l = \{x \in C_T : x(0) = 0, |x(t) - x(s)| < l|t - s| \text{ for } t, s \in [0, T]\}$ with the metric $\rho(x, y) = \sup\{|x(t) - y(t)|/t : t \in [0, T]\}$, and show that the operator F associated to equation (4) via formula (14) acts in E_l and is contractive. **QED**

Thus, weakening of condition (9) ($k = \sum_{i=1}^{r} b_i < 1$) requires the imposition of additional restrictions on h_i and on the dependence of the function f on the first $n + 1$ arguments. The solution obtained in this case is smoother then earlier; its uniqueness is asserted only in this smoothness class. Below we describe a generalization of this result.

Consider the equation

$$x'(t) = f_3(t, x_{g_1(t)}, \dots, x_{g_r(t)}, x'_{h_1(t)}, \dots, x'_{h_r(t)}), \tag{18}$$

and suppose that instead of (15)–(17) the following conditions are satisfied for $t \in [0, T]$:

$$|f_3(t_1, u_1^1, \dots, u_r^1, v_1^1, \dots, v_r^1) - f_3(t_2, u_1^2, \dots, u_r^2, v_1^2, \dots, v_r^2)|$$

$$\le p|t_1 - t_2|^\alpha + \sum_{i=1}^{r} a_i \|u_i^1 - u_i^2\|_t + \sum_{i=1}^{r} b_i \|v_i^1 - v_i^2\|_t \quad (\alpha \in (0, 1]); \tag{19}$$

$$\sum_{i=1}^{r} b_i \sup_{0 \le t \le T} [D^* h_i(t)]^\alpha = k < 1, \tag{20}$$

where $D^* h(t)$ is the right upper derivative of the function $h(t)$.

4.3.9. Existence and uniqueness theorem for equation (18). *Suppose h_i, g_i are continuous, $h_i(0) = g_i(0)$ $(i = 1, \dots, r)$ and conditions (17), (19), and (20) are satisfied.*

Then the Cauchy problem for equation (18) *with null initial condition has a unique solution on* $[0,T]$, *whose derivative satisfies the Hölder condition with exponent* α.

The **proof** is analogous to that of Theorem 4.3.8, with the difference that here instead of E_l one considers the space $E_l^\alpha = \{x \in C_T : x(0) = 0, |x(t) - x(s)| \le l|t-s|^\alpha\}$, with the metric $\rho(x,y) = \sup\{|x(t) - y(t)|/t^\alpha : t \in [0,T]\}$. **QED**

Now consider the equation (3) with $g(t,x) \equiv g(t)$. We introduce the following conditions:

the functions f_1, g, h *are continuous,* $f_1(0,0,0) = 0$ *and for any* $R > 0$ *there is a constant* m_R *such that*

$$|f_1(t,x,y) - f_1(t,0,0)| \le m_R(t + |x|) \text{ for } x, y \in \mathbf{R}^m,\ |x| < R; \tag{21}$$

$|f_1(t,x,y_1) - f_1(t,x,y_2)| \le \omega(|y_1 - y_2|)$, *where the nondecreasing function* $\omega: [0,\infty) \to [0,\infty)$ *is such that* $\omega(h(t)) \le kt$, *with* $k < 1$, *and the inequality*

$$z(t) \le \omega[z(\overline{h}(t))] \quad (\text{with } \overline{h}(t) = \max_{\tau\in[0,t]} h(\tau)) \tag{22}$$

has only the zero solution in the class of nondecreasing nonnegative functions $z(t)$ *that satisfy the condition* $z(t) \le M_z t$.

4.3.10. Existence theorem for equation (3). *Suppose conditions* (21) *and* (22) *are satisfied. Then there exist* $T, l > 0$ *such that equation* (3) *with the null initial condition has at least one solution* x *on the segment* $[0,T]$, *which is such that* $|x'(t)| \le lt$, $t \in [0,T]$.

Proof. Define an MNC ψ in the space $\tilde{C}_0 = \{x \in C_T : x(0) = 0\}$ by the rule

$$[\psi(\Omega)](t) = \chi_t(\Omega_t).$$

Suppose $\Omega \subset \tilde{C}_0$ and $\psi(\Omega) \le \psi(F\Omega)$, where the operator F is associated to equation (3) via formula (14). Fix $t \in (0,T]$ and let $\{y_1, \dots, y_k\}$ be an $(a+\varepsilon)$-net of the set $\Omega_{\overline{h}(t)}$, where $a = [\psi(\Omega)](\overline{h}(t))$. It is readily seen that the set $\Omega^1 = \{z \in C([0,t], \mathbf{R}^m) : z(\tau) = f_1(\tau, (Jy)(g(\tau)), y_i(h(\tau))),\ \tau \in [0,t],\ y \in \Omega,\ i = 1, \dots, k\}$ is relatively compact in $\tilde{C}_0$. Next, for any $x \in (F\Omega)_t$ there is an y_i such that $|y(h(\tau)) - y_i(h(\tau))| \le a + \varepsilon$, where $(Fy)_t = x$ and $\tau \in [0,t]$. Set $z(\tau) = f_1(\tau, (Jy)(g(\tau)), y_i(h(\tau)))$. Then $z \in \Omega^1$ and

$$|z(\tau) - x(\tau)| \le \omega(|y_i(h(\tau)) - y(h(\tau))|) \le \omega(a + \varepsilon).$$

Using the continuity of ω, we obtain

$$\chi_t[(F\Omega)_t] \le \omega[\chi_{\overline{h}(t)}(\Omega_{\overline{h}(t)})]$$

and, by the choice of Ω, $[\psi(\Omega)](t) \leq \omega([\psi(\Omega)](\overline{h}(t)))$. It remains to observe that if

$$\Omega \subset \tilde{C}_0^1 = \{x \in C_T \colon \|x\|_{\tilde{C}_0^1} = \sup_{0<t\leq T} |t^{-1}x(t)| < \infty\}$$

and the norms of the elements of Ω are bounded by some constant M, then $[\psi(\Omega)](t) \leq Mt$. It follows that $[\psi(\Omega)](t) = 0$ for all $t \in (0, T]$, and so F is ψ-condensing on bounded subsets of the space $\tilde{C}_0^1$. Finally, it remains to compute the relative index of the corresponding operator on the intersection of some ball in $\tilde{C}_0$ and a ball in $\tilde{C}_0^1$ and verify that it is different from zero (see [35]). **QED**

4.3.11. Examples. (a) As examples of functions $\omega(t)$ and $h(t)$ one can take $\omega(t) = kt$ and any function $h(t)$ satisfying $0 \leq h(t) \leq rt$, where $kr < 1$, as well as others for which the function $\alpha(t) = \omega(h(t))$ has the following properties: $\alpha^n(t) \to 0$ when $n \to \infty$, $\omega(\alpha(t)) \leq \alpha(\omega(t))$, and $t^{-1}\omega(t)$ is nonincreasing.

(b) The condition $\omega(\overline{h}(t)) \leq kt, k < 1$ is essential for global solvability. It cannot be replaced even by the requirement that the iterates of the function $\alpha(t) = \omega(\overline{h}(t))$ tend to zero. Indeed, consider the problem

$$x'(t) = t + x'(\sin t),\ t \in [0, \pi/2],\ x(0) = 0.$$

In the present case $\omega(t) = t$, $h(t) = \overline{h}(t) = \sin t$. If this problem has a smooth solution, then

$$x'(t) = t + x'(\sin t) = \ldots = t + \sum_{n=1}^{k} \sin^n t + x'(\sin^{k+1} t),$$

but the series $\sum_{n=1}^{\infty} \sin^n t$ diverges for $t = \pi/2$.

(c) In Theorem 4.3.10 it is assumed that the function f grows no faster than linearly near zero in the first argument. The following example shows that this requirement is essential: if the problem

$$x'(t) = 2\sqrt{t} + 2|x'(t/2)|,\ x(0) = 0$$

has a smooth solution x on $[0, \varepsilon]$, then $x'(t) \geq 0$; hence

$$x'(t) = 2\sqrt{t} + 2x'(t/4) = \ldots = 2n\sqrt{t} + 2^n x'(2^{-2n}t) \to \infty \text{ when } n \to \infty.$$

4.3.12. Generalized solutions. When the "sewing" condition (5) is not satisfied the Cauchy problem for equation (1) is not solvable in the classical sense. However, one can pose the problem of the existence of solutions in some generalized sense. One of the possible definitions of a generalized solution goes as follows. Consider equation (6) instead of (1). Arguing as in Lemma 4.3.4 one can readily establish that for any $\eta \in (0, \infty)$ the set of solutions of problem (6), (2) is nonempty and compact in C_T^1 for some $T > 0$. We call *generalized solution of problem* (1), (2) the limit, in the space C_T, of a sequence $\{x_{\eta_k}\}$ of solutions of problem (6), (2) when $\eta_k \to 0$. From the foregoing discussion it easily follows that if conditions (7)–(9) are satisfied, then problem (1), (2) has at least one generalized solution on $(-\infty, T]$ (for some $T > 0$), whose derivative satisfies the Lipschitz condition.

Considering now equation (3) (or (4)), it makes sense to speak about solutions analogous to Carathéodory's generalized solutions. Below we shall discuss the solvability of equation (3) in the space $W_p^1([-a, T], \mathbf{R}^m)$. Recall that $W_p^1([a, b], \mathbf{R}^m)$ designates the space of all absolutely continuous functions on $[a, b]$ whose first derivatives belong to $L_p([a, b], \mathbf{R}^m)$. Let Σ denote the σ-algebra of Lebesgue-measurable subsets of the segment $[-a, T]$.

Consider the following conditions:

$$h^{-1}(e) \in \Sigma \textit{ whenever } e \in \Sigma, \textit{ and } \sup_{e \in \Sigma,\ \mathrm{mes}\, e > 0} [\mathrm{mes}\, h^{-1}(e)]/\mathrm{mes}\, e < \infty; \tag{23}$$

the function $g(t, x)$ is measurable in t for all $x \in \mathbf{R}^m$ and is continuous in x for almost all $t \in [0, T]$; moreover, $g(t, x) \in [t-a, t]$ for all $t \in [0, T]$, $x \in \mathbf{R}^m$; (24)

for any $R > 0$ there exists a function $m_R \in L_p[0, T]$ such that $|f_1(t, x, 0)| \le m_R(t)$ for all $x \in \mathbf{R}^m$, $|x| \le R$; (25)

$f_1(t, x, y)$ is measurable in t for all $x, y \in \mathbf{R}^m$, is continuous in x for all $y \in \mathbf{R}^m$ and almost all $t \in [0, T]$, and satisfies the Lipschitz condition in y with a constant k such that $kr^{1/p} < 1$, where $r = \sup_{e \subset h^{-1}[0,T],\ \mathrm{mes}\, e > 0} [\mathrm{mes}\, h^{-1}(e)]/\mathrm{mes}\, e$. (26)

4.3.13. Theorem on the existence of a solution in W_p^1. *Suppose conditions (23)–(26) are satisfied. Then problem (3), (2) with an arbitrary initial function $\phi \in W_p^1([-a, 0], \mathbf{R}^m)$ has at least one solution $x \in W_p^1([-a, T], \mathbf{R}^m)$ for some $T > 0$.*

The **proof** uses analogues of Lemmas 4.3.3 and 4.3.4 in the space $W_p^1([-a, T], \mathbf{R}^m)$. We note that the constant $r^{1/p}$ is the norm of the composition operator A, $(Ax)(t) = x(h(t))$, in $L_p([-a, T], \mathbf{R}^m)$.

4.3.14. Example. Let us show that the condition $kr^{1/p} < 1$ in Theorem 4.3.13 is essential. Equation (3) may fail to have a solution for k as small as one wishes if $kr^{1/p} = 1$.

Indeed, consider the problem

$$x'(t) = \alpha(t) + x'(t/2)/2, \quad x(0) = 0$$

where

$$\alpha(t) = 2^n/[n(n+1)], \ t \in [2^{-n}, 2^{-n+1}].$$

Clearly, for $p = 1$ all conditions for the existence of a solution, except for $kr < 1$, are satisfied (here $kr = 1$). Suppose there exists a solution $x \in W_1^1[0, \varepsilon]$. Then for $t \in [0, \varepsilon]$,

$$x(t) = x(t/2) + \int_0^t \alpha(s)ds,$$

and for n such that $2^{-n} \in [0, \varepsilon]$ we have

$$x(2^{-n}) = \int_0^{2^{-n}} \alpha(s)ds + x(2^{-n-1}) = \ldots = \sum_{i=0}^{m} \sum_{j=n+i+1}^{\infty} \int_{2^{-j}}^{2^{-j+1}} \alpha(s)ds + x(2^{-n-m})$$

$$= \sum_{i=0}^{m} \sum_{j=n+i+1}^{\infty} \frac{1}{j(j+1)} + x(2^{-n-m}).$$

Since $x(2^{-n-m}) \to 0$ when $m \to \infty$ and $\sum_{j=n+i+1}^{\infty}(j(j+1))^{-1} = (n+i+1)^{-1}$, we reached a contradiction.

4.3.15. Remark. From the discussion in **4.3.12** it follows that under conditions (7)–(9), (23), and (24) the right-hand side of equation (3) can be regarded as being defined on the space $W_\infty^1([-a,T], \mathbf{R}^m)$ (the space of absolutely continuous functions whose derivatives are essentially bounded) and problem (3), (2) has a local generalized solution, which lies in $W_\infty^1([-a,T], \mathbf{R}^m)$. One can show that a function $x \in W_\infty^1([-a,T], \mathbf{R}^m)$ is a generalized solution of problem (3), (2) if and only if x satisfies equation (3) almost everywhere.

4.3.16. Continuous dependence of the solutions on a parameter. Let us address the problem of the continuous dependence of the solutions of the equations

$$x'(t) = f(t, x_t, x'_t, \mu), \ t \geq 0 \tag{27}$$

and

$$x'(t) = f(t, x_t, x'_t, \mu) + \nu_\eta(t)[x'_-(0) - f(0, x_0, x'_0, \mu)], \ t \geq 0 \tag{28}$$

on the parameter $\mu \in \mathrm{M}$, where M is a metric space with metric d.

Consider the following set of conditions:

$f:[0,T]\times C_0\times C_0\times \mathrm{M}\to \mathbf{R}^m$, the operator f is jointly continuous and satisfies the Lipschitz condition with constant $k<1$ in the third variable. (29)

Theorem. *Suppose* (29) *holds and for $\mu=\mu_0, \eta=\eta_0$ problem* (28), (2) *has a unique solution $x^*(t)$ on the interval $(-\infty,T]$. Then for any $\varepsilon>0$ there is a $\delta>0$ such that for $d(\mu,\mu_0)<\delta$ and $\eta=\eta_0$ problem* (28), (2) *has a solution $x(t)$ on $(-\infty,T]$, which satisfies*

$$\|x-x^*\|_T^1<\varepsilon. \tag{30}$$

Proof. Suppose the contrary holds: there exist ε_0 and $\mu_n\to\mu_0$ such that for $\mu=\mu_n,\eta=\eta_0$ problem (28), (2) has no solution satisfying satisfying the estimate (30) with $\varepsilon=\varepsilon_0$. Let $K=\{\mu_n: n=0,1,2,\dots\}$ and let $F_{\eta_0}: K\times C_T\to C_T$ be the operator associated with equation (28) via formula (14). It then readily follows from Theorem 1.5.7 that F_{η_0} is jointly condensing with constant k with respect to the MNC χ. Proceeding as in the proof of Theorem 4.3.6 it is not difficult to show that

$$\mathrm{ind}_R(F_{\eta_0}(\mu_0,\cdot),U)=1,$$

where $U=\{y\in C_T: y_0=\phi_0'\}$, $R=\{y\in C_T: \|y-x^*\|_T\le r\}$. Consequently, there exists $N(r)$ such that for $n\ge N(r)$ the operator $F_{\eta_0}(\mu_n,\cdot)$ has a fixed point y_n in $\overline{U}_R$. Then $x_n=Jy_n$ is a solution of problem (28), (2) and $\|x_n-x^*\|_T^1<(1+T)r$. Choosing $r<(1+T)^{-1}\varepsilon_0$ we reach a contradiction. **QED**

4.3.17. Notes on the references. Our exposition of Lemmas 4.3.3, 4.3.4 and Theorems 4.3.5, 4.3.6 follows the paper by A. E. Rodkina and B. N. Sadovskiĭ [150]. Theorem 4.3.9 is taken from the paper of A. E. Rodkina [144], and Theorems 4.3.10, 4.3.13 and Examples 4.3.11, 4.3.14—from the works of Yu. A. Dyadchenko [35, 36].

The definition of a generalized solution given in **4.3.12** is borrowed from the paper of R. R. Akhmerov [1].

Theorem 4.3.16 is taken from the paper of M. I. Kamenskiĭ [73]. Close results on the continuous dependence on a parameter for continuously differentiable, as well as generalized solutions of problem (27), (2) can be found in A. E. Rodkina's paper [145]. Therein it is also shown that any solution of problem (1), (2) can be continued to a noncontinuable solution. The theorems on the continuability and uniqueness of the solutions of problem (1), (2) (see [145]) were obtained under the assumption that

$$|f(t,x_t,0)|\le L(t,\|x\|_t) \tag{31}$$

or, respectively,

$$|f(t, x_t, u_t) - f(t, y_t, u_t)| \leq L(t, \|x - y\|_t).$$

These theorems are direct generalizations of results of the theory of ordinary differential equations. The estimate functions L used in them are of Osgood type, for example, $L(t,u) = Ku$, $Ku|\ln u|$, and so on. More precisely, to guarantee continuability it suffices to require that all solutions of the inequality

$$z'(t) \leq (1-k)^{-1} L(t, z(t)),\ t \geq 0 \tag{32}$$

be bounded on $[0,T]$, while to guarantee uniqueness it suffices to assume that inequality (32) with the zero initial condition has no nonzero solutions.

The continuability and uniqueness of generalized solutions were studied by Yu. A. Dyadchenko and A. E. Rodkina [37] and by A. E. Rodkina and B. N. Sadovskiĭ [152]. A sufficient condition for continuability is that estimate (31) be fulfilled. However, in the study of the uniqueness of a generalized solution the situation changes. Specifically, even under the assumption that the right-hand side of the equation is very well behaved (namely, that it satisfies the Lipschitz condition with respect to x_t and x'_t, with the Lipschitz constant with respect to x_t being less than 1) there may be many generalized solutions (see [152]). This "anomaly" does not occur if one assumes that the "memory" of equation (1) (with respect to the derivative) is "short" near the singular point $t = 0$.

In her study of problem (1),(2) (see [146, 147]) A. E. Rodkina showed that this problem is equivalent to a certain Cauchy problem for an equation with delay that is simpler than an equation of neutral type with regard to existence theorems. Therein an essential role was played by the invertibility of the operator $I - f'_3(\cdot, 0, 0)$, where f'_3 is the Fréchet derivative of the function f with respect to the third argument.

4.4. PERIODIC SOLUTIONS OF AN EQUATION OF NEUTRAL TYPE WITH SMALL DELAY

The main results of this section are two theorems. The first asserts the existence of a periodic solution of a functional-differential equation of neutral type with right-hand side periodic in t and a small delay, that is close to a periodic solution of the corresponding equation with null delay. The second is an analogous assertion for an autonomous equation.

In this and the following two sections we reduce the problem of the periodic solutions of an equation of neutral type to various operator equations in order to exhibit a variety of equivalent equations with condensing operators.

The study of periodic solutions of an equation of neutral type with small delay will be continued in Section 4.6, where we address the question of their stability.

4.4.1. Description of the equation. We consider an equation of neutral type with small delay of the type

$$x'(t) = f(\varepsilon, t, W(\varepsilon)x_t, W(\varepsilon)x'_t), \tag{1}$$

where

$$f: [0,1] \times \mathbf{R} \times C([-h,0], \mathbf{R}^m) \times C([-h,0], \mathbf{R}^m) \to \mathbf{R}^m,$$

$x_t, x'_t \in C([-h,0], \mathbf{R}^m)$ denote, as in the preceding section, the prehistory of x and x': $x_t(s) = x(t+s)$, $x'_t(s) = x'(t+s)$ $(s \in [-h,0])$, $W(\varepsilon)$, $\varepsilon \in [0,1]$, is the operator acting in $C([-h,0], \mathbf{R}^m)$ according to the rule

$$[W(\varepsilon)u](s) = (u\varepsilon) \quad (s \in [-h,0]),$$

and, finally, ε is a small parameter that characterizes the magnitude of the delay in equation (1): for a given $\varepsilon \in [0,1]$, the value of the derivative of the unknown function x at time t depends on the behavior of x on the segment $[t - \varepsilon h, t]$.

For $\varepsilon = 0$ equation (1) obviously turns into an ordinary differential equation that is not solved with respect to the highest derivative:

$$x'(t) = f(0, t, I_h x(t), I_h x'(t)) \tag{2}$$

(here I_h denotes the canonical embedding of $\mathbf{R}^m$ in $C([-h,0], \mathbf{R}^m)$).

4.4.2. Formulation of the problem. From now on we shall assume that the right-hand side of equation (1) satisfies the following requirements:

the operator f is jointly continuous, is T-periodic in the second argument, and satisfies the Lipschitz condition with constant $k < 1$ in the last argument; (3)

equation (2) has an isolated T-periodic solution ϕ. (4)

We are interested in the existence, for small $\varepsilon > 0$, of T-periodic solutions of equation (1) that are close to ϕ.

4.4.3. Reduction to an operator equation. Consider the equation of neutral type

$$x'(t) = g(t, x_t, x'_t), \tag{5}$$

with $g: \mathbf{R} \times C([-h,0],\mathbf{R}^m) \times C([-h,0],\mathbf{R}^m) \to \mathbf{R}^m$. In the space C_T^1 of continuously differentiable T-periodic $\mathbf{R}^m$-valued functions, equipped with the norm

$$\|x\|_{C_T^1} = \|x(0)\|_{\mathbf{R}^m} + \|x'\|_C,$$

define the operator Φ by the formula

$$(\Phi x)(t) = x(0) + \int_0^t g(s, x_s, x_s')ds - \left(\frac{t}{T} - \frac{1}{2}\right)\int_0^T g(s, x_s, x_s')ds.$$

Lemma. *If the map g is T-periodic in t and continuous, then the operator Φ acts from C_T^1 into C_T^1. The fixed points of Φ are precisely the T-periodic solutions of equation* (5).

Proof. The fact that for any function $x \in C_T^1$ the function Φx is again continuously differentiable is obvious. Furthermore,

$$(\Phi x)(t+T) = x(0) + \int_0^{t+T} g(s, x_s, x_s')ds - \left(\frac{t+T}{T} - \frac{1}{2}\right)\int_0^T g(s, x_s, x_s')ds$$

$$= x(0) + \int_0^t g(s, x_s, x_s')ds - \left(\frac{t}{T} - \frac{1}{2}\right)\int_0^T g(s, x_s, x_s')ds$$

$$+ \int_t^{t+T} g(s, x_s, x_s')ds - \int_0^T g(s, x_s, x_s')ds = (\Phi x)(t).$$

Thus, Φ acts from C_T^1 into itself.

Now suppose the function x is a T-periodic solution of equation (5). Then obviously

$$\int_0^T g(s, x_s, x_s')ds = \int_0^T x'(s)ds = x(T) - x(0) = 0.$$

Consequently,

$$x(t) = x(0) + \int_0^t x'(s)ds = x(0) + \int_0^t g(s, x_s, x_s')ds - \left(\frac{t}{T} - \frac{1}{2}\right)\int_0^T g(s, x_s, x_s')ds = (\Phi x)(t).$$

Conversely, if x is a fixed point of the operator Φ in C_T^1, then

$$x(0) = x(T) = (\Phi x)(T) = x(0) + \frac{1}{2}\int_0^T g(s, x_s, x_s')ds.$$

Therefore, $\int_0^T g(s, x_s, x_s')ds = 0$, and so

$$x(t) \equiv x(0) + \int_0^t g(s, x_s, x_s')ds,$$

which upon differentiation with respect to t yields the needed conclusion. **QED**

Thus, by this lemma, the problem of the T-periodic solutions of equation (1) is equivalent to that of the fixed points of the operator Φ_ε given by

$$(\Phi_\varepsilon x)(t) = x(0) + \int_0^t f(\varepsilon, s, W(\varepsilon)x_s, W(\varepsilon)x'_s)ds - \left(\frac{t}{T} - \frac{1}{2}\right) \int_0^T f(\varepsilon, s, W(\varepsilon)x_s, W(\varepsilon)x'_s)ds$$

in the space C^1_T.

4.4.4. Lemma. *Suppose conditions* (3) *are satisfied. Then the family of operators* $\{\Phi_\varepsilon : \varepsilon \in [0,1]\}$ *is jointly condensing with constant* k *with respect to the Hausdorff* MNC.

Proof. Let C_T denote the space of T-periodic continuous $\mathbf{R}^m$-valued functions, equipped with the sup-norm. Notice that the Hausdorff MNC $\chi_{C^1_T}$ in C^1_T is connected with the Hausdorff MNC χ_{C_T} in C_T through the relation

$$\chi_{C^1_T}(\Omega) = \chi_{C_T}(\Omega'), \tag{6}$$

where $\Omega' = \{x' : x \in \Omega\}$.

The right-hand side of equation (1) can be regarded as an operator F_ε acting from C^1_T into C_T:

$$(F_\varepsilon x)(t) = f(\varepsilon, t, W(\varepsilon)x_t, W(\varepsilon)x'_t).$$

Let us show that the family of operators $\{F_\varepsilon : \varepsilon \in [0,1]\}$ is $(k, \chi_{C^1_T}, \chi_{C_T})$-bounded. To this end we use Theorem 1.5.7. The operator F_ε admits a diagonal representation through the operator $G_\varepsilon : C^1_T \times C^1_T \to C_T$,

$$[G_\varepsilon(x,y)](t) = f(\varepsilon, t, W(\varepsilon)x_t, W(\varepsilon)y'_t).$$

The fact that the set $\{G_\varepsilon(\Omega, y) : \varepsilon \in [0,1]\}$ is totally bounded for any bounded subset $\Omega \subset C^1_T$ and any $y \in C^1_T$ follows from the continuity of the operators G_ε and the total boundedness of the set of functions $\{t \mapsto W(\varepsilon)x_t : \varepsilon \in [0,1], x \in \Omega\}$ in the space $C(\mathbf{R}, C([-h,0], \mathbf{R}^m))$. That the Lipschitz condition is satisfied for $G_\varepsilon(x, \cdot)$ is obvious:

$$\|G_\varepsilon(x,y_1) - G_\varepsilon(x,y_2)\|_{C_T} \le k \sup_{t \in \mathbf{R}} \|W(\varepsilon)(y'_1)_t - W(\varepsilon)(y'_2)_t\|_C = k\|y'_1 - y'_2\|_{C_T} \le k\|y_1 - y_2\|_{C^1_T}.$$

Therefore,

$$\chi_{C_T}[\Phi(\Omega)] \le k\chi_{C^1_T}(\Omega) \tag{7}$$

for any bounded set $\Omega \subset C^1_T$ (here and below we denote $\Phi(\Omega) = \bigcup_{\varepsilon \in [0,1]} \Phi_\varepsilon(\Omega)$).

Now let H_ε denote the operator acting from C_T^1 into C_T according to the rule

$$(H_\varepsilon x)(t) = \frac{1}{T}\int_0^T f(\varepsilon, s, W(\varepsilon)x_s, W(\varepsilon)x'_s)ds.$$

Clearly, H_ε maps any ball into a compact set. Moreover,

$$(\Phi_\varepsilon x)' = F_\varepsilon x - H_\varepsilon x. \tag{8}$$

Using relations (6)–(8) and properties of the Hausdorff MNC we obtain

$$\chi_{C_T^1}[\Phi(\Omega)] = \chi_{C_T}([\Phi(\Omega)]') \leq \chi_{C_T}[F(\Omega) - H(\Omega)] = \chi_{C_T}[F(\Omega)] \leq k\chi_{C_T^1}(\Omega). \quad \textbf{QED}$$

4.4.5. Lemma. *If conditions* (3) *are satisfied, then the mapping* $(\varepsilon, x) \mapsto \Phi_\varepsilon(x)$ *is continuous from* $[0,1] \times C_T^1$ *into* C_T^1.

Proof. If $\varepsilon_n \to \varepsilon_0$ and $\|x_n - x_0\|_{C_T^1} \to 0$ when $n \to \infty$, then, as is readily seen, the set

$$\{\varepsilon_n\} \times [0,T] \times \{W(\varepsilon_n)(x_n)_t : t \in [0,T]\} \times \{W(\varepsilon_n)(x'_n)_t : t \in [0,T]\}$$

is compact in $[0,1] \times [0,T] \times C([-h,0], \mathbf{R}^m) \times C([-h,0], \mathbf{R}^m)$. Therefore, the needed relation:

$$\|\Phi_{\varepsilon_n} x_n - \Phi_{\varepsilon_0} x_0\|_{C_T^1} \to 0 \quad \text{when } n \to \infty,$$

follows from the periodicity of f in t and the uniform continuity of f on compact sets. **QED**

We wish to emphasize that, generally speaking, the family $\{\Phi_\varepsilon\}$ is not continuous in ε in the uniform operator topology.

The first main result of the present section is the following assertion.

4.4.6. Theorem. *Suppose the (isolated) solution* ϕ *of the equation* (2) *has a nonzero index in the sense that* ϕ *is an isolated point of index* $\neq 0$ *of the operator* Φ_0. *Then there are positive numbers* ε_0 *and* r_0 *such that for any* $\varepsilon \in [0, \varepsilon_0]$ *the set* $\mathfrak{N}(\varepsilon)$ *of* T*-periodic solutions of the equation* (1) *that lie in the ball of center* ϕ *and radius* r_0 *in the space* C_T^1 *is nonempty and, in addition,*

$$\lim_{\varepsilon \to 0} \sup_{x \in \mathfrak{N}(\varepsilon)} \|x - \phi\|_{C_T^1} = 0. \tag{9}$$

Proof. Let $B = B(\phi, r_0)$ be a ball in C_T^1 that contains no solutions of equation (2) other than ϕ. By Theorem 3.4.4 on the local constancy of the index, there is an $\varepsilon_0 > 0$ such that

$$\operatorname{ind}(\Phi_\varepsilon, B) = \operatorname{ind}(\Phi_0, B)$$

for all $\varepsilon \in [0, \varepsilon_0]$. By the hypothesis of the theorem, $\operatorname{ind}(\Phi_0, B) \neq 0$. Now the fact that $\mathfrak{N}(\varepsilon) \neq \emptyset$ follows from property 5° of the index (see **3.1.1**) and Lemma 4.4.3.

Let us prove that (9) holds. Assuming the contrary, there exist $\delta \in (0, r_0]$, $\varepsilon_n \to 0$, and $x_n \in \mathfrak{N}(\varepsilon)$ such that

$$\delta \leq \|x_n - \phi\|_{C_T^1} (\leq r_0). \tag{10}$$

Since $x_n = \Phi_{\varepsilon_n} x_n$ and the family $\{\Phi_\varepsilon\}$ is χ-condensing, the set $\{x_n\}$ is relatively compact: otherwise,

$$\chi(\{x_n\}) = \chi(\{\Phi_{\varepsilon_n} x_n\}) \leq k\chi(\{x_n\}).$$

Hence, with no loss of generality we can assume that $x_n \to x_0 \in B_0$ when $n \to \infty$. Letting $n \to \infty$ in the equality $x_n = \Phi_{\varepsilon_n} x_n$, we get $x_0 = \Phi_0 x_0$; moreover, by (10), $x_0 \neq \phi$, which contradicts the choice of r_0. **QED**

4.4.7. Remark. The nonzero-index condition in Theorem 4.4.6 is satisfied, for example, in the case where the right-hand side of equation (2) is continuously differentiable with respect to the space variables and the corresponding variational equation,

$$x'(t) = a(t)x(t) + b(t)x'(t),$$

where

$$a(t) = \left[\frac{\partial}{\partial u} f(0, t, u, v)\Big|_{u=I_h\phi(t),\ v=I_h\phi'(t)}\right] I_h,$$

$$b(t) = \left[\frac{\partial}{\partial v} f(0, t, u, v)\Big|_{u=I_h\phi(t),\ v=I_h\phi'(t)}\right] I_h,$$

has no nonzero T-periodic solutions. In fact, in this case the operator Φ_0 is differentiable at the point ϕ and 1 is not an eigenvalue of $\Phi_0'(\phi)$. By Theorem 3.3.5,

$$\operatorname{ind}(\Phi_0, B) = \operatorname{ind}(\Phi_0'(\phi), B(0,1)).$$

The right-hand side of this equality is different from zero by Theorem 3.3.1.

Another case in which the index of ϕ is different from zero is when the solution ϕ of equation (2) is exponentially stable. One way of proving this is as follows. Let U be the operator of translation by one period along the trajectories of the ordinary differential equation (2). The exponential stability of ϕ guarantees that for N large enough the iterate

U^N maps the ball $B_1 \subset \mathbf{R}^m$ with center $\phi(0)$ and sufficiently small radius r into itself. Consequently, $\operatorname{ind}(U^N, B_1) = 1$. By a theorem of P. P. Zabreĭko and M. A. Krasnosel'skiĭ [92] (see **3.9.2**), if N is prime then

$$\operatorname{ind}(U, B_1) \equiv \operatorname{ind}(U^N, B_1) \,(\operatorname{mod} N).$$

The equality $\operatorname{ind}(\Phi_0, B) = 1$ follows from the principle of relatedness (see [92]).

4.4.8. Autonomous equations. Let us consider the autonomous version of equation (1):

$$x'(t) = f(\varepsilon, W(\varepsilon)x_t, W(\varepsilon)x_t') \tag{11}$$

and the corresponding ordinary differential equation

$$x'(t) = f(0, I_h x(t), I_h x'(t)). \tag{12}$$

It is known that the investigation of periodic solutions of autonomous differential equations encounters difficulties that arise mainly due the following reasons. Firstly, the period of the solution is not known beforehand. Secondly, a periodic solution of an autonomous equation is not isolated, except, possibly, for the case when it is stationary (all translates of the solution along the t-axis are again periodic solutions). And thirdly, the index of the set of fixed points of the operators which give periodic solutions of an autonomous equation that are generated by translates of a single solution, is, as a rule, equal to zero. One of the consequences of these circumstances is that in general for equation (11) the conditions of Theorem 4.4.6 cannot be satisfied.

As it turns out, a very effective tool for handling the present case is the *method of functionalizing the parameter*, proposed by M. A. Krasnolsel'skiĭ [92, 16].

We begin with an abstract fixed point theorem for operators that depend on a parameter, from which the main result will follow.

4.4.9. General constructions. Let E be a Banach space, $x_0 \in E$, $B(x_0, r)$ a ball in E, $T_0 \in \mathbf{R}$, $\tau > 0$. Suppose

$$\begin{gathered}\Psi\colon [T_0 - \tau, T_0 + \tau] \times [0,1] \times B(x_0, r) \to E \textit{ is a} \\ \textit{continuous and jointly } (k,\chi)\textit{-condensing operator.}\end{gathered} \tag{13}$$

Denote the operator $(T, x) \mapsto \Psi(T, 0, x)$ by F. We shall assume that

$$F(T_0, x_0) = x_0; \tag{14}$$

$$\textit{the operator } F \textit{ is Fréchet differentiable at the point } (T_0, x_0); \tag{15}$$

$$1 \textit{ is a simple eigenvalue of the operator } F'_x(T_0, x_0). \tag{16}$$

By Theorem 1.5.9, the operator $F'_x(T_0, x_0)$ is (k, χ)-condensing, and so (see Chapter 2) 1 is an isolated eigenvalue of $F'_x(T_0, x_0)$. Let g_0 be a normalized eigenvector of $F'_x(T_0, x_0)$ with eigenvalue 1. Since this eigenvalue is isolated and simple, there exists a closed invariant subspace $E_0 \subset E$ of $F'_x(T_0, x_0)$ that is complementary to the subspace spanned by g_0. Finally, we shall assume that

$$e_0 = F'_T(T_0, x_0) \notin E_0. \tag{17}$$

4.4.10. Lemma. *There exist a ball $B(x_0, \rho)$ and a linear functional $T: B(x_0, \rho) \to [T_0-\tau, T_0+\tau]$ such that $T(x_0) = T_0$ and the index $\operatorname{ind}(G, B(x_0, \rho))$, where $Gx = F(T(x), x)$, is defined and different from zero.*

Proof. Let $l \in E^*$ be such that $l(E_0) = 0$ and $l(e_0) = 1$. Set

$$T(x) = T_0 + l(x - x_0).$$

Denote $x - x_0 = h$. It is readily verified that

$$x - Gx = x - F(T(x), x) = x_0 + h - F(l(h) + T_0, x_0 + h)$$

$$= h - F'_T(T_0, x_0)l(h) - F'_x(T_0, h_0)h + q(h), \tag{18}$$

where $q(h)/\|h\| \to 0$ when $h \to 0$. Now suppose $L \in E^*$ annihilates E_0 and $L(g_0) = 1$, and denote by P the projector onto E_0 parallel to g_0: $Ph = h - L(h)g_0$. Then obviously

$$h - F'_x(T_0, x_0)h = [I - F'_x(T_0, x_0)]Ph \in E_0. \tag{19}$$

Substituting (19) in (18) and denoting $[I - F'_x(T_0, x_0)]Ph - l(h)e_0$ by Ah, we get

$$x - Gx = Ah + q(h). \tag{20}$$

Further, since 1 is a simple eigenvalue of $F'_x(T_0, x_0)$, the choice of the subspace E_0 ensures the existence of a constant $C_1 > 0$ such that

$$\|[I - F'_x(T_0, x_0)]Ph\| \geq C_1\|Ph\|. \tag{21}$$

Put $C_2 = |l(g_0)|$. Then

$$|l(h)| = |l(Ph + (I - P)h)| = |l((I - P)h)|$$

$$= |l(L(h)g_0)| = |l(g_0)|\,|L(h)| = C_2\|L(h)g_0\| = C_2\|(I-P)h\|. \tag{22}$$

By (17), there is a constant $C_3 > 0$ such that

$$\|Ah\| \geq C_3(\|[I - F'_x(T_0, x_0)]Ph\| + \|l(h)e_0\|). \tag{23}$$

Finally, relations (21)–(23) imply the existence of a positive constant C such that

$$\|Ah\| \geq C\|h\|. \tag{24}$$

Now choose $\rho > 0$ such that

$$\|q(h)\|/\|h\| < C \quad \text{and} \quad T_0 - \tau \leq T(x_0 + h) \leq T_0 + \tau \tag{25}$$

whenever $\|h\| \leq \rho$. Then one can readily see that from relations (20), (24) and (25) it follows that the operator G has no fixed points on the boundary of the ball $B(x_0, \rho)$, and consequently its index $\operatorname{ind}(G, B(x_0, \rho))$ is defined; moreover, by Theorem 3.3.2,

$$\operatorname{ind}(G, B(x_0, \rho)) = \operatorname{ind}(I - A, B(0, \rho)).$$

By (24) and Theorem 3.3.1, the second index is different from zero. **QED**

It is not difficult to see that for any $\delta < \rho$ one has

$$\operatorname{ind}(G, B(x_0, \delta)) \neq 0.$$

Define the operator $\Psi_0\colon [0,1] \times B(x_0, \rho) \to E$ by the equation $\Psi_0(\varepsilon, x) = \Psi(T(x), \varepsilon, x)$.

4.4.11. Theorem. *Suppose the operator Ψ satisfies the conditions (13)–(17). Then for any $\delta > 0$ there exists an $\varepsilon(\delta) > 0$ such that for $\varepsilon \in [0, \varepsilon(\delta)]$ the equation*

$$x = \Psi_0(\varepsilon, x)$$

has at least one solution in $B(x_0, \delta)$.

Proof. Fix $\delta > 0$. With no loss of generality we may assume that $\delta < \rho$. If we can show that the operator Ψ_0 is jointly condensing, then by Theorem 3.4.1 its continuity will imply that, for small $\varepsilon > 0$,

$$\operatorname{ind}(\Psi_0(\varepsilon, \cdot), B(x_0, \delta)) = \operatorname{ind}(G, B(x_0, \delta)),$$

and consequently

$$\operatorname{ind}(\Psi_0(\varepsilon, \cdot), B(x_0, \delta)) \neq 0,$$

which yields the desired conclusion.

Thus, it remains to show that Ψ_0 is condensing on the ball $B(x_0, \rho)$ with respect to the Hausdorff MNC. To this end we remark that

$$\Omega_1 = \Psi_0([0,1], \Omega) \subset \Psi([T_0 - \tau, T_0 + \tau], [0,1], \Omega) = \Omega_2.$$

Hence, since Ψ is (k, χ)-condensing and χ is monotone,

$$\chi[\Psi_0([0,1], \Omega)] = \chi(\Omega_1) \leq \chi(\Omega_2) \leq k\chi(\Omega). \quad \textbf{QED}$$

4.4.12. The basic conditions. We shall assume that the right-hand side of equation (11) satisfies the following requirements:

f is jointly continuous and satisfies the Lipschitz condition with constant $k < 1$ in the third variable; (26)

equation (2) has a nonstationary T_0-periodic solution ϕ (27)

(the simpler case of a stationary solution will be considered below). Let q denote the map $\mathbf{R}^m \times \mathbf{R}^m \to \mathbf{R}^m$ defined as

$$q(x,y) = f(0, I_h x, I_h y).$$

We shall assume that

q is continuously differentiable. (28)

Denote $a(t) = \partial q(x,y)/\partial x|_{x=\phi(t),\, y=\phi'(t)}$, $b(t) = \partial q(x,y)/\partial y|_{x=\phi(t),\, y=\phi'(t)}$, and let $W(t,s)$ designate the operator of translation from time s to time t (see, e.g., [86]) along the trajectories of the equation

$$x'(t) = a(t)x(t) + b(t)x'(t). \tag{29}$$

Since the norm of the operator $b(t)$ is not larger than k for all $t \in \mathbf{R}$ (thanks to condition (26)) and $k < 1$, system (29) is equivalent to the system

$$x'(t) = [I - b(t)]^{-1} a(t)x(t), \tag{30}$$

and so the operator $W(t,s)$ is indeed defined.

Clearly, the function $\psi(t) = \phi'(t) \not\equiv 0$ is a T_0-periodic solution of equations (29) and (30) (this is verified by substituting ϕ in the equation and differentiating the resulting identity with respect to t). Therefore, 1 is an eigenvalue of the operator $W(T_0, 0)$. We shall assume that

1 is a simple eigenvalue of $W(T_0, 0)$. (31)

Our goal is to show that under the conditions listed above equation (11) has a periodic solution that is C^1-close to ϕ and has period close to T_0, i.e., to prove the following result.

4.4.13. Theorem. *Suppose that conditions (26)–(28) and (31) are satisfied. Then for any $\delta > 0$ there is an $\varepsilon_0 > 0$ such that for $\varepsilon \in [0, \varepsilon_0]$ equation (11) has a T-periodic solution x such that $\|x - \phi\|_{C^1([0,T_0],\mathbf{R}^m)} \leq \delta$ and $|T - T_0| \leq \delta$.*

Plan of the proof. Searching for T-periodic solutions of equation (11) is equivalent to searching for T_0-periodic solutions of the equation

$$y'(\tau) = \frac{T}{T_0} f\Big(\varepsilon, W\Big(\frac{\varepsilon T_0}{T}\Big) y_\tau, \frac{T_0}{T} W\Big(\frac{\varepsilon T_0}{T}\Big) y'_\tau\Big). \tag{32}$$

This is readily seen by performing the change of variables $\tau = T_0 t/T$, $y(\tau) = x(T\tau/T_0)$ in equation (11).

In its turn, the problem of the T_0-periodic solutions of equation (32) reduces to the problem of the fixed points of a certain condensing operator, and we will show that the operator in question satisfies the conditions of Theorem 4.4.11.

4.4.14. Reduction to an operator equation. Let us go back to equation (5). Suppose that g satisfies the conditions of Lemma 4.4.3. The equivalent operator equation will be considered in the space $\mathbf{R}^m \times C_T$.

In equation (5) we denote $x'(t)$ by $w(t)$ and $x(0)$ by λ. Then x can be expressed in terms of w and λ as

$$x(t) = \lambda + \int_0^t w(s)ds.$$

Since we are interested only in T-periodic solutions of equation (5), we may replace the above expression for x by

$$x(t) = \lambda + \int_0^t w(s)ds - \frac{t}{T}\int_0^T w(s)ds$$

(the last term is actually equal to zero, because $\int_0^T w(s)ds = x(T) - x(0) = 0$). Denote $\frac{1}{T}\int_0^T w(s)ds = M(w)$. Let J be the operator defined by the formula

$$(Jw)(t) = \int_0^t w(s)ds - tM(w).$$

Then equation (5) takes on the form

$$w(t) = g(t, \lambda + (Jw)_t, w_t). \tag{33}$$

Notice that we also have the equality

$$\lambda = \lambda - M(w). \tag{34}$$

Combining (33) and (34) in a system, we obtain the following operator equation in $\mathbf{R}^m \times C_T$:

$$(\lambda, w) = H(\lambda, w),$$

where

$$H(\lambda, w) = (\lambda - M(w), g(t, \lambda + (Jw)_t, w_t)).$$

Finally, consider the operator $\bar{H}$, the first component of which, $\bar{H}_1$, has the same form as H, while the second component, $\bar{H}_2$, is defined by the equation

$$[\bar{H}_2(\lambda, w)](t) = [H_2(\lambda, w)](t) + r(t)M(w),$$

where $r(t)$ is an arbitrary continuous T-periodic $m \times m$ matrix-valued function of t. In our constructions we will choose $r(t)$ so that

$$r(t)\phi'(0) = [I - b(t)]\left[\phi'(0) - \frac{1}{T_0}\phi'(t)\right] - \phi''(t).$$

Let ξ_i be a coordinate of the vector $\phi'(0)$ that is not equal to zero. Then for $r(t)$ one can take, for example, the matrix in which the i-th column is equal to

$$\xi_i^{-1}[I - b(t)]\left[\phi'(0) - \frac{1}{T_0}\phi'(t)\right] - \xi_i^{-1}\phi''(t),$$

while the remaining columns are equal to zero.

The indicated choice of the matrix $r(t)$ is connected with the fact that condition (31) of Theorem 4.4.13 is formulated in terms of the translation operator. The presence of the additional term involving $r(t)$ makes the verification of conditions (16) and (17) technically easier.

The following assertion is established with no difficulty.

Lemma. *The operator $\bar{H}$ acts continuously from $\mathbf{R}^m \times C_T$ into $\mathbf{R}^m \times C_T$. If x is a T-periodic solution of equation (5), then the pair $(x(0), x')$ is a fixed point of $\bar{H}$. Conversely, if (λ, w) is a fixed point of $\bar{H}$, then the function $x(t) \equiv \lambda + \int_0^t w(s)ds$ is a T-periodic solution of equation (5).*

By this lemma, the problem of the T-periodic solutions of equation (32) is equivalent to the operator equation

$$x = \Psi(T, \varepsilon, x)$$

in the space $E = \mathbf{R}^m \times C_{T_0}$, where $x = (\lambda, w)$,

$$\Psi(T, \varepsilon, x) = \bar{H}(T, \varepsilon, \lambda, w)$$

$$= \left(\lambda - M(w), \frac{T}{T_0} f\left(\varepsilon, \lambda + W\left(\frac{\varepsilon T_0}{T}\right)(Jw)_t, \frac{T_0}{T} W\left(\frac{\varepsilon T_0}{T}\right) w_t\right) + r(t) M(w)\right).$$

We shall assume that $\varepsilon T_0 / T \leq 1$.

4.4.15. Lemma. *Suppose condition* (26) *is satisfied. The the operator* Ψ *is continuous and jointly condensing with respect to the Hausdorff* MNC *in* $\mathbf{R}^m \times C_{T_0}$.

Proof. The continuity of the first component of the operator $\bar{H}$ is obvious. To prove the continuity of the second component H_2, we proceed by reductio ad absurdum: suppose $\lambda_n \to \lambda_0$, $\varepsilon_n \to \varepsilon_0$, $T_n \to T_1$ when $n \to \infty$, $w_n \in C_{T_0}$, $\|w_n - w_0\|_C \to 0$ when $n \to \infty$, and $\delta > 0$ are such that

$$\|H_2(T_n, \varepsilon_n, \lambda_n, w_n) - H_2(T_1, \varepsilon_0, \lambda_0, w_0)\|_C \geq \delta.$$

This means that there exists a convergent sequence $t_n \in [0, T_0]$ such that

$$\left\| \frac{T_n}{T} \left[f\left(\varepsilon_n, \lambda_n + W\left(\frac{\varepsilon_n T_0}{T_n}\right)(Jw_n)_{t_n}, \frac{T_0}{T_n} W\left(\frac{\varepsilon_n T_0}{T_n}\right)(w_n)_{t_n}\right) - \right.\right.$$

$$\left.\left. f\left(\varepsilon_0, \lambda_0 + W\left(\frac{\varepsilon_0 T_0}{T_1}\right)(Jw_0)_{t_n}, \frac{T_0}{T_1} W\left(\frac{\varepsilon_0 T_0}{T_1}\right)(w_0)_{t_n}\right)\right] \right\| \geq \delta.$$

Taking into account the continuity of f, J, W and letting $n \to \infty$, we reach a contradiction.

To show that Ψ is condensing, it clearly suffices to verify that the operator $\bar{H}_2 \colon [T_0 - \tau, T_0 + \tau] \times [0,1] \times \mathbf{R}^m \times C_{T_0} \to C_{T_0}$ is jointly $\chi_{C_{T_0}}$-condensing on bounded subsets. To this end we use Theorem 1.5.7. $\bar{H}_2$ admits a diagonal representation of the form

$$\bar{H}_2(T, \varepsilon, \lambda, w) = G_2(T, \varepsilon, \lambda, w, w),$$

where

$$[G_2(T, \varepsilon, \lambda, u, v)](t) = \frac{T}{T_0} f\left(\varepsilon, \lambda + W\left(\frac{\varepsilon T_0}{T}\right)(Ju)_t, W\left(\frac{\varepsilon T_0}{T}\right) v_t\right) + r(t) M(v).$$

The total boundedness of the set $G_2([T_0 - \tau, T_0 + \tau], [0,1], \Omega_1, \Omega_2, \{v\})$ for arbitrary bounded subsets $\Omega_1 \subset \mathbf{R}^m$ and $\Omega_2 \subset C_{T_0}$ follows from the continuity of the operators G_2 and $W(\varepsilon)$, the total boundedness of Ω_1 and the compactness of the operator J. Finally, the fact that G_2 satisfies the Lipschitz condition in the last argument follows from the Lipschitz condition for f and the obvious inequality $\|W(\varepsilon)\| \leq 1$ $(\varepsilon \in [0,1])$. **QED**

4.4.16. Verification of the conditions of Theorem 4.4.11 for the operator Ψ. Condition (13) is exactly the assertion of Lemma 4.4.15. Let us take for x_0 the pair $(\phi(0), \phi')$. Then condition (14) follows from (27) and the lemma in **4.4.14**, while condition (15) is a consequence of (28); moreover,

$$[F'_x(T_0, x_0)](\mu, y) = [\Psi'_x(T_0, 0, x_0)](\mu, y)$$

$$= (\mu - M(y), a(t)[\mu + (Jy)(t)] + b(t)y(t)) + r(t)M(y). \tag{35}$$

We next turn to condition (16). By condition (31), equation (29) has the nonzero T_0-periodic solution ϕ', and no others that are linearly independent of it. Hence, by Lemma 4.3.14, the operator $F'_x(T_0, x_0)$ has 1 as an eigenvalue; a corresponding eigenvector is the pair $g_0 = (\phi'(0), \phi'')$, and there are no other linearly independent eigenvectors. Now let us show that there are no (root) vectors associated to g_0. In fact, assuming the contrary, there is a pair $(\nu, \kappa) \in \mathbf{R}^m \times C_{T_0}$ such that

$$\nu + \phi'(0) = \nu - M(\kappa),$$

$$\phi''(t) + \kappa(t) = a(t)[\nu + J(\kappa)](t) + b(t)\kappa(t) + r(t)M(\kappa).$$

This gives

$$\phi''(t) + \kappa(t) = a(t)\Big[\nu + \int_0^t \kappa(s)ds + t\phi'(0)\Big] + b(t)\kappa(t) - r(t)\phi'(0).$$

Denote $\nu + \int_0^t \kappa(s)ds + t\phi'(0)$ by $z_0(t)$. By the choice of $r(t)$ (see **4.4.13**), the function $z_0(t)$ will be a T_0 periodic solution of the equation

$$\frac{1}{T_0}[I - b(t)]\phi'(t) = [I - b(t)]z_0'(t) - a(t)z_0(t).$$

Then, obviously,

$$z_0(t) = W(t,0)z_0(0) + \frac{1}{T_0}\int_0^t W(t,s)\phi'(s)ds. \tag{36}$$

Next, since ϕ' is a solution of the linearized equation (29),

$$\phi'(t) = W(t,0)\phi'(0).$$

Substituting this expression for ϕ' in (36), we obtain

$$z_0(t) = W(t,0)z_0(0) + \frac{1}{T_0}\int_0^t W(t,0)\phi'(0)ds = W(t,0)\nu + \frac{t}{T_0}W(t,0)\phi'(0).$$

Clearly, $z_0(T_0) = z_0(0)$, because $M(z_0') = 0$. Hence, $z_0(0) = W(T_0,0)\nu + W(T_0,0)\phi'(0)$. Since ϕ' is a solution of equation (29), $\phi'(0) = W(T_0,0)\phi'(0)$. Then

$$W(T_0,0)\nu + \phi'(0) = \nu,$$

i.e., ν is root vector of the operator $W(T_0,0)$ belonging to the eigenvalue 1, which contradicts condition (31).

Thus, it remains to verify that condition (17) is fulfilled, i.e., that $e_0 = F_T'(T_0,x_0) \notin E_0$, where E_0 is an invariant subspace of the operator $F_x'(T_0,x_0)$ that complements the subspace spanned by the eigenvector g_0. Since $[I - F_x'(T_0,x_0)]E = E_0$, it suffices to show that $e_0 \notin [I - F_x'(T_0,x_0)]E$. It is readily seen that

$$e_0 = \Big(0, \frac{1}{T_0}[I - b(t)]\phi'(t)\Big). \tag{37}$$

Proceeding again by reductio ad absurdum, suppose there exist a $\mu_0 \in \mathbf{R}^m$ and a $y_0 \in C_{T_0}$ such that $e_0 = (\mu_0, y_0) - [F_x'(T_0,x_0)](\mu_0,y_0)$. By relations (35) and (37), this yields

$$\Big(0, \frac{1}{T_0}[I - b(t)]\phi'(t)\Big) = (\mu_0,y_0) - (\mu_0 - M(y_0), a(t)(\mu_0 + (Jy_0)(t)) + b(t)y_0(t) + r(t)M(y_0)).$$

We thus get the following system of equations for μ_0 and y_0:

$$0 = \mu_0 - \mu_0 + M(y_0),$$

$$\frac{1}{T_0}[I - b(t)]\phi'(t) = y_0(t) - a(t)(\mu_0 + (Jy_0)(t)) - b(t)y_0(t) - r(t)M(y_0).$$

The first equation gives $M(y_0) = 0$. Denoting $\mu_0 + \int_0^t y_0(s)ds$ by $z_0(t)$, the second equation implies that z_0 is a T_0-periodic solution of the equation

$$\frac{1}{T_0}[I - b(t)]\phi'(t) = [I - b(t)]z_0'(t) - a(t)z_0(t),$$

i.e., we arrived at the same contradiction as in the proof that the eigenvalue 1 is simple. **QED**

In the case where ϕ is a stationary solution of equation (12) one can readily establish the following assertion.

4.4.17. Theorem. *Suppose equation (12) has a stationary solution $\phi(t) \equiv p_0$. Suppose further that the operator $g\colon [0,1] \times \mathbf{R}^m \to \mathbf{R}^m$, defined by the rule $g(\varepsilon,p) = f(\varepsilon, I_h p, I_h 0)$, is continuous and differentiable at the point p_0 for $\varepsilon = 0$, and that the operator $g_p'(0,p_0)$ is nondegenerate. Then for small ε equation (11) has a stationary solution close to ϕ.*

Proof. It suffices to apply the implicit function theorem to the equation

$$g(\varepsilon, p) = 0.$$

Note that the assumption that the derivative $g'_p(0, p_0)$ exists and is nondegenerate can be replaced by the less restrictive requirement that p_0 be an isolated point of index zero of the operator $g(0, \cdot)$.

4.4.18. Notes on the references. The exposition in this section followed works of R. R. Akhmerov, M. I. Kamenskiĭ, A. S. Potapov, and B. N. Sadovskiĭ [10] and of R. R. Akhmerov, M. I. Kamenskiĭ, V. S. Kozyakin, and A. V. Sobolev [9]. The operators Φ and Ψ are taken from the papers of B. N. Sadovskiĭ [159, 160].

4.5. THE AVERAGING PRINCIPLE FOR EQUATIONS OF NEUTRAL TYPE

In this section we prove theorems on averaging over a bounded interval and in the class of periodic solutions of equations of neutral type. The theory of MNCs and condensing operators proves here useful in establishing the existence of solutions of the nonaveraged equation as well as in investigating the stability of solutions (see Section 4.6).

4.5.1. Formulation of the problem. We consider an equation of neutral type with a small parameter multiplying the right-hand side:

$$x'(t) = \varepsilon f(t, x_t, x'_t), \tag{1}$$

where, as earlier, $f: \mathbf{R} \times C([-h, 0], \mathbf{R}^m) \times C([-h, 0], \mathbf{R}^m) \to \mathbf{R}^m$ and x_t and x'_t are the prehistory of the unknown function x and of its derivative x', respectively. Alongside with equation (1) we consider the *averaged equation*

$$x'(t) = \varepsilon f_0(x(t)), \tag{2}$$

where

$$f_0(x) = \lim_{t \to \infty} \frac{1}{t} \int_0^t f(s, I_h x, 0) ds \quad (x \in \mathbf{R}^m); \tag{3}$$

recall that I_h denotes the canonical embedding of $\mathbf{R}^m$ in $C([-h, 0], \mathbf{R}^m)$.

We want to elucidate under what conditions and in what sense the solutions of the original equation (1) are close, for small ε, to the "corresponding" solutions of the averaged equation (2) on intervals that grow when ε is decreased, or on an unbounded interval.

4.5.2. Basic conditions. Let us list conditions (referred to below as *basic*) that the operator f will satisfy in all theorems of the present section:

$$f \text{ is continuous in } t; \tag{4}$$

f is continuous in the second and third variables, uniformly in all variables:

$$\|f(t,u_1,v_1) - f(t,u_2,v_2)\| \le \alpha(\|u_1 - u_2\|_C) + \beta(\|v_1 - v_2\|_C),$$
$$\text{where } \alpha(\xi), \beta(\xi) \to 0 \text{ when } \xi \to 0; \tag{5}$$

$$\text{the limit (3) exists for any } x \in \mathbf{R}^m; \tag{6}$$

$$f \text{ is bounded on its whole domain.} \tag{7}$$

From now on we shall assume that $\varepsilon \in (0, 1]$.

4.5.3. Passing to compressed functions. In the investigation of equation (1) one uses the change of variables

$$x = W(\varepsilon)y, \tag{8}$$

where $W(\varepsilon)$ acts as $[W(\varepsilon)y](t) = y(\varepsilon t)$. Notice that $[W(\varepsilon)y]'(t) = \varepsilon[W(\varepsilon)y'](t)$. Hence, this change of variables brings equation (1) to the form

$$[W(\varepsilon)y'](t) = f(t, [W(\varepsilon)y]_t, \varepsilon[W(\varepsilon)y']_t).$$

After simple transformations we arrive at the equation

$$y'(t) = f(\varepsilon^{-1}t, W(\varepsilon)y_{t,\varepsilon}, \varepsilon W(\varepsilon)y'_{t,\varepsilon}), \tag{9}$$

where $y_{t,\varepsilon}$ and $y'_{t,\varepsilon}$ denote the restrictions of the functions y_t and y'_t, respectively, to the segment $[-\varepsilon h, 0]$.

In view of (8), the new unknown function y is expressible through x as

$$y(t) = ([W(\varepsilon)]^{-1}x)(t) = [W(\varepsilon^{-1})x](t) = x(\varepsilon^{-1}t),$$

i.e., the graph of y is obtained by compressing the graph of x along the t-axis "ε^{-1} times". Accordingly, (9) will be referred to as the *equation in compressed functions*. In the literature this passage is often treated as passage to a new, "slow" time εt with preservation of the same unknown function; however, we shall not use here that interpretation.

In compressed functions the averaged equation (2) becomes

$$y'(t) = f_0(y(t)). \tag{10}$$

Clearly, for small ε the right-hand sides of equations (9) and (10) are close in some integral sense, and so one can expect that their solutions will also be close. Giving a rigorous meaning to this last assertion is the content of the basic averaging theorems.

We shall need the following easy consequence of the Krasnosel'skiĭ-Kreĭn theorem [88, 32].

4.5.4. Lemma. *Suppose that the operator $g: \mathbf{R} \times \mathbf{R}^m \to \mathbf{R}^m$ is continuous, the continuity in the second argument being jointly uniform, and that the limit*

$$g_0(x) = \lim_{t \to \infty} \frac{1}{t} \int_0^t g(s, x) ds$$

exists for all $x \in \mathbf{R}^m$. Let $x^0 \in C([0,T], \mathbf{R}^m)$. Then for any $t \in [0,T]$,

$$\lim_{\varepsilon \to 0} \int_0^t g(\varepsilon^{-1}s, x^0(s)) ds = \int_0^t g_0(x^0(s)) ds.$$

Using this lemma one establishes the following result.

4.5.5. Theorem on passing to the limit under the integral sign. *Suppose f satisfies the basic conditions and the functions $x^\varepsilon \in C^1([-h,T], \mathbf{R}^m)$ ($\varepsilon \in [0,1]$) are such that*

$$\|x^\varepsilon - x^0\|_C \to 0 \text{ when } \varepsilon \to 0 \tag{11}$$

and their derivatives are uniformly bounded:

$$\|(x^\varepsilon)'\|_C \le M. \tag{12}$$

Then for any $t \in [0,T]$,

$$\lim_{\varepsilon \to 0} \int_0^t f(\varepsilon^{-1}s, W(\varepsilon)x^\varepsilon_{s,\varepsilon}, \varepsilon W(\varepsilon)(x^\varepsilon)'_{s,\varepsilon}) ds = \int_0^t f_0(x^0(s)) ds.$$

Proof. By assumptions (11) and (12),

$$\lim_{\varepsilon \to 0} \sup_{s \in [0,T]} \|W(\varepsilon)x^\varepsilon_{s,\varepsilon} - I_h x^0(s)\|_C = 0$$

and

$$\lim_{\varepsilon\to 0} \sup_{s\in[0,T]} \|\varepsilon W(\varepsilon)(x^\varepsilon)'_{s,\varepsilon}\|_C = 0.$$

Hence, by the uniform continuity of f,

$$f(\varepsilon^{-1}s, W(\varepsilon)x^\varepsilon_{s,\varepsilon}, \varepsilon W(\varepsilon)(x^\varepsilon)'_{s,\varepsilon}) - f(\varepsilon^{-1}s, I_h x^0(s), 0) \to 0 \quad \text{when} \quad \varepsilon \to 0$$

uniformly in $s \in [0,T]$, and consequently

$$\lim_{\varepsilon\to 0}\int_0^t f(\varepsilon^{-1}s, W(\varepsilon)x^\varepsilon_{s,\varepsilon}, \varepsilon W(\varepsilon)(x^\varepsilon)'_{s,\varepsilon})ds = \lim_{\varepsilon\to 0}\int_0^t f(\varepsilon^{-1}s, I_h x^0(s), 0)ds.$$

Now notice that the operator $g: \mathbf{R}\times\mathbf{R}^m \to \mathbf{R}^m$, defined by the formula

$$g(t,x) = f(t, I_h x, 0),$$

satisfies the conditions of Lemma 4.5.4, and hence

$$\lim_{\varepsilon\to 0}\int_0^t f(\varepsilon^{-1}s, I_h x^0(s), 0)ds = \int_0^t f_0(x^0(s))ds,$$

which in conjunction with the above relation between limits proves the assertion of the theorem. **QED**

4.5.6. Remark. In the proof of Theorem 4.5.5 we made no use of the assumption that the operator f is bounded on its whole domain. However, in the sequel we shall use the fact that, under the boundedness assumption on f, condition (12) can be relaxed to

$$\|(x^\varepsilon)'\|_{C([\delta(\varepsilon),T],\mathbf{R}^m} \le M, \quad \text{where } \delta(\varepsilon)\to 0 \text{ when } \varepsilon \to 0.$$

In fact, we can redefine x^ε left of $\delta(\varepsilon)$ so that Theorem 4.5.5 applies to the resulting functions $\tilde{x}^\varepsilon$. But then its conclusion is also valid for the original functions x^ε, because, by the boundedness of f,

$$\lim_{\varepsilon\to 0}\int_0^t \left[f(\varepsilon^{-1}s, W(\varepsilon)x^\varepsilon_{s,\varepsilon}, \varepsilon W(\varepsilon)(x^\varepsilon)'_{s,\varepsilon}) - f(\varepsilon^{-1}s, W(\varepsilon)\tilde{x}^\varepsilon_{s,\varepsilon}, \varepsilon W(\varepsilon)(\tilde{x}^\varepsilon)'_{s,\varepsilon})\right]ds =$$

$$\lim_{\varepsilon\to 0}\int_0^{\delta(\varepsilon)+\varepsilon h} \left[f(\varepsilon^{-1}s, W(\varepsilon)x^\varepsilon_{s,\varepsilon}, \varepsilon W(\varepsilon)(x^\varepsilon)'_{s,\varepsilon}) - f(\varepsilon^{-1}s, W(\varepsilon)\tilde{x}^\varepsilon_{s,\varepsilon}, \varepsilon W(\varepsilon)(\tilde{x}^\varepsilon)'_{s,\varepsilon})\right]ds = 0.$$

4.5.7. Remark. In the arguments of the two preceding subsections the operator f may be allowed to depend also on ε, provided only that $f(\varepsilon,t,u,v) \to f(0,t,u,v)$ uniformly in t, u, v as $\varepsilon \to 0$ and the operator $(t, u, v) \mapsto f(0, t, u, v)$ satisfies the basic conditions. To prove this it suffices to notice that

$$\int_0^t [f(\varepsilon, \varepsilon^{-1}s, W(\varepsilon)x^\varepsilon_{s,\varepsilon}, \varepsilon W(\varepsilon)(x^\varepsilon)'_{s,\varepsilon}) - f(0, \varepsilon^{-1}s, W(\varepsilon)x^\varepsilon_{s,\varepsilon}, \varepsilon W(\varepsilon)(x^\varepsilon)'_{s,\varepsilon})]\, ds \to 0$$

when $\varepsilon \to 0$.

4.5.8. The averaging problem on a bounded interval. Here we are interested in the closeness of the generalized solutions of the nonaveraged equation (1) and solutions of the corresponding averaged equation (2) for small ε on an interval of length of order ε^{-1}.

We first consider equations for which the sewing condition does not come into play and which are used to define generalized solutions (see **4.3.12**):

$$x'(t) = \varepsilon f(t, x_t, x'_t) + [x'_-(0) - \varepsilon f(0, x_0, x'_0)]\nu_\mu(t); \tag{13}$$

here ν_μ is a piecewise-linear function that is equal to 0 on $[\mu, \infty)$ and to 1 at $t = 0$. We shall consider solutions of equation (13) with the initial condition

$$x(t) = \phi(t) \quad (t \in [-h, 0]), \tag{14}$$

and solutions of the averaged equation (2) with the initial condition

$$x(0) = \phi(0). \tag{15}$$

In addition, we shall consider the following initial conditions for equation (2):

$$x(0) = \phi(0) + \frac{\mu}{2}\phi'(0) \tag{16}$$

(these will be needed in the next subsection). As usual, we shall assume that

$$\phi \in C^1([-h, 0], \mathbf{R}^m). \tag{17}$$

We will also assume that the following condition is satisfied:

for $\varepsilon = 1$ the averaged problem (2), (15) *has a unique solution y^1 on some interval* $[0, T]$. (18)

Then, obviously, for $\varepsilon \in (0, 1]$ the averaged problem (2), (15) has a unique solution y^ε on the segment $[0, \varepsilon^{-1}T]$, given by the formula $y^\varepsilon(t) = y^1(\varepsilon t)$.

Let $\mathfrak{M}(\varepsilon,\mu)$ denote the set of all solutions of problem (13), (14) on the segment $[-h,\varepsilon^{-1}T]$, where T is the number appearing in condition (18). We are asking whether it is true that

$$\sup_{x\in\mathfrak{M}(\varepsilon,\mu)} ||y^{\varepsilon}-x|| \to 0 \text{ when } \varepsilon,\mu\to 0, \tag{19}$$

where $||\cdot||$ is the norm in the space $C([0,\varepsilon^{-1}T],\mathbf{R}^m)$. Without assuming beforehand that the set $\mathfrak{M}(\varepsilon,\mu)$ is nonempty, we make the customary convention that $\sup\emptyset=-\infty$. Thus, assertion (19) means, in particular, that for ε and μ small enough,

$$\mathfrak{M}(\varepsilon,\mu)\neq\emptyset.$$

In order to formulate a weaker assertion about the closeness of solutions under the assumption of their existence, we introduce the notation

$$\mathfrak{N}(\varepsilon,\mu)=\mathfrak{M}(\varepsilon,\mu)\cup\{y^{\varepsilon}\}.$$

Then the conditional variant of assertion (19) takes on the form

$$\sup_{x\in\mathfrak{N}(\varepsilon,\mu)} ||y^{\varepsilon}-x|| \to 0 \text{ when } \varepsilon,\mu\to 0. \tag{20}$$

First we prove assertion (20), and then we will show that with the help of the theorems on the solvability of the Cauchy problem it implies the unconditional assertion (19).

4.5.9. Theorem. *Let the operator f satisfy the basic conditions and let the requirements* (17), (18) *be fulfilled. Then relation* (20) *holds true.*

Proof. Suppose (20) is not true, i.e., there exist solutions x^k of the nonaveraged problem (13), (14) for $\varepsilon=\varepsilon_k$, $\mu=\mu_k$, where $\varepsilon_k\to 0$ and $\mu_k\to 0$ when $k\to\infty$, such that

$$||x^k-y^{\varepsilon_k}||\geq\delta>0. \tag{21}$$

Consider the compressed functions $z^k=W(\varepsilon_k^{-1})x^k$. They satisfy the identities

$$(z^k)'(t)=f(\varepsilon_k^{-1}t,W(\varepsilon_k)z^k_{t,\varepsilon_k},\varepsilon_kW(\varepsilon_k)(z^k)'_{t,\varepsilon_k})+\varepsilon_k^{-1}[\phi'_-(0)-\varepsilon_kf(0,\phi,\phi')]\nu_{\mu_k}(\varepsilon_k^{-1}t), \tag{22}$$

or, equivalently,

$$z^k(t)=\phi(0)+\int_0^t f(\varepsilon_k^{-1}s,W(\varepsilon_k)z^k_{s,\varepsilon_k},\varepsilon_kW(\varepsilon_k)(z^k)'_{s,\varepsilon_k})ds$$

$$+[\phi'_-(0) - \varepsilon_k f(0,\phi,\phi')]\varepsilon_k^{-1}\int_0^t \nu_{\mu_k}(\varepsilon_k^{-1}s)ds. \tag{23}$$

It is readily seen that the last term in the right-hand side converges to zero when $k \to \infty$, uniformly in $t \in [0,T]$; as for the first two terms, they form a uniformly bounded and equicontinuous family of functions thanks to the boundedness of f. Hence, by Arzelà's theorem, we may assume, with no loss of generality, that the sequence $\{z^k\}$ converges in $C([0,T],\mathbf{R}^m)$ to some function z. In the present situation we can apply the theorem on passing to the limit under the integral sign, or rather Remark 4.5.6 to that theorem; indeed, (22) implies the existence of a constant M such that

$$\|(z^k)'\|_{C([\varepsilon_k\mu_k,T],\mathbf{R}^m)} \leq M.$$

Letting $k \to \infty$ in (23), we obtain

$$z(t) = \phi(0) + \int_0^t f_0(z(s))ds, \quad t \in [0,T].$$

By condition (18), this further implies that $z = y^1$. However, inequality (21) yields

$$\|z^k - y^1\|_C = \|z^k - z\|_C \geq \delta > 0,$$

which contradicts the convergence of z^k to z. **QED**

4.5.10. Remark. Under the hypotheses of the preceding theorem one can assert that

$$\sup_{x\in\mathfrak{N}(\varepsilon,\mu)} \|(y^\varepsilon)' - x'\| \to 0 \text{ when } \varepsilon,\mu \to 0,$$

where $\|\cdot\|$ is the norm in the space $C^1([\mu,\varepsilon^{-1}T],\mathbf{R}^m)$. In fact, from equations (2), (13), and the boundedness of f it immediately follows that on the segment $[\mu,\varepsilon^{-1}T]$ the derivatives of the solutions of the averaged and nonaveraged equations are of order ε.

In the next subsection we will need an assertion about the closeness of the solutions of problem (13), (14) to the solutions of the averaged equation (2) with initial conditions (16). For this we shall allow the operator f in the right-hand side of equation (1) to depend on ε, i.e., $f = f(\varepsilon,t,u,v)$. As a replacement for (18) we shall assume that

for $\varepsilon = 1$ the averaged equation (2) *has a unique solution on the segment* $[0,T]$ *for arbitrary initial conditions.* (24)

In this case for any $\varepsilon \in (0,1]$ the Cauchy problem (2), (16) has a unique solution on the segment $[0,\varepsilon^{-1}T]$, say, $y^{\varepsilon,\mu}$. Denote $\mathfrak{R}(\varepsilon,\mu) = \mathfrak{M}(\varepsilon,\mu) \cup \{y^{\varepsilon,\mu}\}$. The question we wish to answer is whether

$$\sup_{x\in\mathfrak{R}(\varepsilon,\mu)} \|y^{\varepsilon,\mu} - x\|_{C^1([\mu,\varepsilon^{-1}T],\mathbf{R}^m)} \to 0 \ \text{ when } \varepsilon \to 0, \tag{25}$$

uniformly in μ.

4.5.11. Theorem. *Suppose $f(\varepsilon,t,u,v) \to f(0,t,u,v)$ when $\varepsilon \to 0$, uniformly in t,u,v, and the map $(t,u,v) \mapsto f(0,t,u,v)$ satisfies the basic conditions. Finally, suppose that requirements* (17) *and* (24) *are met. Then relation* (25) *holds.*

Proof. As in the proof of Theorem 4.5.9, if one assumes that (25) fails, then there are solutions x^k of the nonaveraged problem (13), (14) for $\varepsilon = \varepsilon_k$, $\mu = \mu_k$, with $\varepsilon_k \to 0$ when $k \to \infty$ and $\mu_k \in (0,1]$, such that

$$||x^k - y^{\varepsilon_k,\mu_k}|| \geq \delta > 0.$$

With no loss of generality we may assume that $\mu_k \to \mu_0 \in [0,1]$ when $k \to \infty$. From this point on the arguments differ from the proof of Theorem 4.5.9 at two steps. Namely, in passing to the limit in relation (23), first, the convergence of the second term in the right-hand side to $\int_0^t f_0(z(s))ds$ is guaranteed by Remark 4.5.6 rather than Theorem 4.5.5; second, the last term in (23) converges to $(\mu_0/2)\phi'(0)$, and not to zero. The other arguments remain unchanged. **QED**

In the next subsection we consider problem (1), (14). Let $\mathfrak{M}(\varepsilon)$ denote the set of generalized solutions of this problem on the segment $[-h, \varepsilon^{-1}T]$ and, as above, set $\mathfrak{N}(\varepsilon) = \mathfrak{M}(\varepsilon) \cup \{y^\varepsilon\}$. We shall prove the relation

$$\sup_{x\in\mathfrak{N}(\varepsilon)} ||y^\varepsilon - x|| \to 0 \text{ when } \varepsilon \to 0; \tag{26}$$

here $||\cdot||$ is the norm in $W^1_\infty([0,\varepsilon^{-1}T], \mathbf{R}^m)$.

4.5.12. Theorem. *Under the assumptions of Theorem 4.5.9 relation* (26) *holds.*

Proof. First let us estimate, in the metric of the space C, the distance between the generalized solution u^ε of problem (1), (14) and the solution y^ε. For the generalized solution u^ε choose a function $x^\varepsilon \in \mathfrak{N}(\varepsilon, \mu(\varepsilon))$ such that $||u^\varepsilon - x^\varepsilon||_C \leq \varepsilon$ and $\mu(\varepsilon) \leq \varepsilon$. Then

$$||u^\varepsilon - y^\varepsilon||_C \leq ||u^\varepsilon - x^\varepsilon||_C + ||x^\varepsilon - y^\varepsilon||_C \leq \varepsilon + \sup_{x\in\mathfrak{N}(\varepsilon,\mu(\varepsilon))} ||y^\varepsilon - x||_C.$$

Consequently,

$$\sup_{x\in\mathfrak{N}(\varepsilon)} ||y^\varepsilon - x||_C \to 0 \text{ when } \varepsilon \to 0. \tag{27}$$

As we already remarked, the derivative of the solution of problem (13), (14) on the segment $[\mu, \varepsilon^{-1}T]$ is of order ε, and so the solution itself satisfies the Lipschitz condition on this segment with a Lipschitz constant of the same order. The Lipschitz condition remains valid under a uniform passage to the limit with preservation of the Lipschitz constant. Consequently, the derivative of the generalized solution of problem (1), (14) (there, where it exists) has order ε. Finally, the same is true for the derivative of the solution y^ε of the averaged problem. In conjunction with (27), this yields the desired conclusion. **QED**

4.5.13. Unconditional variants of the averaging principle. As we already remarked, the theorems proved to this point have a conditional character: they assert that if the solutions of the nonaveraged problem exist on the segment $[-h, \varepsilon^{-1}T]$, then they are close to the solutions y^ε. Unconditional assertions can be obtained with no difficulty by incorporating in the respective theorems some hypotheses that guarantee the existence of solutions. As implied by the results of Section 4.3, one such hypothesis is as follows:

$$f \textit{ satisfies the Lipschitz condition in the last argument,} \tag{28}$$

i.e., the function β in (5) is linear: $\beta(\xi) = k\xi$.

4.5.14. Theorem. *Suppose the conditions of Theorem 4.5.9 and condition* (28) *are satisfied. Then relations* (19) *and* (26) *hold.*

Proof. For sufficiently small ε the operator εf satisfies the Lipschitz condition in the last argument, with Lipschitz constant smaller than unity. Hence, the basic conditions guarantee that the set $\mathfrak{M}(\varepsilon, \mu)$ is nonempty.

Relation (19) now follows from (20). Finally, (19) and the fact that $\mathfrak{M}(\varepsilon, \mu) \neq \emptyset$ imply that $\mathfrak{M}(\varepsilon) \neq \emptyset$ for small ε. **QED**

4.5.15. Periodic solutions in the averaging principle. In the remaining part of this section we consider T-periodic solutions of equation (1), assuming that the following additional condition is satisfied:

$$\textit{the operator } f \textit{ is periodic in } t \textit{ with period } T. \tag{29}$$

The main restriction on the averaged equation imposed below is that equation (2) have an isolated singular point $x^*(t) \equiv x_*$. More precisely, let $r_0 > 0$ be such that

$$\begin{aligned}&\textit{in the ball of radius } r_0 \textit{ centered at } x_0 \textit{ the equation}\\ &f_0(x) = 0 \textit{ has a unique solution } x_*.\end{aligned} \tag{30}$$

The problem we are concerned with can be stated as follows: let $\mathfrak{M}(\varepsilon)$ denote the set of all T-periodic solutions of equation (1) that satisfy the inequality

$$\|x(0) - x_*\| \leq r_0. \tag{31}$$

Can one assert that

$$\sup_{x \in \mathfrak{M}(\varepsilon)} \|x^* - x\|_{C_T^1} \to 0 \ \text{ when } \varepsilon \to 0\,? \tag{32}$$

As above, this question admits both a conditional and an unconditional interpretation. Relation (32) represents the unconditional variant; the conditional variant is

$$\sup_{x \in \mathfrak{N}(\varepsilon)} \|x^* - x\|_{C_T^1} \to 0 \ \text{ when } \varepsilon \to 0, \tag{33}$$

where $\mathfrak{N}(\varepsilon) = \mathfrak{M}(\varepsilon) \cup \{x^*\}$. Correspondingly, we shall prove two theorems on averaging in the class of periodic solutions.

We remark that the periodicity assumption (29) on f guarantees the existence of the averaged operator f_0, i.e., the fulfillment of condition (6). Moreover, in the present case

$$f_0(x) = \frac{1}{T} \int_0^T f(s, I_h x, 0) ds.$$

4.5.16. Theorem. *Suppose the operator f satisfies the basic conditions and that conditions* (29), (30) *are fulfilled. Then* (33) *holds.*

Proof. Essentially we have to show that (33) holds in the metric of C, since the derivatives of the solutions of (1) are of order ε. Suppose, then, that (33) does not hold in the metric of C, i.e., there are $\delta > 0$, $\varepsilon_k \to 0$ $(k \to \infty)$, and $x^k \in \mathfrak{M}(\varepsilon_k)$, such that $\|x^k - x^*\|_C \geq \delta$. Taking into account that the derivatives of the functions x^k converge uniformly to zero, the functions themselves are T-periodic, and their values at zero are bounded by a common constant, we may assume with no loss of generality that the sequence $\{x^k\}$ converges uniformly to a constant function z. Obviously,

$$r_0 \geq \|z - x^*\|_C \geq \delta. \tag{34}$$

The functions $z^k = W(\varepsilon_k^{-1})x^k$ also converge uniformly to z. Moreover, their derivatives are uniformly bounded, because

$$(z^k)'(t) = f(\varepsilon_k^{-1} t, W(\varepsilon_k) z^k_{t,\varepsilon_k}, \varepsilon_k W(\varepsilon_k)(z^k)'_{t,\varepsilon_k}).$$

Consequently, Theorem 4.5.5 on passing to the limit under the integral sign applies to the equality

$$z^k(t) = z^k(0) + \int_0^t f(\varepsilon_k^{-1} s, W(\varepsilon_k) z^k_{s,\varepsilon_k}, \varepsilon_k W(\varepsilon_k)(z^k)'_{s,\varepsilon_k}) ds,$$

yielding

$$z(t) = z(0) + \int_0^t f_0(z(s))ds.$$

Since z is constant, this implies $f_0(z(0)) = 0$, which in conjunction with (34) contradicts condition (30). **QED**

4.5.17. Theorem. *Supplement the hypotheses of the preceding theorem by condition* (28) *and the following condition:*

the index of the singular point x_* of the finite-dimensional vector field $-f_0$ is different from zero. (35)

Then relation (32) *holds.*

Notice that a simple test for condition (35) to be satisfied is the existence and nondegeneracy of the derivative $f_0'(x_*)$.

Proof. By the preceding theorem, is suffices to establish that $\mathfrak{M}(\varepsilon) \neq \emptyset$ for sufficiently small ε. As shown in **4.4.3**, the T-periodic solutions of equation (1) are precisely the fixed points of the operator Φ_ε,

$$[\Phi_\varepsilon(x)](t) = x(0) + \varepsilon \int_0^t f(s, x_s, x_s')ds - \varepsilon\Big(\frac{t}{T} - \frac{1}{2}\Big) \int_0^T f(s, x_s, x_s')ds. \qquad (36)$$

This operator acts in the space C_T^1 of continuously-differentiable T-periodic $\mathbf{R}^m$-valued functions, equipped with the norm $||x||_{C_T^1} = ||x(0)|| + ||x'||_C$. By Lemma 4.4.4, for ε small enough the operator Φ_ε is $\chi_{C_T^1}$-condensing. Next, by Lemma 4.4.3, the T-periodic solutions of the averaged equation (2) coincide precisely with the fixed points of the compact operator $\Phi_{0\varepsilon}$,

$$[\Phi_{0\varepsilon}(x)](t) = x(0) + \varepsilon \int_0^t f_0(x(s))ds - \varepsilon\Big(\frac{t}{T} - \frac{1}{2}\Big) \int_0^T f_0(x(s))ds, \qquad (37)$$

acting in the same space. In Lemma 4.5.19 below it will be established that on small spheres in C_T^1 centered at x^* the vector fields $I - \Phi_\varepsilon$ and $I - \Phi_{0\varepsilon}$ do not point in opposite directions for sufficiently small ε and so, by Theorem 3.2.5, the fixed-point indices of the operators Φ_ε and $\Phi_{0\varepsilon}$ on small balls coincide. In Lemma 4.5.18 it will be shown that the fixed-point index of $\Phi_{0\varepsilon}$ coincides with the index of the singular point x_* of the finite-dimensional vector field $-f_0$, and so, by condition (35), it is different from zero. Thus, the fixed-point index of the operator Φ_ε is different from zero for small ε. By property 5° (see **3.1.1**), Φ_ε has at least one fixed point near x^*—a T-periodic solution of equation (1).

In this manner the proof is reduced to the following two lemmas.

4.5.18. Lemma. *Under the assumptions of Theorem 4.5.17, the fixed-point index* $\operatorname{ind}(\Phi_{0\varepsilon}, B)$ *of the operator* $\Phi_{0\varepsilon}$ (see (37)) *on any ball* B *in* C^1_T *centered at* x^* *and of radius smaller than* r_0 (see condition (30)) *is defined and equals the index of the singular point* x_* *of the vector field* $-f_0$.

Proof. Let us show that for small ε the family of operators (with parameter λ) $G(\cdot, \varepsilon, \lambda)$, defined as

$$[G(x, \varepsilon, \lambda)](t) = x(0) + \lambda\varepsilon \int_0^t f_0(x(s))ds$$

$$-\lambda\varepsilon\left(\frac{t}{T} - \frac{1}{2}\right) \int_0^t f_0(x(s))ds + (1-\lambda)\varepsilon \int_0^T f_0(x(s))ds \quad (\lambda \in [0,1]),$$

effects a compact homotopy on the ball B from $\Phi_{0\varepsilon}$ to the operator $\tilde{\Phi}_{0\varepsilon}$, where

$$[\tilde{\Phi}_{0\varepsilon}(x)](t) = x(0) + \varepsilon \int_0^T f_0(x(s))ds.$$

To this end it obviously suffices to verify that $G(x, \varepsilon, \lambda) \neq x$ for small ε and all $\lambda \in [0,1]$ and $x \in S$, where $S = \partial B$. Suppose this is not the case. Then there are sequences ε_k, with $\varepsilon_k \to 0$ as $k \to \infty$, $\lambda_k \in [0,1]$, and $x^k \in S$, such that

$$x^k(t) = x^k(0) + \lambda_k\varepsilon_k \int_0^t f_0(x^k(s))ds$$

$$-\lambda_k\varepsilon_k\left(\frac{t}{T} - \frac{1}{2}\right) \int_0^T f_0(x^k(s))ds + (1-\lambda_k)\varepsilon_k \int_0^T f_0(x^k(s))ds. \tag{38}$$

The sequence $\{x^k\}$ is compact in C^1_T, since it lies on the sphere S and the derivatives of the functions x^k are of order ε. Therefore, with no loss of generality we may assume that $\{x_k\}$ converges to a function $x \in S$; then clearly x is a constant function. Setting $t = T$ in (38) and taking into account the fact that $1 - \lambda_k/2$ does not vanish we obtain, after obvious simplifications,

$$\int_0^T f_0(x^k(s))ds = 0.$$

Letting $k \to \infty$ gives

$$\int_0^T f_0(x(s))ds = 0.$$

Since x is a constant, $f_0(x(0)) = 0$. This contradicts condition (30), because $r_0 \geq ||x(0) - x_*|| > 0$.

Thus, for small ε the operators $\Phi_{0\varepsilon}$ and $\tilde{\Phi}_{0\varepsilon}$ are homotopic on B. This implies that for small ε the index $\operatorname{ind}(\Phi_{0\varepsilon}, B)$ is defined and equals $\operatorname{ind}(\tilde{\Phi}_{0\varepsilon}, B)$.

Now notice that the range of the operator $\tilde{\Phi}_{0\varepsilon}$ lies in the subspace E of C_T^1 consisting of the constant functions. Using the restriction principle (see **3.4.4**) we get

$$\operatorname{ind}(\tilde{\Phi}_{0,\varepsilon}, B) = \operatorname{ind}(\tilde{\Phi}_{0\varepsilon}, B \cap E). \tag{39}$$

Since

$$[\tilde{\Phi}_{0\varepsilon}(x)](t) \equiv x(0) + \varepsilon T f_0(x(0))$$

whenever $x \in E$, the right-hand side of (39) is equal to the index of the singular point x_* of the finite-dimensional vector field $-f_0$. **QED**

4.5.19. Lemma. *Suppose the conditions of Theorem 4.5.17 are satisfied. Let S be a sphere of center x^* and radius $\leq r_0$ in C_T^1. Then for small ε the condensing vector fields $I - \Phi_{0\varepsilon}$ and $I - \Phi_\varepsilon$ do not point in opposite directions on S.*

Proof. Assuming the contrary, there are sequences $\varepsilon_k > 0$, $x^k \in S$, and $\mu_k, \lambda_k \geq 0$, such that $\varepsilon_k \to 0$ when $k \to \infty$, $\mu_k + \lambda_k > 0$, and

$$\mu_k(I - \Phi_{\varepsilon_k})x^k = -\lambda_k(I - \Phi_{0\varepsilon_k})x^k. \tag{40}$$

By the preceding lemma, $I - \Phi_{0\varepsilon_k}$ does not vanish in S for sufficiently large k. Therefore, we may assume that $\mu_k \neq 0$. Denoting $l_k = \lambda_k \mu_k^{-1}$, we rewrite (40) in the form

$$(1 + l_k)x^k(t) = (1 + l_k)x^k(0) + \varepsilon_k \int_0^t f(s, x_s^k, (x^k)'_s)ds + l_k\varepsilon_k \int_0^t f_0(x^k(s))ds$$

$$-\Big(\frac{t}{T} - \frac{1}{2}\Big)\Big[\varepsilon_k \int_0^T f(s, x_s^k, (x^k)'_s)ds + l_k\varepsilon_k \int_0^T f_0(x^k(s))ds\Big]. \tag{41}$$

Setting here $t = T$, we get

$$\int_0^T f(s, x_s^k, (x^k)'_s)ds + l_k \int_0^T f_0(x^k(s))ds = 0,$$

and so we can recast (41) as

$$x^k(t) = x^k(0) + (1 + l_k)^{-1}\varepsilon_k \int_0^t f(s, x_s^k, (x^k)'_s)ds + (1 + l_k)^{-1} l_k\varepsilon_k \int_0^t f_0(x^k(s))ds. \tag{42}$$

This shows, in particular, that the derivatives of x^k have order ε_k. Since, in addition, the functions x^k themselves lie in S, i.e., they are bounded by a common constant and T-periodic, Arzelà's theorem allows us to assume, with no loss of generality, that the sequence $\{x^k\}$ converges to a function $y \in C_T$. Furthermore, it is clear that the function y is constant and

$$r_0 \geq ||y(0) - x_*|| > 0. \tag{43}$$

Now let us make the standard change of variables $y^k = W(\varepsilon_k^{-1})x^k$ in (41):

$$y^k(t) = y^k(0) + (1 + l_k)^{-1} \int_0^t f(\varepsilon_k^{-1} s, W(\varepsilon_k) y^k_{s,\varepsilon_k}, \varepsilon_k W(\varepsilon_k)(y^k)'_{s,\varepsilon_k}) ds$$

$$+(1 + l_k)^{-1} l_k \int_0^t f_0(y^k(s)) ds.$$

Clearly, the functions y^k also converge in C to y, and their derivatives are bounded by a common constant. Hence, we can apply Theorem 4.5.5 on passing to the limit under the integral sign (we assume, with no loss of generality, that the sequence l_k converges to a finite or infinite limit $l \geq 0$) to obtain:

$$y(t) = y(0) + (1 + l)^{-1} \int_0^t f_0(y(s)) ds + (1 + l)^{-1} l \int_0^t f_0(y(s)) ds.$$

If $l = \infty$, then the coefficient in front of the first [resp. second] integral is equal to zero [resp. one]. Thus, whatever the value of l,

$$y(t) = y(0) + \int_0^t f_0(y(s)) ds.$$

Since the function y is constant, we obtain $f_0(y(0)) = 0$, which in conjunction with (43) contradicts condition (30). The lemma is proved and the proof of the unconditional variant of the averaging principle in the class of periodic solutions is thus complete. **QED**

4.5.20. Notes on the references. The results concerning the averaging principle for the Cauchy problem are taken from the paper of R. R. Akhmerov [1], while those concerning the averaging principle in the class of periodic solutions are taken from papers of R. R. Akhmerov and M. I. Kamenskiĭ [4] and of R. R. Akhmerov [3].

4.6. ON THE STABILITY OF SOLUTIONS OF EQUATIONS OF NEUTRAL TYPE

In this section we describe a tool that allows one to study the stability of periodic solutions of equations of neutral type with small delay and periodic solutions of equations of neutral type in the setting of the averaging principle.

4.6.1. Preliminary remarks. Let us recall a number of notions and features of stability theory for equations of neutral type. Consider the equation

$$x'(t) = g(t, x_t, x'_t). \tag{1}$$

Here, as usual, $g: \mathbf{R} \times C([-h, 0], \mathbf{R}^m) \times C([-h, 0], \mathbf{R}^m) \to \mathbf{R}^m$ and x_t and x'_t denote the prehistory of the unknown function and of its derivative, respectively. Suppose that equation (1) admits the trivial solution $x(t) \equiv 0$, i.e.,

$$g(t, 0, 0) \equiv 0. \tag{2}$$

Alongside with (1), consider the equation with additional term

$$x'(t) = g(t, x_t, x'_t) + [x'_-(0) - g(0, x_0, x'_0)]\nu_\mu(t), \tag{3}$$

where as above ν_μ is a piecewise-linear function that is equal to 0 on $[\mu, \infty)$ and to 1 at zero (see **4.3.12**). For equation (3) the sewing condition at $t = 0$ is satisfied for any initial function $\phi \in C^1([-h, 0], \mathbf{R}^m)$. Suppose that for any such function ϕ the solution $x(t)$ of equation (3) with the initial condition

$$x(t) = \phi(t), \quad t \in [-h, 0], \tag{4}$$

exists and is unique on the semiaxis $[-h, \infty)$. This is the case, for example, when (see [145, 150])

the operator g is continuous and satisfies the
Lipschitz condition in the second and third arguments
with respective constants K and k, and $k < 1$. (5)

Furthermore (see [151]), condition (5) implies the existence of a continuous function $M: [0, \infty) \to [0, \infty)$ such that

$$||x_t||_{C^1} \leq M(t)||\phi||_{C^1}, \; t \geq 0; \tag{6}$$

here x is the solution of problem (3), (4) and $\mu \in (0, 1]$.

From the estimate (6) it follows that any generalized solution x of the Cauchy problem (1), (4) (see **4.3.12**) can be continued to $[-h, \infty)$ and

$$||x_t||_{W^1_\infty} \leq M(t)||\phi||_{C^1}.$$

4.6.2. Definition of exponential stability. We say that the zero solution of equation (1) is *exponentially stable* if there exist positive numbers r, M and γ such that any generalized solution x of the Cauchy problem (1), (4) can be continued to $[-h, \infty)$ and satisfies the estimate

$$||x_t||_{W^1_\infty} \leq Me^{-\gamma t}||\phi||_{C^1}, \ \ t \geq 0,$$

whenever $||\phi||_{C^1} \leq r$.

4.6.3. Definition. The zero solution of equation (1) is said to be *unstable* if there exist a positive number δ, a sequence $\{t_n\}$ $(t_n \geq 0)$, and a sequence $\{x^n\}$ of generalized solutions of (1), with x_n defined on $[-h, t_n]$, such that

$$||x_0^n||_{C^1} \to 0 \text{ when } n \to \infty \quad \text{ and } \quad ||x_{t_n}^n||_{W^1_\infty} \geq \delta.$$

4.6.4. Stability by the first approximation. Suppose that, in addition to conditions (2) and (5), the operator g satisfies the following two requirements:

g can be written in the form $g(t, u, v) = A(t)u + B(t)v + q(t, u, v)$, where $A(t)$ and $B(t)$ are continuous functions of t with values linear operators acting from $C([-h, 0], \mathbf{R}^m)$ to $\mathbf{R}^m$, and $q(t, u, v)/(||u||_C + ||v||_C) \to 0$ when $||u||_C, ||v||_C \to 0$, uniformly in t; (7)

g is periodic in t with period T (8)

(with no loss of generality one can assume that $h < T$).

Let $\mathcal{V}_{\mu,t}$ denote the *operator of translation from time* 0 *to time t along the trajectories of equation* (3). By definition, $\mathcal{V}_{\mu,t}$ acts in the space $C^1([-h, 0], \mathbf{R}^m)$ according to the rule

$$[\mathcal{V}_{\mu,t}(\phi)](s) = x(t + s), \ \ s \in [-h, 0],$$

where x is the solution of the Cauchy problem (3), (4). In [151] it is shown that under conditions (5), (7) the operator $\mathcal{V}_{\mu,t}$ is differentiable at zero for any μ, t, and its derivative $\mathcal{U}_{\mu,t}$ is the operator of translation from time 0 to time t along the trajectories of the linearized equation

$$x'(t) = A(t)x_t + B(t)x_t' + [x_-'(0) - A(0)x_0 - B(0)x_0']\nu_\mu(t). \tag{9}$$

Let $E \subset C^1([-h, 0], \mathbf{R}^m)$ denote the subspace consisting of the functions that satisfy the sewing condition for the linearized equation (1):

$$x'(t) = A(t)x_t + B(t)x_t',$$

i.e., the set of all $\phi \in C^1([-h,0],\mathbf{R}^m)$ such that $\phi'_-(0) = A(0)\phi + B(0)\phi'$. Denote $\mathcal{U}_\mu = \mathcal{U}_{\mu,T}$. Then, first, $\mathcal{U}_\mu$ maps $C^1([-h,0],\mathbf{R}^m)$ into E and, second, $\mathcal{U}_\mu|_E$ does not depend on μ. Hence, in particular, the spectrum of the operator $\mathcal{U}_\mu$ does not depend on μ.

We shall use the following analogue of Lyapunov's theorem on stability by the first approximation for equations of neutral type (see [7, 8]).

4.6.5. Theorem. *Suppose conditions* (5), (7) *and* (8) *are satisfied. If the spectral radius of the operators* $\mathcal{U}_\mu$ *is smaller than* 1, *then the zero solutions of equation* (1) *is exponentially stable. If the operators* $\mathcal{U}_\mu$ *have at least one eigenvalue of modulus larger than* 1, *then the zero solution of equation* (1) *is unstable.*

4.6.6. Equations with small delay. Let us return to the study of the equation of neutral type with small delay considered in Section 4.4,

$$x'(t) = f(\varepsilon, t, W(\varepsilon)x_t, W(\varepsilon)x'_t). \tag{10}$$

From here on we shall assume that conditions (3) and (4) of Section 4.4 are fulfilled. In addition, we shall assume that

the operator f is jointly continuously differentiable in the third and fourth arguments. (11)

Let ϕ be a T-periodic solution of the equation

$$x'(t) = f(0, t, I_h x(t), I_h x'(t)), \tag{12}$$

and let x^ε be a T-periodic solution of equation (10), the existence of which is guaranteed by Theorem 4.4.6. Linearization of (12) along the solution x^0 yields the equation

$$x'(t) = a(t)x(t) + b(t)x'(t), \tag{13}$$

where

$$a(t) = \frac{\partial}{\partial x} f(0, t, I_h x, I_h(x^0)'(t))|_{x=x^0(t)}, b(t) = \frac{\partial}{\partial x} f(0, t, I_h x^0(t), I_h y)|_{y=(x^0)'(t)}.$$

Let U_0 denote the operator of translation by the period T along the trajectories of the (ordinary differential) equation (13). Recall (see. e.g., [86]) that if all the multipliers of system (13) (i.e., the eigenvalues of U_0) are smaller than 1 in modulus, then the solution x^0 of equation (12) is exponentially stable, whereas if system (13) has at least one multiplier of modulus larger than 1, then the solution x^0 of equation (12) is unstable.

4.6.7. Theorem. *If all the multipliers of system* (13) *are smaller than* 1 *in modulus, then for small* ε *the solutions* x^ε *of equation* (10) *are exponentially stable. If system* (13) *has at least one multiplier of modulus larger than* 1*, then for small* ε *the solutions* x^ε *of equation* (10) *are unstable.*

Proof. Fix $\mu < T$ and for arbitrary $\varepsilon_0 \in (0, T/h]$ denote by V_ε ($\varepsilon \in [0, \varepsilon_0]$) the operator of translation from time 0 to time T along the trajectories of the equation

$$x'(t) = f(\varepsilon, t, W(\varepsilon)x_t, W(\varepsilon)x'_t) + [x'_-(0) - f(\varepsilon, 0, W(\varepsilon)x_0, W(\varepsilon)x'_0)]\nu_\mu(t), \qquad (14)$$

which acts in the space $C^1([-\varepsilon_0 h, 0], \mathbf{R}^m)$. We show below (Lemma 4.6.8) that for $\varepsilon_0 > 0$ small enough the family of operators $\{V_\varepsilon : \varepsilon \in [0, \varepsilon_0]\}$ is (q, χ)-condensing on bounded subsets, where χ is the Hausdorff MNC in $C^1([-\varepsilon_0 h, 0], \mathbf{R}^m)$ (here we consider that this space is equipped with the norm $||x||_{C^1} = ||x(-\varepsilon_0 h)|| + ||x'||_C$).

As discussed above, the operator V_ε is continuously differentiable and its derivative at the point x_0^ε, U_ε, is the operator of translation from time 0 to time T along the trajectories of the equation

$$x'(t) = A_\varepsilon(t)W(\varepsilon)x_t + B_\varepsilon(t)W(\varepsilon)x'_t$$

$$+[x'_-(0) - A_\varepsilon(0)W(\varepsilon)x_0 - B_\varepsilon(0)W(\varepsilon)x'_0]\nu_\mu(t), \qquad (15)$$

where $A_\varepsilon(t)$ is the derivative of the operator $f(\varepsilon, t, \cdot, W(\varepsilon)(x^\varepsilon)'_t)$ at the point $W(\varepsilon)x_t^\varepsilon$ and $B_\varepsilon(t)$ is the derivative of the operator $f(\varepsilon, t, W(\varepsilon)x_t^\varepsilon, \cdot)$ at the point $W(\varepsilon)(x^\varepsilon)'_t$. By Theorem 1.5.9, the family of operators $\{U_\varepsilon : \varepsilon \in [0, \varepsilon_0]\}$ is (q, χ)-bounded.

It follows from the results of the paper [73] that the family $\{U_\varepsilon\}$ is strongly continuous in ε at zero, i.e., $U_\varepsilon x \to U_0 x$ for all x when $\varepsilon \to 0$. Hence, Theorem 2.7.4 applies to any sequence $\{U_{\varepsilon_n}\}$ such that $\varepsilon_n \to 0$ when $n \to \infty$.

Let us show that if all multipliers of system (13) are smaller than 1 in modulus, then for small ε the spectral radius of the operators U_ε is also smaller than 1, and hence, by Theorem 4.6.5, the solutions x^ε are exponentially stable. Indeed, suppose this is not the case. Then there is a sequence $\{\varepsilon_n\}$, $\varepsilon_n \to 0$ when $n \to \infty$, such that the operators U_{ε_n} have spectral radius ≥ 1. By Theorem 2.7.4, the spectral radius of the operator U_0 will be then ≥ 1, which contradicts the assumptions of the theorem.

Now suppose that the system (13) has a multiplier λ_0 such that $|\lambda_0| > 1$. Then for small ε the operators U_ε have spectral points outside the unit disk (by Theorem 2.6.11, these points are necessarily eigenvalues of finite multiplicity). Indeed, assuming the contrary one can find a sequence $\varepsilon_n \to 0$, $n \to \infty$, for which the spectrum $\sigma(U_{\varepsilon_n})$ of U_{ε_n} does not intersect the exterior C of the closed unit disk. By Theorem 2.7.4, this implies

$\sigma(U_0) \cap C = \emptyset$, which contradicts the existence of λ_0. Thus, for small ε the set $\sigma(U_\varepsilon) \cap C$ is nonempty, and consequently, by Theorem 4.6.5, the solutions x^ε are unstable.

Thus, the proof of Theorem 4.6.7 is reduces to that of the following assertion.

4.6.8. Lemma. *There exists an $\varepsilon_0 > 0$ such that the family of operators $\{V_\varepsilon : \varepsilon \in [0, \varepsilon_0]\}$ is (q, χ)-condensing for some $q < 1$.*

Proof. For $\varepsilon \in [0, \varepsilon_0]$ the right-hand side of equation (14) can be regarded as an operator G_ε acting from $C^1([a - \varepsilon_0 h, b], \mathbf{R}^m)$ into $C([a, b], \mathbf{R}^m)$ for all a, b with $a < b$. First, let us show that

$$\chi_C[G(\Omega)] \leq k\chi_{C^1}(\Omega) \tag{16}$$

for all bounded sets $\Omega \subset C^1([a - \varepsilon_0 h, b], \mathbf{R}^m)$; here $G(\Omega) = \bigcup_{\varepsilon \in [0,\varepsilon_0]} G_\varepsilon(\Omega)$, and χ_C and χ_{C^1} are the Hausdorff MNCs in the spaces $C([a, b], \mathbf{R}^m)$ and $C^1([a - \varepsilon_0 h, b], \mathbf{R}^m)$, respectively (we take $||x||_{C^1} = ||x(a - \varepsilon_0 h)|| + ||x'||_C$, and accordingly (see **4.4.4**) $\chi_{C^1}(\Omega) = \chi_{C([a-\varepsilon_0 h, b], \mathbf{R}^m)}(\Omega')$, where $\Omega' = \{x' : x \in \Omega\}$). Represent the operator G_ε as a sum $G_\varepsilon^1 + G_\varepsilon^2$, where G_ε^1 and G_ε^2 correspond to the first and respectively to the second term in the right-hand side of equation (14). Then clearly the operator G_ε^2 is compact, and G_ε^1 admits the diagonal representation $G_\varepsilon^1(x) = H_\varepsilon(x, x)$ through the operator H_ε,

$$[H_\varepsilon(x, y)](t) = f(\varepsilon, t, W(\varepsilon)x_t, W(\varepsilon)y_t').$$

Inequality (16) is now seen to follow from Theorem 1.5.7.

Let us go back to the translation operator. We subject ε_0 to the inequality $n\varepsilon_0 h < T$; the positive integer n will be specified below. Given an arbitrary bounded set $\Omega \subset C^1([-\varepsilon_0, h], \mathbf{R}^m)$, we let Δ denote the set of all solutions of equation (14) ($\varepsilon \in [0, \varepsilon_0]$) that are defined on the segment $[-\varepsilon_0 h, T]$ and have initial conditions $\phi \in \Omega$. It is readily seen that for $0 \leq j \leq n + 1$,

$$\Delta'|_{[T - j\varepsilon_0 h, T]} \subset G(\Delta|_{[T-(j+1)\varepsilon_0 h, T]}). \tag{17}$$

By the definition of the translation operator,

$$V(\Omega) = \Delta|_{[T - \varepsilon_0 h, T]},$$

and so

$$\chi_{C^1}[V(\Omega)] = \chi_{C^1}(\Delta|_{[T-\varepsilon_0 h, T]}) = \chi_C(\Delta'|_{[T-\varepsilon_0 h, T]}). \tag{18}$$

Using inclusion (17) and inequality (16) for $j = 1, \ldots, n$, we derive from (18) the inequality:

$$\chi_{C^1}[V(\Omega)] \leq k^n \chi_{C^1}(\Delta|_{[T-(n+1)\varepsilon_0 h, T]}),$$

or, recalling that $n\varepsilon_0 h < T$,

$$\chi_{C^1}[V(\Omega)] \leq k^n \chi_{C^1}(\Delta). \tag{19}$$

Now notice that the MNC χ_C is semi-additive with respect to partitioning of a segment in the following sense:

$$\chi_{C([a,b],\mathbf{R}^m)}(\Lambda) \leq \chi_{C([a,c],\mathbf{R}^m)}(\Lambda|_{[a,c]}) + \chi_{C([c,b],\mathbf{R}^m)}(\Lambda|_{[c,b]})$$

for any $\Lambda \subset C([a,b],\mathbf{R}^m)$ and $c \in [a,b]$. Consequently,

$$\chi_{C^1}(\Delta) \leq \chi_{C^1}(\Delta|_{[\varepsilon_0 h,0]}) + \chi_{C^1}(\Delta|_{[0,T]}) \leq \chi_{C^1}(\Omega) + k\chi_{C^1}(\Delta)$$

(in the last inequality we used again (17) and (16)).

Thus,

$$\chi_{C^1}(\Delta) \leq \frac{1}{1-k}\chi_{C^1}(\Omega). \tag{20}$$

Now (19) and (20) imply the inequality

$$\chi_{C^1}[V(\Omega)] \leq \frac{k^n}{1-k}\chi_{C^1}(\Omega),$$

which shows that for sufficiently large n (i.e., for sufficiently small ε_0) the family of operators $\{V_\varepsilon : \varepsilon \in [0, \varepsilon_0]\}$ is (q,χ)-bounded, as needed. This completes the proof of the lemma and, together with it, that of Theorem 4.6.7. **QED**

4.6.9. Stability of periodic solutions in the averaging principle. Let us consider the equation

$$x'(t) = \varepsilon f(t, x_t, x'_t) \tag{21}$$

(see Section 4.5). We shall assume that conditions (4), (5), (7), and (28)–(30) of Section 4.5 are satisfied. Then, by Theorem 4.5.17, for small ε equation (21) has a T-periodic solution x^ε that is close to the stationary solution of the averaged equation. Our next objective is to show that the stability properties of the solutions x^ε are determined by the stability properties of the solution $x^0 = x^*$ of the averaged equation

$$x'(t) = f_0(x(t)). \tag{22}$$

We shall furthermore assume that

the operator f is jointly continuously differentiable in the second and third arguments. (23)

As one can readily see, in this case the operator f_0 is also continuously differentiable. Let D denote its derivative at the point x_* (recall that x_* is an isolated solution of the equation $f_0(x) = 0$). We shall assume that

$$\textit{the operator } D \textit{ has no eigenvalues on the imaginary axis.} \tag{24}$$

In particular, this condition guarantees (see **4.5.17**) that condition (35) of Section 4.5 is satisfied.

4.6.10. Theorem. *If the eigenvalues of the operator D lie in the left half-plane (i.e., the solution x^0 of the averaged equation (22) is exponentially stable), then for small ε the solutions x^ε of equation (21) are exponentially stable. If, however, D has at least one eigenvalue with positive real part (i.e., the solution x^0 of equation (22) is unstable), then the solutions x^ε of equation (21) are unstable.*

Proof. Let $A_\varepsilon(t)$ [resp. $B_\varepsilon(t)$] denote the derivative of the operator $f(t, \cdot, (x^\varepsilon)'_t)$ [resp. $f(t, x^\varepsilon_t, \cdot)$] at the point x^ε_t [resp. $(x^\varepsilon)'_t$]. Next, let V_ε ($\varepsilon > 0$) denote the operator of translation from time 0 to time $T[\varepsilon^{-1}]$ ($[a]$ denotes the integer part of the number a) along the trajectories of the equation

$$x'(t) = \varepsilon A_\varepsilon(t)x_t + \varepsilon B_\varepsilon(t)x'_t + [x'_-(0) - A_\varepsilon(0)x_0 - B_\varepsilon(0)x'_0]\nu_\mu(t); \tag{25}$$

here it is assumed that μ is a sufficiently small positive number. Now let V_0 denote the operator that acts in $C^1([-h, 0], \mathbf{R}^m)$ according to the rule

$$V_0(\phi) = I_h\left[e^D\left(\phi(0) + \frac{\mu}{2}\phi'(0)\right)\right]$$

(here I_h is the canonical embedding of $\mathbf{R}^m$ in $C^1([-h, 0], \mathbf{R}^m)$).

As above, it suffices to show that, for small $\varepsilon > 0$, in the first case the operators V_ε have spectral radius smaller than 1, while in the second case they have eigenvalues of modulus larger than 1. Let us check that the family of operators $\{V_\varepsilon : \varepsilon \in [0, \varepsilon_0]\}$ is strongly continuous in the parameter ε at $\varepsilon = 0$ and is condensing with respect to a nice MNC (namely, χ_{C^1}).

The first of these assertions follows from Theorem 4.5.11. Indeed, let y^ε be the solution of the Cauchy problem

$$x' = \varepsilon Dx, \quad x(0) = \phi(0) + \frac{\mu}{2}\phi'(0),$$

i.e.,

$$y^\varepsilon(t) = e^{\varepsilon tD}\left[\phi(0) + \frac{\mu}{2}\phi'(0)\right].$$

Then

$$||V_\varepsilon(\phi) - V_0(\phi)||_{C^1} \le \left|\left|V_\varepsilon(\phi) - y^\varepsilon_{T[\varepsilon^{-1}]}\right|\right|_{C^1} + \left|\left|y^\varepsilon_{T[\varepsilon^{-1}]} - I_h\left[e^D\left(\phi(0) + \frac{\mu}{2}\phi'(0)\right)\right]\right|\right|_{C^1}.$$

The first term in the right-hand side of this inequality tends to zero as $\varepsilon \to 0$ by virtue of Theorem 4.5.11; that the second term also tends to zero is obvious. Thus,

$$V_\varepsilon(\phi) \to V_0(\phi) \ \text{ when } \varepsilon \to 0$$

for all $\phi \in C^1([-h, 0], \mathbf{R}^m)$. In a similar manner on proves that

$$\tilde{V}_\varepsilon(\phi) \to \tilde{V}_0(\phi) \ \text{ when } \varepsilon \to 0, \tag{26}$$

where $\tilde{V}_\varepsilon$ is the "operator of translation" from time 0 to time $T[\varepsilon^{-1}]$ along the trajectories of equation (25), acting from $C^1([-h, 0], \mathbf{R}^m)$ into $C^1([T[\varepsilon^{-1}] - 2h, T[\varepsilon^{-1}]], \mathbf{R}^m)$ as

$$[\tilde{V}_\varepsilon(\phi)](s) = x(s), \quad s \in [T[\varepsilon^{-1}] - 2h, T[\varepsilon^{-1}]]$$

(here x is the solution of equation (25) with initial function ϕ). From the linearity and continuity of the operators $\tilde{V}_\varepsilon$, relation (26), and the uniform boundedness principle it follows that, for sufficiently small $\varepsilon_1 > 0$, the norms $||\tilde{V}_\varepsilon||$ ($\varepsilon \in [0, \varepsilon_1]$) are bounded by a common constant M. Next, taking into account that for small $\varepsilon_2 > 0$ we have

$$||A_\varepsilon(t)|| \le N, \ \ ||B_\varepsilon(t)|| \le N \quad (t \in \mathbf{R}, \ \varepsilon \in [0, \varepsilon_2]),$$

we obtain

$$||(V_\varepsilon(\phi))'||_C \le 2\varepsilon N M ||\phi||_{C^1}.$$

If we now take $\varepsilon_0 < \min\{\varepsilon_1, \varepsilon_2, (2NM)^{-1}\}$, then

$$\chi_{C^1}[V(B)] = \chi_{C^1}\Big[\bigcup_{\varepsilon \in [0,\varepsilon_0]} V_\varepsilon(B)\Big] = \chi_C\Big[\bigcup_{\varepsilon \in [0,\varepsilon_0]} (V_\varepsilon(B))'\Big] \le 2\varepsilon_0 N M < 1,$$

where B is the unit ball in $C^1([-h, 0], \mathbf{R}^m)$. By the linearity of V_ε, this implies that the family of operators $\{V_\varepsilon : \varepsilon \in [0, \varepsilon_0]\}$ is χ_{C^1}-condensing with a constant smaller than 1.

Thus, Theorem 2.7.4 applies to any sequence $\{V_{\varepsilon_n}\}$ ($\varepsilon_n \to 0$ when $n \to \infty$).

From here on one can argue much in the same manner as in the proof of Theorem 4.6.7. In fact, if all eigenvalues of the operator D lie in the open left half-plane, then, as one can readily verify, for sufficiently small $\mu > 0$ the spectrum of the operator V_0 lies in the unit disk. But then, by Theorem 2.7.4, for small $\varepsilon < 1$ the spectral radius of the operators V_ε is smaller than 1, and consequently the solutions x^ε of equation (21) are exponentially stable. If, however, D has at least one eigenvalue in the right half-plane, then V_0 has an eigenvalue of modulus larger than 1. Therefore, the operator V_ε also has such an eigenvalue when ε is small. By Theorem 4.6.5, the latter means that the solutions x^ε are unstable. **QED**

4.6.11. Notes on the references. The results of this section are taken from the papers of R. R. Akhmerov and M. I. Kamenskiĭ [5–7], and the treatment follows the paper of M. I. Kamenskiĭ [77]. The operator of translation along the trajectories of an equation of neutral type is described in the paper of B. N. Sadovskiĭ [159] (see also the paper of R. R. Akhmerov and M. I. Kamenskiĭ [8]). In [74] M. I. Kamenskiĭ studies the stability of periodic solutions of autonomous equations of neutral type with small delay and proves an analogue of the Andronov-Witt theorem for equations of neutral type. Condensing operators also prove useful in the problem of absolute stability of equations of neutral type (see R. R. Akhmerov, N. G. Kazakova, and A. V. Pokrovskiĭ [11]).

4.7. FLOQUET THEORY FOR EQUATIONS OF NEUTRAL TYPE

In this section and the next one we study the structure of the set of solutions of a homogeneous system of linear differential-difference equations of neutral type with periodic coefficients. The exposition of the main results of Sections 4.7 and 4.8 follows the papers of M. I. Kamenskiĭ [75] and respectively the papers of M. I. Kamenskiĭ [77] and of R. R. Akhmerov, M. I. Kamenskiĭ, A. S. Potapov, and B. N. Sadovskiĭ (see [10, p. 203]).

4.7.1. Formulation of the problem. The system under consideration has the form

$$x'(t) = A(t)x(t) + B(t)x(t-h) + C(t)x'(t-h). \tag{1}$$

The unknown function x of real argument t takes values in the n-dimensional complex space $\mathbf{C}^n$, the matrix functions $A(t), B(t)$, and $C(t)$ are continuous on $\mathbf{R}$ and T-periodic, and the delay $h > 0$ is constant. We shall assume that $T > h$; the case $T < h$ is considered in [75]. If the initial function $\phi\colon [-h, 0] \to \mathbf{C}^m$ is continuously differentiable and satisfies the sewing condition

$$\phi'(0) = A(0)\phi(0) + B(0)\phi(-h) + C(0)\phi'(-h), \tag{2}$$

then the problem of the existence and uniqueness of the solution of equation (1) with the initial condition

$$x(t) = \phi(t), \quad t \in [-h, 0] \tag{3}$$

is solved with no difficulty. In fact, on the segment $[0,h]$ the function $x(t)$ is the solution of the linear nonhomogeneous ordinary differential equation

$$x'(t) = A(t)x(t) + f(t), \tag{4}$$

where $f(t) = B(t)\phi(t-h) + C(t)\phi'(t-h)$, with the initial condition $x(0) = \phi(0)$.

Similarly, if the solution of equation (1) on the segment $[(m-1)h, mh]$, denoted x_m, is already known, then on the segment $[mh,(m+1)h]$ it can be continued by the solution of equation (4), in which the function f is defined by the formula

$$f(t) = B(t)x_m(t-h) + C(t)x'_m(t-h),$$

and the initial condition is

$$x(mh) = x_m(mh).$$

The method for solving equation (1) described above is called the *step method* (see [111]). It is readily observed that, for any function $\phi \in C^1([-h,0],\mathbf{C}^n)$ (= the space of continuously differentiable $\mathbf{C}^n$-valued functions defined on $[-h,0]$) that satisfies the sewing condition (2), the solution x_ϕ of equation (1) with initial condition (3) is defined on any segment $[-h,\tau]$ $(\tau > 0)$ and depends continuously on ϕ in the norm of $C^1([-h,0],\mathbf{C}^n)$.

Let C_0^1 denote the subspace of $C^1([-h,0],\mathbf{C}^n)$ consisting of the functions ϕ that satisfy condition (2). Recall that the operator of translation by time T along the trajectories of equation (1) acts from C_0^1 into itself according to the rule

$$(V\phi)(s) = x_\phi(T+s), \quad s \in [-h,0].$$

4.7.2. Floquet solutions. Let λ be a nonzero eigenvalue of the operator V, $\phi^0 \in C_0^1$ be a corresponding eigenfunction: $V\phi^0 = \lambda\phi^0$, and $\phi^1,\dots,\phi^m \in C_0^1$ be associated functions, i.e., $V\phi^p = \lambda\phi^p + \phi^{p-1}$, $p = 1,\dots,m$. Proceeding as in the case of ordinary differential equations or equations with delay (see [111]) one can easily show that the solution x^0 with the initial condition $x^0(t) = \phi^0(t), t \in [-h,0]$, has the form

$$x^0(t) = \exp\Big(\frac{t}{T}\ln\lambda\Big)y^0(t), \tag{5}$$

where y^0 is a T-periodic function. Also, the solutions x^i with the initial conditions $x^i(t) = \phi^i(t)$, $t \in [-h,0]$ have the form

$$x^i(t) = \exp\Big(\frac{t}{T}\ln\lambda\Big)\sum_{j=0}^{i} P_{ij}(t)y^j(t), \tag{6}$$

where the P_{ij} are polynomials of degree $i-j$ and the y^j are T-periodic functions.

Solutions of the form (5) and (6) are usually referred to (see [111]) as *Floquet solutions* and *associated Floquet solutions*, respectively, and the numbers $T^{-1}\ln\lambda$ are called *Floquet exponents*.

4.7.3. A special MNC. Let us define an MNC in the space C_0^1, with respect to which the translation operator will be condensing on any bounded subset of C_0^1. We wish to emphasize that the constructions carried out below are in many respects similar to the corresponding reasoning in the preceding section (see **4.6.8**). However, utilization of the MNC α enables us to obtain the stronger assertion that is needed in the present situation.

Let Ω be a bounded subset of $C^1([-h,0],\mathbf{C}^n)$. Set

$$\psi(\Omega)=\alpha(\Omega'), \tag{7}$$

where $\Omega'=\{x' : x\in\Omega\}$ and α is the Kuratowski MNC in the space $C([-h,0],\mathbf{C}^n)$.

4.7.4. Remark. It is readily seen that ψ is a monotone, semi-additive, algebraically semi-additive, semi-homogeneous, regular MNC. All these properties follow from the corresponding properties of the Kuratowski MNC (see **1.1.4**). Therefore, ψ generates a normal MNC $\tilde{\psi}$ by the rule $\tilde{\psi}(X)=\psi(\{x_n\})$, where $\{x_n\}$ is the set of elements of the sequence X.

4.7.5. Lemma. *Let $\Omega\subset C([a,b],\mathbf{C}^n)$ be a bounded subset. Then for any $c\in[a,b]$,*

$$\alpha_{[a,b]}(\Omega)=\max\left\{\alpha_{[a,c]}(\Omega),\alpha_{[c,b]}(\Omega)\right\}, \tag{8}$$

where the notations $\alpha_{[a,c]}(\Omega),\alpha_{[c,b]}(\Omega)$, and $\alpha_{[a,b]}(\Omega)$ are self-evident.

Proof. The inequality

$$\alpha_{[a,b]}(\Omega)\geq\max\left\{\alpha_{[a,c]}(\Omega),\alpha_{[c,b]}(\Omega)\right\}$$

is plain. Let us establish the opposite inequality. Let $\{X_i\colon i=1,\dots,q\}$ and $\{Y_j\colon j=1,\dots,p\}$ be partitions of the sets $\Omega|_{[a,c]}$ and $\Omega|_{[c,b]}$, respectively. Define a partition $\{Z_{ij}\colon i=1,\dots,q; j=1,\dots,p\}$ of Ω by the following rule:

$$Z_{ij}=\{x\colon\ x\in\Omega,\ x|_{[a,c]}\in X_i,\ x|_{[c,b]}\in Y_j\}.$$

Then clearly $\operatorname{diam} Z_{ij}\leq\max\{\operatorname{diam} X_i,\operatorname{diam} Y_j\}$, which completes the proof of the lemma. **QED**

4.7.6. Theorem. *The translation operator V is (k,ψ)-bounded on any bounded subset of the space C_0^1, where the constant k is such that* $\max_t ||C(t)|| \leq k$.

Proof. First let us consider the auxiliary operator $F: C^1([-h,T], \mathbf{C}^n) \to C([0,T], \mathbf{C}^n)$ defined as

$$(Fz)(t) = A(t)z(t) + B(t)z(t-h) + C(t)z'(t-h).$$

Since the first two terms are compact operators and the third is a contractive operator,

$$\alpha_{[0,T]}(F\Delta) \leq k\alpha_{[-h,T]}(\Delta').$$

Now let Ω be an arbitrary bounded subset of C_0^1. Then

$$\psi[V(\Omega)] = \alpha_{[-h,0]}([V(\Omega]') = \alpha_{[T-h,T]}(\Delta'), \tag{9}$$

where $\Delta = \{x_\phi : \phi \in \Omega\}$, x_ϕ being the solution of equation (1) with initial condition (3). Clearly,

$$\alpha_{[T-h,T]}(\Delta') \leq \alpha_{[0,T]}(\Delta'). \tag{10}$$

Further, $\Delta'|_{[0,T]} = F\Delta$, because the functions from the set $\Delta|_{[0,T]}$ are restrictions of solutions of (1) to the segment $[0,T]$; therefore,

$$\alpha_{[0,T]}(\Delta') \leq k \max\left\{\alpha_{[-h,0]}(\Omega'), \alpha_{[0,T]}(\Delta')\right\}.$$

If the maximum in the right-hand side of the last inequality is attained by the second term, then $\alpha_{[0,T]}(\Delta') = 0$ and the inequality

$$\psi[V(\Omega)] \leq k\psi(\Omega)$$

is obviously satisfied. In the opposite case,

$$\alpha_{[0,T]}(\Delta') \leq k\alpha_{[-h,0]}(\Omega') = k\psi(\Omega).$$

Consequently, (9) and (10) imply

$$\psi[V(\Omega)] \leq k\psi(\Omega). \quad \textbf{QED}$$

4.7.7. Remark. For the normal MNC $\tilde{\psi}$ one has

$$||V||^{\tilde{\psi}} \leq k.$$

This is readily checked with the help of Theorem 4.7.6.

4.7.8. Theorem. *For any* $\varepsilon > 0$ *the solution* x *of equation* (1) *can be represented as a sum of Floquet solutions, associated Floquet solutions, and solutions that decay faster than* $\exp[(t\ln(k+\varepsilon))/T]$.

Proof. As indicated in Remark 4.7.7, the operator V is $(k, \tilde{\psi}, \tilde{\psi})$-bounded. Hence, by Theorem 2.6.11, for any given $\varepsilon > 0$, V can have only finitely many points of its spectrum outside the disk of radius $k+\varepsilon/2$ centered at zero , and those points $\lambda_1, \dots, \lambda_p$ are necesarily eigenvalues of finite multiplicity. Let $\phi_1^0, \dots, \phi_p^0$ be eigenvectors corresponding to these eigenvalues, and let ϕ_q^j, $q = 1, \dots, p$, $j = 1, \dots, m_q$ be associated vectors. Decompose the space C_0^1 into the direct sum of the finite-dimensional subspace L spanned by the vectors $\{\phi_q^j\colon\, q = 1, \dots, p,\ j = 0, 1, \dots, m_q\}$ and the V-invariant subspace M corresponding to the part of the spectrum lying inside the disk of radius $k+\varepsilon/2$. Then any solution x of equation (1) can be written as a sum of solutions y and z such that the functions $y_0, z_0 \in C_0^1$, defined as

$$y_0 = y|_{[-h,0]}, \quad z_0 = z|_{[-h,0]},$$

belong to the subspaces L and M, respectively. But y can be represented as a linear combination of solutions with initial functions ϕ_q^j, $q = 1, \dots, q$, $j = 0, 1, \dots, m_q$, which, as we remarked in **4.7.2**, are Floquet solutions or associated Floquet solutions. Notice that the spectral radius of the operator $V|_M$ is smaller than $k+\varepsilon/2$. Using this and proceeding as in the case of ordinary differential equations (see, e.g., [86]), it is not hard to show that the solution z satisfies the estimate

$$||z(t)|| + ||z'(t)|| \leq C\exp\Big(\frac{t}{T}\ln(k + 3\varepsilon/4)\Big),$$

which completes the proof of the theorem. **QED**

4.8. CONTINUOUS DEPENDENCE OF THE FLOQUET EXPONENTS ON THE DELAY

In this section, as in the preceding one, we consider the equation of neutral type

$$x'(t) = A(t)x(t) + B(t)x(t-h) + C(t)x'(t-h), \tag{1}$$

where $x(t) \in \mathbf{C}^n, A(t), B(t), C(t)$ are linear operators on $\mathbf{C}^n$ that depend continuously (in the operator norm) and T-periodically on t. However, in contrast to Section 4.7, the delay $h \leq h_0$ is a varying parameter. We shall also assume that $||C(t)|| \leq k < 1$.

In the case where A, B, C are constant, the Floquet exponents coincide with the roots of the characteristic quasi-polynomial (see [41])

$$\det \left| \lambda - A - Be^{-\lambda h} - \lambda C e^{-\lambda h} \right|$$

and, being roots of an analytic function, they depend continuously on h. In this section we use Theorem 2.7.4 to establish an analogous fact for the group of the Floquet exponents of equation (1) that are "responsible for stability."

4.8.1. An auxiliary equation. As we already did several times, alongside with (1) we consider the equation

$$x'(t) = A(t)x(t) + B(t)x(t-h) + C(t)x'(t-h) +$$

$$[x'_-(0) - A(0)x(0) - B(0)x(-h) - C(0)x'(-h)]\nu_\mu(t), \tag{2}$$

where ν_μ is a piecewise-linear function, equal to 1 for $t = 0$ and to 0 for $t \geq \mu > 0$. We shall always assume that the parameter μ satisfies the constraint $\mu < T$.

It is readily seen that for equation (2) the sewing condition is satisfied for any initial function $\phi \in C^1([-h_0, 0], \mathbf{C}^n)$. The solution of equation (2) with arbitrary initial function $\phi \in C^1([-h_0, 0], \mathbf{C}^n)$ can be found by the step method, as described in **4.7.1**.

Let $W_\tau(h)$ denote the operator of translation by time τ along the trajectories of equation (2), and let $V_\tau(h)$ denote the operator of translation by time τ along the trajectories of equation (1), defined on functions that satisfy the sewing condition. For $\tau > \mu$, $W_\tau(h)$ maps the space $C^1([-h_0, 0], \mathbf{C}^n)$ into the set of functions satisfying the sewing condition. Consequently, the eigenfunctions of the operator $W_T(h)$ corresponding to nonzero eigenvalues will satisfy the sewing condition. This implies right away that the eigenvalues of the operators $W_T(h)$ and $V_T(h)$ coincide.

We shall need the fact that the translation operator $W_\tau(h)\phi$ is jointly condensing in the variables h, ϕ, with a constant $k < 1$, with respect to the MNC given by the rule

$$\psi(\Omega) = \chi_C(\Omega'). \tag{3}$$

The difference between the present problem and the one considered in **4.6.8** is that here one has to deal with a family of operators depending on a parameter h. As it turns out, here, too, it suffices that the time τ be large compared with h.

4.8.2. Lemma on the translation operator. *The operator $W_\tau(h)\phi$ is jointly condensing in the variables h, ϕ with constant $k^n/(1-k)$, where $n = [\tau/h_0]$, with respect to the MNC ψ given by formula* (3).

The **proof** is esentially a word-for-word repetition of the arguments of **4.6.7**, and will be omitted.

4.8.3. Remark. Let us use Lemma 4.8.2 to find an upper bound for the Fredholm radius of the operator $W_T(h)$. By the lemma and the monotonicity of the Hausdorff MNC, one has the estimate

$$\chi[W_{mT}(h)B] \leq k^n(1-k)^{-1},$$

where χ is the Hausdorff MNC in the space $C^1([-h,0],\mathbf{C}^n)$ with the norm $||x||_{C^1} = ||x(-h)|| + ||x'||_C$, B is the unit ball in this space, and $n = [mT/h]$. Hence, relations (1) and (2) of Section 2.4. imply

$$||W_{mT}(h)||^\psi \leq k^n(1-k)^{-1};$$

here ψ denotes the normal measure of noncompactness generated by χ. Next, from the formula for $R_\Phi[W_T(h)]$ (see Theorem 2.6.11) it obviously follows that

$$R_\Phi[W_T(h)] = \lim_{m\to\infty}(||W_T^m(h)||^\psi)^{1/m} = \lim_{m\to\infty}(||W_{mT}(h)||^\psi)^{1/m} \leq \lim_{m\to\infty} k^{[mT/h]/m}(1-k)^{-1/m}.$$

This yields the inequality

$$R_\Phi[W_T(h)] \leq k^{T/h}.$$

It follows that for any $\varepsilon > 0$ outside the disk of radius $k^{T/h}+\varepsilon$ there can be only finitely many points of the spectrum of the operator $W_T(h)$, which are necessarily eigenvalues of finite multiplicity. Notice also that if $|\lambda| > k^{T/h}+\varepsilon$, then for sufficiently large m we have

$$|\lambda|^m > k^{[mT/h]}(1-k)^{-1} \tag{4}$$

Before turning to the main theorem of this section, let us prove an elementary result.

4.8.4. Remark on distinguishing powers. *Let $\sigma = \{\lambda_1, \dots, \lambda_N\}$ be a set of distinct complex numbers. Then there exist arbitrarily large numbers m such that $\lambda_1^m, \dots, \lambda_N^m$ are also distinct.* In this case we say that *the m-th power distinguishes the points of σ.*

Proof. Let $\lambda_j = \rho_j e^{2i\pi\alpha_j}$, $|\alpha_j| < 1$. Notice that if $\rho_p \neq \rho_q$ or if $\alpha_p - \alpha_q$ is irrational, then the numbers λ_p^m and λ_q^m are distinct for all m. Indeed, in the first case $|\lambda_p^m| \neq |\lambda_q^m|$.

In the second case, if one assumes that $\lambda_p^m = \lambda_q^m$ for some m, then $2\pi\alpha_p m = 2\pi\alpha_q m + 2\pi r$, where r is an integer, which implies $\alpha_p - \alpha_q = r/m$, contradicting the irrationality of $\alpha_p - \alpha_q$.

Let Λ denote the subset of σ consisting of all $\lambda_j \in \sigma$ with the property that there is a $\lambda_l \in \sigma$ such that $\rho_j = \rho_l$ and the difference $\alpha_j - \alpha_l \neq 0$ is rational. Let Q be the set of denominators of all such irreducible fractions $\alpha_j - \alpha_l$. Consider numbers of the form $m = (r \cdot \prod_{q \in Q} q) + 1$, where r is an arbitrary integer. We claim that the numbers $\lambda_1^m, \ldots, \lambda_N^m$ are distinct. In fact, if for a pair λ_p and λ_q one has $|\lambda_p| \neq |\lambda_q|$ or the difference $\alpha_p - \alpha_q$ is rational then, as we observed above, $\lambda_p^m \neq \lambda_q^m$. If now $|\lambda_p| = |\lambda_q|$ and $\alpha_p - \alpha_q \neq 0$ is rational, then $\lambda_p, \lambda_q \in \Lambda$ and $\alpha_p - \alpha_q = l/q$, where $q \in Q$ and the fraction l/q is irreducible. Hence,

$$m\alpha_p - m\alpha_q = r_1 + lq,$$

where r_1 is an integer and $l/q \neq 0$. Consequently, $\arg \lambda_p^m \neq \arg \lambda_p^m$, which completes the proof.

No let us return to the study of the behavior of the Floquet exponents for equation (1).

4.8.5. Theorem on the continuous dependence of the Floquet exponents on the delay. *For any $\delta > 0$, the group of the Floquet exponents $\mu(h)$ of equation* (1) *that satisfy the inequality*

$$\operatorname{Re} \mu(h) \geq \frac{1}{h_0} \ln k + \delta, \tag{5}$$

depends continuously on the magnitude of the delay h.

Proof. As it follows from **4.7.2**, the Floquet exponents satisfying the estimate (5) correspond to the eigenvalues of the operator $W_T(h)$ that lie outside the disk of radius $k^{T/h_0} + \varepsilon$, where $\varepsilon > 0$ depends on δ ($\varepsilon \to 0$ when $\delta \to 0$). Now take a seqeuence h_n such that $h_n \to h_\infty \in [0, h_0]$ when $n \to \infty$.

Let σ_n denote the part of the spectrum of the operator $W_T(h_n)$ lying outside the disk of radius $k^{T/h_0} + \varepsilon$. We claim that

$$\operatorname{dist}(\sigma_n, \sigma_\infty) \to 0 \text{ when } n \to \infty,$$

where dist denotes the Hausdorff distance.

Assume the contrary. Then either there exists a sequence $\lambda_n \in \sigma_n$ with the property that $\inf_{\lambda \in \sigma_\infty} |\lambda_n - \lambda|$ does not converge to zero when $n \to \infty$, or there exists an eigenvalue $\overline{\lambda}$ of the operator $W_T(h_\infty)$ such that $\inf_{\lambda \in \sigma_n} |\overline{\lambda} - \lambda|$ does not converge to zero when $n \to \infty$.

Let us examine the first case. With no loss of generality we may assume that $\lambda_n \to \lambda_0$. Pick a number m such that the m-th power distinguishes the points of $\{\lambda_0\} \cup \sigma_\infty$ (see **4.8.4**) and inequality (4) holds for all $\lambda \in \{\lambda_0\} \cup \sigma_\infty$. By **4.8.2** and **2.7.1**, the operators $A_n = W_{mT}(h_n)$ satisfy the conditions of Theorem 2.7.4. Consequently, assertions (a) and (b) of that theorem and the spectral mapping theorem (see [34]) imply

$$\operatorname{dist}(\sigma_n^m, \sigma_\infty^m) \to 0 \quad \text{when } n \to \infty. \tag{6}$$

Here for a set $\sigma \subset \mathbf{C}$ one denotes $\sigma^m = \{\lambda^m\colon \lambda \in \sigma\}$. But the sequence $\lambda_n^m \in \sigma_n^m$ converges to $\lambda_0^m \in \sigma_\infty^m$ when $n \to \infty$, which contradicts (6).

Thus, we showed that the limit points of a sequence of eigenvalues $\{\lambda_n\}$ of the operators $W_T(h_n)$ such that $\lambda_n \in \sigma_n$, lie necessarily in σ_∞.

To complete the proof it remains to exclude the second possibility. Choose a number m such that the m-th power distinguishes the points of σ_∞ and inequality (4) holds for all $\lambda \in \sigma_\infty$. Then proceeding exactly as above one again establishes relation (6). Let U be a neighborhood of the set $\sigma_\infty^m \setminus \{\bar{\lambda}_m\}$ such that $\bar{\lambda}^m \notin \bar{U}$. Then there exists a neighborhood V of the set $\sigma_\infty \setminus \{\bar{\lambda}\}$ such that $\bar{\lambda} \notin \bar{V}$ and $V^m \subset U$. By the assumption made and what we proved above, $\sigma_n \subset V$ for n large enough. But then $\sigma_n^m \subset U$, which contradicts (6), thereby completing the proof. **QED**

4.9. MEASURES OF NONCOMPACTNESS AND CONDENSING OPERATORS IN SPACES OF INTEGRABLE FUNCTIONS

In this section we study the connection between the MNCs χ and μ in the spaces L_p and study operators of Hammerstein type that are condensing with respect to these MNCs.

4.9.1. Notation and auxiliary facts. Let V be a fixed set of finite Lebesgue measure in a finite-dimensional space. Recall that, for any $p \in [1, \infty)$, L_p denotes the space of equivalence classes x of measurable functions $\xi\colon V \to \mathbf{R}$ for which the norm

$$\|x\|_p = \left(\int_V |\xi(t)|^p dt\right)^{1/p} < \infty,$$

and L_∞ denotes the space of equivalence classes of measurable functions $\xi\colon V \to \mathbf{R}$ for which the norm

$$\|x\|_\infty = \operatorname{vrai\,sup}_{t \in V} |x(t)| = \inf_{\xi \in x} \sup_{t \in V} |\xi(t)| < \infty.$$

Throughout this section we shall identify an equivalence class x and its representative ξ, and refer to x as a function. Whenever we are interested in the domain V of a function from L_p, we shall write $L_p(V)$.

Let $\kappa_D(t)$ denote the *indicator of the set* $D \subset V$:

$$\kappa_D(t) = \begin{cases} 1, & \text{if } t \in D, \\ 0, & \text{if } t \notin D, \end{cases}$$

and let P_D denote the operator of multiplication by κ_D:

$$(P_D x)(t) = \kappa_D(t) x(t).$$

Let us recall some facts of the theory of L_p spaces and operators in such spaces (for details the reader is referred, for example, to the book [93]).

The functions forming a set $\Omega \subset L_p$ are said to have *equi-absolutely continuous norms* if

$$\|P_D x\|_p \to 0 \quad \text{when} \quad \operatorname{mes} d \to 0$$

uniformly in $x \in \Omega$.

We shall use the following compactness criterion: *a set $\Omega \subset L_p$ is compact if and only it is compact in measure and the functions that form Ω have equi-absolutely continuous norms.* Here *compactness in measure* means compactness in the normed space S of all measurable, almost everywhere finite functions x, equipped with the norm

$$\|x\| = \inf_{s>0} \left\{ s + \operatorname{mes}\{t \colon |x(t)| \geq s\} \right\}.$$

An operator $A \colon L_p \to L_q$ $(1 \leq p, q \leq \infty)$ is said to be *positive* if it takes nonnegative functions into nonnegative ones. A linear operator $A \colon L_p \to L_q$ is said to be *regular* if it can be written as the difference of two positive operators.

The linear integral operator A,

$$(Ax)(t) = \int_V K(t,s) x(s) ds \tag{1}$$

is regular from L_p into L_q if the operator $|A|$, defined by the rule

$$(|A|x)(t) = \int_V |K(t,s)| x(s) ds,$$

acts from L_p into L_q. We shall need the following assertion: *any linear regular integral operator acting from L_∞ into L_q $(1 \leq q < \infty)$ is compact* (see [93]).

Let the function $f: V \times \mathbf{R} \to \mathbf{R}$ satisfy the *Carathéodory conditions*, i.e., the function $t \mapsto f(t,u)$ is measurable for any $u \in \mathbf{R}$ and the function $u \mapsto f(t,u)$ is continuous for almost all $t \in V$. The operator F,

$$[F(x)](t) = f(t, x(t)),$$

is called a *superposition operator* (or *Nemytskiĭ operator*).

It is known that *if a superposition operator acts from L_p into L_q $(q < \infty)$, then it is continuous. Moreover, if $p < \infty$, then*

$$|f(t,u)| \leq a|u|^{p/q} + b(t), \tag{2}$$

where $b(t) \in L_q$.

4.9.2. The MNC μ. First let us study the connection between the MNCs χ and μ in the spaces L_p. Recall (see **1.8.2**) that the MNC μ in L_p is defined by the rule

$$\mu(\Omega) = \lim_{\text{mes}\, D \to 0} \sup_{x \in \Omega} ||P_D x||_p.$$

It is readily verified that the function μ is an MNC in the sense of the general definition, and has all properties enumerated in **1.1.4**, except for regularity. It vanishes on any set Ω whose elements have equi-absolutely continuous norms. At the same time, the system of Rademacher functions in $L_p[0,1]$ (see, e.g., [94]),

$$x_m(t) = \operatorname{sign} \sin(2^{m-1}\pi t),$$

is not relatively compact for any p, whereas $\mu(\{x_m\}_{m=1}^{\infty}) = 0$, as trivial estimates show.

It is also easy to see that

$$\mu[B(x_0, r)] = r. \tag{3}$$

4.9.3. Theorem. *For any bounded set Ω in L_p,*

$$\mu(\Omega) \leq \chi(\Omega); \tag{4}$$

moreover, if Ω is compact in measure, then

$$\mu(\Omega) = \chi(\Omega). \tag{5}$$

Proof. Inequality (4) is readily established: if $\{x_1, \dots, x_M\}$ is a $[\chi(\Omega)+\varepsilon]$-net of Ω, then

$$\Omega \subset \bigcup_{i=1}^{m} [x_i + (\chi(\Omega)+\varepsilon)B],$$

where B is the unit ball in L_p. But then

$$\mu(\Omega) \le \mu\Big(\bigcup_{i=1}^{m}[x_i+(\chi(\Omega)+\varepsilon)B]\Big) = \max_{1\le i\le m} \mu[x_i+(\chi(\Omega)+\varepsilon)B] = (\chi(\Omega)+\varepsilon)\mu(B) = \chi(\Omega)+\varepsilon,$$

which by the arbitrariness of $\varepsilon > 0$ yields (4).

Now let us prove (5). For any $a \ge 0$ and $x \in L_p$ we put

$$D(x,a) = \{t \in V\colon |x(t)| \ge a\}.$$

Then it is readily seen that the boundedness of Ω implies

$$\lim_{a\to\infty} \sup_{x\in\Omega} \operatorname{mes} D(x,a) = 0,$$

and so

$$\sup_{x\in\Omega} \|P_{D(x,a)}x\|_p \le \mu(\Omega) + \varepsilon(a), \tag{6}$$

where $\varepsilon(a) \to 0$ when $a \to 0$.

Let $[x]_a$ denote the "cut-off" of the function x by the number a:

$$[x]_a = \begin{cases} x(t), \text{ if } |x(t)| \le a, \\ a \cdot \operatorname{sign} x(t), \text{ if } |x(t)| > a. \end{cases}$$

Set $[\Omega]_a = \{[x]_a\colon x \in \Omega\}$. Notice that $[\Omega]_a$ is compact in measure together with Ω. Furthermore, since the norms of the functions from $[\Omega]_a$ are equi-absolutely continuous, $[\Omega]_a$ is compact, and hence

$$\chi([\Omega]_a) = 0.$$

This readily implies the equality

$$\chi(\Omega) = \chi(\{x - [x]_a\colon x \in \Omega\}). \tag{7}$$

Further, we obviously have

$$|(x - [x]_a)(t)| \le |(P_{D(x,a)}x)(t)|,$$

and so, by inequality (6),

$$\|x - [x]_a\|_p \le \|P_{D(x,a)}x\|_p \le \mu(\Omega) + \varepsilon.$$

Taking account of (7) and the obvious estimate of the MNC χ of an arbitrary set through the supremum of the norms of the elements in that set, we get $\chi(\Omega) \leq \mu(\Omega)+\varepsilon(a)$. Letting here $a \to \infty$, we obtain the inequality $\chi(\Omega) \leq \mu(\Omega)$, which in conjunction to (4) yields (5). **QED**

4.9.4. Remark. Therorem 4.9.3 proves useful in the following situation. Suppose we were able to establish the (k, μ, χ)-boundeness of an operator $A: L_p \to L_q$, i.e., that the inequality

$$\chi_q[A(\Omega)] \leq k\mu_p(\Omega)$$

holds for any bounded $\Omega \subset L_p$ (here the indices appended to the MNCs indicate the space where they are considered). Then, by Theorem 4.9.3,

$$\mu_q[A(\Omega)] \leq \chi_q[A(\Omega)] \leq k\mu_p(\Omega) \leq k\chi_p(\Omega).$$

Thus, the operator A turns out to be both (k, μ_p, μ_q)-bounded and (k, χ_p, χ_q)-bounded. It is exactly this kind of situation that arises in the next subsection, where we prove a theorem that establishes, in particular, a new property of Hammerstein integral operators. It is known that, as a rule, such operators are compact, but they can loose this property at boundary points of the L-characteristic of the linear part. The theorem proved below asserts that at those boundary points a Hammerstein operator is locally condensing, with a constant that tends to zero when the neighborhood is contracted to a point.

4.9.5. Theorem. *Let A be a continuous linear operator which acts from L_p into L_q $(1 \leq p < q < \infty)$ and is compact as an operator from L_∞ into L_q. Let $F: L_q \to L_p$ be a superposition operator generated by a function f. Then the operator $A \circ F: L_q \to L_q$ is (k, μ, χ)-bounded on each ball $B(x_0, r)$, with $k = a||A||r^{q/p-1}$, where a is the constant appearing in* (2) *and $||A|| = ||A||_{L_p \to L_q}$.*

Proof. Let $B(x_0, r)$ be an arbitrary ball in L_q and $\Omega \subset B(x_0, r)$. The proof reduces to verifying of the following two inequalities:

$$\chi[(A \circ F)(\Omega)] \leq ||A||\mu[F(\Omega)], \tag{8}$$

$$\mu[F(\Omega)] \leq ar^{q/p-1}\mu(\Omega). \tag{9}$$

To prove (8) we remark that for any bounded $\Omega_1 \subset L_p$ we have

$$\chi[A(\Omega)_1] \leq \overline{\lim_{\text{mes } D \to 0}} \sup_{x \in \Omega_1} ||AP_D x||_q. \tag{10}$$

In fact, let $D(x,a) = \{t \in V\colon |x(t)| \geq a\}$ and $T(x,a) = \{t \in V\colon |x(t)| < a\} = V \setminus D(x,a)$. Then, clearly,

$$\Omega_1 \subset \{P_{D(x,a)}x\colon x \in \Omega_1\} + \{P_{T(x,a)}x\colon x \subset \Omega_1\}.$$

Since $\sup_{x\in\Omega_1} ||P_{T(x,a)}x||_\infty \leq a$ and A is compact as an operator from L_∞ into L_q, the set $A(\{P_{T(x,a)}x\colon x \in \Omega_1\})$ is relatively compact. It follows that

$$\chi[A(\Omega_1)] \leq \chi(\{AP_{D(a,x)}x\colon x \in \Omega_1\}).$$

Taking into account that

$$\lim_{a\to\infty} \sup_{x\in\Omega_1} \operatorname{mes} D(x,a) = 0,$$

we get (10). To complete the proof (8) it now suffices to notice that

$$\chi[(A \circ F)(\Omega)] \leq \overline{\lim_{\operatorname{mes} D\to 0}} \sup_{y\in F(\Omega)} ||AP_D y|| = \overline{\lim_{\operatorname{mes} D\to 0}} \sup_{x\in\Omega} ||AP_D F(x)||$$

$$\leq ||A|| \overline{\lim_{\operatorname{mes} D\to 0}} \sup_{x\in\Omega} ||P_D F(x)|| = ||A||\mu[F(\Omega)].$$

It remains to prove (9). By estimate (2),

$$|f(t,x(t))| \leq a|x(t)|^{q/p} + b(t).$$

Denote $f_1(t,u) = a|u|^{q/p} + b(t)$, $f_2(t,u) = a|u|^{q/p}$, and let F_1 and F_2 be the coresponding superposition operators, acting from L_q into L_p. Then, by the definition of the MNC μ,

$$\mu[F(\Omega)] = \overline{\lim_{\operatorname{mes} D\to 0}} \sup_{x\in\Omega} ||P_D f(\cdot,x(\cdot))||_p \leq \overline{\lim_{\operatorname{mes} D\to 0}} \sup_{x\in\Omega} ||P_D f_1(\cdot,x(\cdot))||_p = \mu[F_1(\Omega)].$$

Further, by the invariance of μ under translations, $\mu[F_1(\Omega)] = \mu[F_2(\Omega)]$. Finally, a simple calculation shows that $\mu[F_2(\Omega)] = a[\mu(\Omega)]^{q/p}$, and so

$$\mu[F(\Omega)] \leq a[\mu(\Omega)]^{q/p-1}\mu(\Omega) \leq ar^{q/p-1}\mu(\Omega). \quad \textbf{QED}$$

4.9.6. Remark. Under the assumptions of Theorem 4.9.5, the operator $A \circ F$ is $(a||A||r^{q/p-1}, \chi)$-bounded on any ball of radius r in L_q—this follows from Remark 4.9.4. However, $A \circ F$ is not necessarily χ-condensing on the entire space L_q. For example, the operators A and F defined as (here $V = (0,1)$)

$$(Ax)(t) = \int_0^1 \frac{x(s)}{|t-s|^\lambda} ds, \quad (Fx)(t) = |x(t)|^{q/p} \operatorname{sign}[x(t)],$$

where $\lambda = 1 + q^{-1} - p^{-1}$ $(1 < p < q < \infty)$ satisfy the conditions of Theorem 4.9.5. Nevertheless, the MNC χ of the image of the unit ball in L_q under the map $A \circ F$ is larger than 1 (see [46]).

4.9.7. Remark. Suppose the derivative $(A \circ F)'_{x_0}$ of the operator $A \circ F$ at the point x_0 exists. Since the constant $a\|A\|r^{q/p-1}$, with which $A \circ F$ is condensing on the ball $B(x_0, r)$, tends to zero when $r \to 0$, it then follows, by Theorem 1.5.9, that the operator $(A \circ F)'_{x_0}$ is χ-condensing with an arbitrarily small constant, and as such it maps bounded subsets into relatively compact ones. Therefore, in particular, the operator described in the preceding subsection also provides an example of a non-χ-condensing operator whose derivative at any point maps bounded subsets into relatively compact ones.

4.9.8. Application to integral equations. As an elementary example of application of Theorem 4.9.5 we consider the problem of the solvability of the integral equation of Hammerstein type

$$x(t) = \int_V K(t,s) f(s, x(s)) ds. \tag{11}$$

Theorem. *Suppose the operator A, given by formula (1), acts from L_p into L_q and is regular, and the superposition operator F generated by the function f acts from L_q into L_p. Suppose further that there is an $r_0 \in (0, a\|A\|^{1/(1-q/p)})$, such that $ar_0^{q/p} + \|b\|_p \leq r_0/\|A\|$. Then equation (11) has at least one solution of class L_q in the ball $B(0, r_0)$.*

Proof. Clearly, A also acts from L_∞ into L_q, and consequently (see **4.9.1**) it is compact from L_∞ into L_q. By Theorem 4.9.5, the operator $A \circ F$ is (k, χ)-bounded on $B(0, r_0)$. By virtue of our choice of r_0, the constant $K = a\|A\|r_0^{q/p-1}$ is smaller than 1, and so $A \circ F$ is (k, χ)-condensing. It remains to observe that the condition $ar_0^{q/p} + \|b\|_p \leq r_0/\|A\|$ guarantees the invariance of the ball $B(0, r_0)$ under $A \circ F$. Now from Theorem 1.5.11 it follows that $A \circ F$ has a fixed point in $B(0, r_0)$, or, equivalently, that equation (11) has a solution. **QED**

4.9.9. Notes on the references. The exposition of the results of this section follows the papers of N. A. Erzakova [44, 46]. Let us give, without proof, a number of assertions concerning the theory of MNCs and condensing operators in L_p-spaces (see [42, 44, 46]).

The following analogue of Theorem 4.9.3 holds true for the MNC β: *for any bounded set $\Omega \subset L_p$,*

$$\beta(\Omega) \geq 2^{1/p}\mu(\Omega);$$

moreover, if Ω is compact in measure, then this inequality becomes an equality. This assertion is not a simple corollary of Theorem 4.9.3. Although from it and Theorem 4.9.3 it indeed follows that β and χ are proportional on sets Ω that are compact in measure: $\beta(\Omega) = 2^{1/p}\chi(\Omega)$, *the* MNCs *$\beta$ and χ are not proportional for $p \neq 2$* (for $p = 2$ they are connected by the relation $\beta(\Omega) = 2^{1/2}\chi(\Omega)$).

An operator $A: L_p \to L_q$ is said to be *partially additive* (see [93]) if for any set of functions $x_1, \ldots, x_n \in L_p$ with disjoint supports one has

$$A(x_1 + \ldots + x_n) = Ax_1 + \ldots + Ax_n - (n-1)A0.$$

For example, the Uryson integral operator A,

$$(Ax)(t) = \int_V K(t, s, x(s))ds \tag{12}$$

is partially additive.

Theorem. *Suppose the continuous operator $A: L_p \to L_q$ ($1 \leq p, q < \infty$) maps bounded sets into sets that are relatively compact in measure and is partially additive. Suppose further that A is continuous as an operator acting from L_∞ into L_q. Then the following conditions are equivalent:*

1) *A is (k, μ)-bounded;*
2) *A is (k, χ)-bounded;*
3) *A is $(2^{1/q-1/p}k, \beta)$-bounded;*
4) *$\mu[A(B(0,r))] \leq kr$ for any r (here $B(0,r)$ is the ball of radius r centered at 0 in L_q);*
5) $\overline{\lim}_{\text{mes } D_1 + \text{mes } D_2 \to 0} \sup_{\|x\|_p \leq r} \left\|P_{D_1} A P_{D_2} x\right\|_q \leq kr$ *for any $r > 0$.*

Theorem. *Let A be the Uryson operator defined by formula* (12). *Suppose the kernel of A satisfies the Carathéodory conditions* (see [93]) *and that A is continuous as an operator $L_p \to L_q$ and compact as an operator $L_\infty \to L_q$. Suppose further that for any fixed $x \in L_p$ the function $(t,s) \mapsto K(t, s, x(s))$ belongs to $L_p(V \times V)$. If, for any $r > 0$,*

$$\overline{\lim_{\text{mes } D \to 0}} \sup_{\|x\|_p \leq r} \left\| \int_D K(t, s, x(s))ds \right\|_q \leq kr,$$

then the operator A is (k, μ, χ)-bounded and hence, by Remark 4.6.4, (k, χ)-bounded.

Theorem. *Let $A: L_p \to L_1$ be a linear operator that also acts from L_p into L_∞ ($1 \leq p < \infty$). Then*

$$\|A\|^{(\beta)} \leq \left[\left(2 - \frac{\|A\|_{p\to 1}}{(\text{mes } V)\|A\|_{p\to\infty}}\right) 2^{-1/p}\right] \|A\|_{p\to 1},$$

where $\|A\|_{p\to q}$ *is the norm of* A *as an operator* $L_p \to L_q$ *and* $\|A\|^{(\beta)}$ *is its* β*-norm (see* **2.4.10***) as an operator* $L_p \to L_1$.

In particular, if $p = 1$, then

$$\|A\|^{(\beta)} < \|A\|,$$

whereas an analogous statement for the χ-norm of a linear operator is in general not true.

For other applications to integral operators the reader may consult the paper of J. Appell, E. de Pascale, and P. P. Zabreĭko [14].

REFERENCES

[1] Akhmerov, R. R., *On the averaging principle for functional-differential equations,* Ukrain. Math. Zh. **25** (1973), 579–588. (Russian). MR **48**#6607

[2] Akhmerov, R. R., *The existence of solutions of the Cauchy problem for a certain class of functional-differential equations of pass-ahead type,* in: *Proceedings of the IXth Scientific Conference of the Faculty of Phys.-Math. and Natural Sciences,* Patrice Lumumba Peoples's Friendship Univ., Moscow, Izd.-vo UDN, 1974, pp. 5–7. (Russian).

[3] Akhmerov, R. R., *A remark on the principle of averaging on the whole line for equations of neutral type,* Differentsial′nye Uravneniya **13** (1977), 1506. (Russian). MR **58**#22920

[4] Akhmerov, R. R. and Kamenskiĭ, M. I., *On the second theorem of N. N. Bogolyubov in the averaging principle for functional-differential equations of neutral type,* Differentsial′nye Uraveniya **10** (1974), 537–540. (Russian) MR **49**#7560

[5] Akhmerov, R. R. and Kamenskiĭ, M. I., *The averaging principle, and the stability of periodic solutions of equations of neutral type,* Voronezh Gos. Univ. Trudy Nauchn.-Issled. Inst. Mat. VGU, Vyp. 15 Teor. Sb. Stateĭ Nelineĭ. Funktsional. Anal., 1974, pp. 9–13. (Russian). MR **57**#6731

[6] Akhmerov, R. R. and Kamenskiĭ, M. I., *On the question of the stability of the equilibrium state of a system of functional-differential equations of neutral type with small deviation of the argument,* Uspekhi Mat. Nauk **30** (1975) no. 2(182), 205–206. (Russian). MR **53**#8615

[7] Akhmerov, R. R. and Kamenskiĭ, M. I., *A certain approach to the study of the stability of periodic solutions in the averaging principle for functional-differential equations of neutral type,* Comment. Math. Univ. Carolinae **16** (1975), 293–313. MR **51**#6093

[8] Akhmerov, R. R. and Kamenskiĭ, M. I., *Stability by the first approximation of ensembles of dynamical systems with applications to neutral type equations,* Nonlinear Anal. **11** (1987), no. 6, 651–664. MR 88g:34118

[9] Akhmerov, R. R., Kamenskiĭ, M. I., Kozyakin, V. S., and Sobolev, A. V., *Periodic solutions of autonomous functional-differential equations of neutral type with small lag,* Differentsial′nye Uravneniya **10** (1974), 1923–1931. (Russian). MR **51**#3644

[10] Akhmerov, R. R., Kamenskiĭ, M. I., Potapov, A. S., and Sadovskiĭ, B. N., *Condensing operators,* Mathematical Analysis, Vol. 18, 185–250, Akad. Nauk SSSR, Vsesoyz. Inst. Nauchn. i Tekhn. Informatsii, Moscow, 1980. (Russian). MR 83e:47039

[11] Akhmerov, R. R., Kazakova, N. G., and Pokrovskiĭ, A. V., *Principle of the absence of bounded solutions and the absolute stability of equations of neutral type,* Serdica **4** (1978), no. 1, 61–69. (Russian). MR 80k:34101

[12] Amann, H., *Fixed points of asymptotically linear maps in ordered Banach spaces,* J. Functional Analysis **14** (1973), 162–171. MR **50**#3019

[13] Ambrosetti, A., *Un teorema di esistenza per le equazioni differenziali negli spazi di Banach,* Rend. Sem. Mat. Univ. Padova **39** (1967), 349–361. MR **36**#5478

[14] Appell, J., de Pascale, E., and Zabreĭko, P. P., *An application of B. N. Sadovskij's fixed point principle to nonlinear singular equations,* Z. Anal. Anwendungen **6** (1987), no. 3, 193–208. MR 88m:47086

[15] Banaś, J. and Goebel, K., *Measures of Noncompactness in Banach Spaces,* Lecture Notes in Pure and Applied Mathematics, 60. Marcel Dekker, Inc., New York, 1980. MR 82f:47066

[16] Bobylev, N. A., and Krasnosel′skiĭ, M. A., *A functionalization of the parameter and a theorem of relatedness for autonomous systems,* Differentsial′nye Uravneniya **6** (1970), 1946–1952. (Russian). MR **42**#8005

[17] Bondarenko, V. A., *The existence of a universal measure of noncompactness,* Problemy Mat. Anal. Slozh. Sistem no. 2 (1968), 18–21. (Russian). MR **46**#749

[18] Borisovich, Yu. G., *On the relative rotation of compact vector fields in linear spaces,* Trudy Sem. Funktsion. Anal. Voronezh. Gos. Univ. **12** (1969), 3–27. (Russian).

[19] Borisovich, Yu. G., Gel′man, B. P., Myshkis, A. D., and Obukhovskiĭ, V. V., *Topological methods in the theory of multivalued mappings,* Uspekhi Mat. Nauk **35** (1980), no. 1(211), 59–126. (Russian). MR 81e:55004

[20] Borisovich, Yu. G. and Obukhovskiĭ, V. V., *Topological properties, theory of rotation, and fixed point theorems for a class of noncompact multivalued mappings,* Voronezh. Gos. Univ., 1980, 34 pages, deposited VINITI no. 5033-80. (Russian).

[21] Borisovich, Yu. G. and Sapronov, Yu. I., *On the topological theory of densifying operators,* Dokl. Akad. Nauk SSSR **183** (1968), 18–20. (Russian). MR **38**#6414

[22] Borisovich, Yu. G. and Sapronov, Yu. I., *On the topological theory of compactly supported mappings,* Trudy Sem. Funktsion. Anal. Voronezh. Gos. Univ. **12** (1969), 43-68. (Russian).

[23] Bourbaki, N., *Elements of Mathematics. General Topology. Part I.*, Hermann, Paris; Addison-Wesley, Reading, Mass.-London-Don Mills, Ont., 1966. MR **34**#5044a

[24] Browder, F. E., *On the fixed point index for continuous mappings of locally connected spaces,* Summa Brasil. Math., no. 4 (1980), 253–293;. MR **26**#4354

[25] Cellina, A., *On the existence of solutions of ordinary differential equations in Banach spaces,* Funkcial. Ekvac. **14** (1971), 129–136. MR **46**#3937

[26] Daneš, J., *Some fixed point theorems in metric and Banach spaces,* Comment. Math. Univ. Carolinae **12** (1971), 37–51. MR **44**#4604

[27] Daneš, J., *On the Istrăţescu's measure of noncompactness,* Bull. Math. Soc. Sci. Math. R. S. Roumanie (N.S.) **16(64)** (1972), no. 4, 403–406. MR **50**#14389

[28] Daneš, J., *On densifying and related mappings and their application in nonlinear functional analysis,* in: *Theory of nonlinear operators, Proc. Summer School, Neuchâtel, 1972),* pp. 15–56, Schr. Zentralinst. Math. Mech. Akad. Wiss. DDR Heft 20, Akademie-Verlag, Berlin, 1974. MR **50**#14388

[29] Darbo, G. *Punti uniti in transformazioni a codominio non compatto,* Rend. Sem. Mat. Univ. Padova **24** (1955), 84–92. MR **16**#1140

[30] De Blasi, F. S., *Compactness gauges and fixed points,* Atti. Acad. Naz. Lincei Rend. Cl. Sci. Fis. Mat. Natur. (8)**57** (1974), no. 3-4, 170–176. MR**53**#1560

[31] De Blasi, F. S., *On a property of the unit sphere in a Banach space,* Bull. Math. Soc. Sci. Math. R. S. Roumanie (N.S.) **21(69)** (1977), no. 3-4, 259–262. MR **58**#2475

[32] Demidovich, B. P., *Lectures on the Mathematical Theory of Stability,* "Nauka", Moscow, 1967. (Russian). MR **37**#1716

[33] Dugundji, J., *An extension of Tietze's theorem,* Pacific J. Math. **1** (1951), 353–367. MR **13**#373

[34] Dunford, N. and Schwartz, J. T., *Linear Operators. I. General Theory.* with the assistance of W. G. Bade and R. G. Bartle, Pure and Appl. Math., Vol. 7, Interscience, New-York, 1958. MR **22**#8302

[35] Dyadchenko, Yu. A., *On a Volterra-type nonlinear operator equation type,* Candidate of Science Dissertation, Voronezh, 1978. (Russian).

[36] Dyadchenko, Yu. A., *On the solvability of a Volterra-type nonlinear operator equation,* in: *Theory of operator equations,* Izd.-vo Voronezh. Gos. Univ., Voronezh, 1979, pp. 22–33. (Russian).

[37] Dyadchenko, Yu. A. and Rodkina, A. E., *On the continuability and the continuous dependence on a parameter of generalized solutions of neutral type equations,* Voronezh. Gos. Univ. Trudy Nauchn.-Issled. Mat. VGU no. 12, Sb. Stateĭ Prikl. Anal. (1974), 28–31. (Russian). MR **57**#10163

[38] Dyadchenko, Yu. A. and Rodkina, A. E., *Generalized solutions of equations of neutral type,* Voronezh. Gos. Univ. Trudy Nauchn.-Issled. Inst. Mat. VGU no. 15, Sb. Stateĭ Nelineĭn. Funktsional. Anal. (1974), 24–32. (Russian). MR **57**#6713

[39] Edmunds, D. E., Potter, A. J. B., and Stuart, C. A., *Non-compact positive operators,* Proc. Roy. Soc. London Ser. A **328** (1972), no. 1572, 67–81. MR **58**#3058

[40] Eisenfeld, J. and Lakshmikantham, V., *On a measure of nonconvexity and applications,* Yokohama Math. J. **24** (1976), no. 1-2, 133–140. MR **54**#13657

[41] El′sgol′ts, L. E. and Norkin, S. B., *Introduction to the Theory of Differential Equations with Deviating Argument,* (second edition), "Nauka", Moscow, 1971. (Russian). MR **50**#5133; English transl.: *Introduction to the Theory and Application of Differential Equations with Deviating Arguments,* Math. in Science and Engineering, Vol. 105, Academic Press, New York, London, 1973. MR **50**#5134

[42] Erzakova, N. A., *On the measure of noncompactness β of linear operators in spaces of summable functions,* Voronezh. Gos. Univ., Voronezh, 1982, 15p; deposited VINITI No. 6133-82. (Russian).

[43] Erzakova, N. A., *A measure of noncompactness,* in: *Approximate Methods for Investigating Differential Equations and their Applications,* Kuĭbyshev. Gos. Univ., Kuĭbyshev, 1982, pp. 58–60. (Russian). MR 85f:47071

[44] Erzakova, N. A., *χ-Condensing operators in spaces of summable functions,* in: *Abstracts of Reports to the All-Union School on Function Theory Dedicated to the 100th Anniversary of Academician N. N. Luzin's Birthday,* Izd.-vo Kemerov. Univ., Kemerovo, 1983, p. 43. (Russian).

[45] Erzakova, N. A. , *On the inner Hausdorff measure of noncompactness,* in: *Theory of operators in function spaces,* Izd.-vo Voronezh. Gos. Univ., Voronezh, 1983, pp. 37–45. (Russian).

[46] Erzakova, N. A., *On condensing operators in L_p-spaces,* Voronezh. Gos. Univ., Voronezh, 1983, 25p; deposited VINITI No. 2850-83. (Russian).

[47] Erzakova, N. A., *On measures of noncompactness in Banach spaces,* Candidate of Science Dissertation, Voronezh, 1983. (Russian).

[48] Fitzpatrick, P. M., *A generalized degree for uniform limits of A-proper mappings,* J. Math. Anal. Appl. **35** (1971), 536–552. MR **43**#6788

[49] Furi, A. and Vignoli, A., *On a property of the unit sphere in a linear normed space,* Bull. Acad. Polon. Sci. Sér. Math. Astronom. Phys. **18** (1970), 333–334. MR **41**#8969

[50] Gatica, J. A., *Fixed point theorems for k-set-contractions and pseudocontractive mappings,* J. Math. Anal. Appl. **46** (1974), 555–564. MR **49**#11328

[51] Gershteĭn, V. M., *On the theory of dissipative differential equations in a Banach space,* Funktsional. Anal. i Prilozhen. **4** (1970), no. 3, 99–100. (Russian). MR **43**#666

[52] Gikhman, I. M. and Skorohod, A. V., *The theory of Stochastic Proccsses,* Vols. I and III, "Nauka", Moscow, 1971 and 1975. (Russian). MR **49**#6287 and MR **58**# 31323a; English transl.: *The Theory of Stochastic Processes,* Vols. I and III, Die Grundlehren der mathematischen Wissenschaften, Band 210 and Band 232, Springer-Verlag, Berlin, New York, 1974 and 1979. MR **49**#11603 and MR **58**#31323b.

[53] Godunov, A. N., *The Peano theorem in Banach spaces,* Funktsional. Anal. i Prilozhen. **9** (1974), no. 1, 59–60. (Russian). MR **51**#1051

[54] Gohberg, I. and Kreĭn, M. G., Fundamental aspects of defect numbers, root numbers, and indexes of linear operators, Uspekhi Mat. Nauk (N.S.) **12** (1957), no. 2(74), 43–118. (Russian). MR **20**#3459

[55] Gol′denshteĭn, L. S., Gohberg, I., and Markus, A. S. *Investigation of some properties of bounded linear operators and of the connection with their q-norm,* Uchen. Zap. Kishinev. Gos. Univ., no. 29 (1957), 29–36. (Russian)

[56] Gol′denshteĭn, L. S. and Markus, A. S., *On the measure of noncompactness of bounded sets and of linear operators,* in: *Studies in Algebra and Mathematical Analysis,* Izd.-vo "Kartya Moldavenyaske", Kishinev, 1965, pp. 45–54. (Russian). MR **35**#789

[57] Gol′tser, Ya. I. and Zverkin, A. M., *The existence and uniqueness of the solutions of differential equations with lag in a Banach space,* Differentsial′nye Uravneniya **12** (1976), no. 8, 1404–1409. (Russian). MR **55**#13035

[58] Grothendieck, A., *Espaces Vectoriels Topologiques,* Instituto di Matemática Pura e Aplicada, Universidade de São Paulo, São Paulo, 1954. MR **17**#1110; English transl.: *Topological Vector Spaces,* Notes in Mathematics and its Applications, Gordon and Breach, New York, London, Paris, 1973. MR **51**#8772

[59] Hadžić, O., *A fixed-point theorem for mappings with a ψ-densifying iteration in locally convex spaces,* Mat. Vesnik **2(15)(30)** (1978), no. 2, 105–109. MR 80f:47050

[60] Hale, J. *Theory of Functional Differential Equations,* Second edition, Applied Math. Sciences, Vol. 3, Springer-Verlag, New York-Heidelberg, 1977. MR **58**#22904

[61] Himmelberg, C. J., Porter, J. R., and Van Vleck, F. S., *Fixed point theorems for condensing multifunctions,* Proc. Amer. Math. Soc. **23** (1969), 635–641. MR **39**#7480

[62] Istrăţescu, A. I. and Istrăţescu, V. I., *A generalization of collectively compact sets of operators.* I., Rev. Roumaine Math. Pures Appl. **17** (1972), 33–37. MR **46**#6092

[63] Istrăţescu, I. I., *Some spectral properties for the I-contraction operators,* Math. Balkanika **4** (1974), 267–269. MR **51**#8870

[64] Istrăţescu, V. I., *On a measure of noncompactness,* Bull. Math. Soc. Sci. Math. R. S. Roumanie (N.S.) **16(64)** (1972), No. 2, 195–197. MR **49**#6180

[65] Istrăţescu, V. I and Istrăţescu, A. I., *On the theory of fixed points for some classes of mappings.* III. Atti. Acad. Naz. Licei Rend. Cl. Sci. Fis. Mat. Natur. (8) **49** (1970), 43–46. MR **50**#10921c

[66] Istrăţescu, V. I and Istrăţescu, A. I., *On the theory of fixed points for some classes of mappings.* II. Rev. Roumaine Math. Pures Appl. **16** (1971), 1073–1076. MR **50**#10921b

[67] Jones, G. S., *A Functional approach to fixed-point analysis of noncompact operators,* Math. Systems Theory **6** (1972/73), 375–382. MR **51**#1501

[68] Kalmykov, A. A., *Fixed points of condensing operators that expand a cone,* Perm. Gos. Univ. Uchen. Zap. No. 309, Mat. Problemy Nelineĭn. Anal. (1974), 29–32. (Russian). MR **52**#1425

[69] Kalmykov, A. A., *A condition for the (k,χ)-boundedness of integral operators in spaces of functions that are continuous and bounded on an infinite interval,* Perm. Gos. Univ, Perm′, 1977, 16p; deposited VINITI No. 1194-77. (Russian).

[70] Kalmykov, A. A., *Projective measures of noncompactness and projectively-condensing operators,* Perm. Gos. Univ, Perm′, 1977, 31p; deposited VINITI No. 1674-77. (Russian).

[71] Kamenskiĭ, G. A., *Existence, uniqueness, and continuous dependence on initial values of the solutions of systems of differential equations with deviating argument of neutral type,* Mat. Sb. (N.S.) **55(97)** (1961), 363–378. (Russian). MR **26**#6536

[72] Kamenskiĭ, M. I., *On the Peano theorem in infinite-dimensional spaces,* Mat. Zametki **11** (1972), 569–576. (Russian). MR **46**#3940

[73] Kamenskiĭ, M. I., *Computation of the index of an isolated solution of the Cauchy problem for a functional-differential equation of neutral type,* in: *Collection of Papers by Postgraduates,* Vol. 1, Voronezh. Gos. Univ., Voronezh, 1973, pp. 6–12. (Russian).

[74] Kamenskiĭ, M. I., *On the Andronov-Witt theorem for functional-differential equations of neutral type*, Voronezh Gos. Univ. Trudy Nauchn.-Issled. Inst. Mat. VGU no. 12, Sb. Stateĭ po Prikl. Anal. (1974), 31–36. (Russian). MR **57**#10165

[75] Kamenskiĭ, M. I., *On the operator of translation along the trajectories of equations of neutral type,* in: *Collection of Papers by Postgraduates,* Vol. 2, Voronezh. Gos. Univ., Voronezh, 1974, pp. 19–22. (Russian).

[76] Kamenskiĭ, M. I., *Measures of noncompactness and the perturbation theory of linear operators,* Tartu Riikl. Ül. Toimetised no. 430 (1977), 112–122. (Russian). MR **58**#7151

[77] Kamenskiĭ, M. I., *The operator of translation along the trajectories of equations of neutral type that depend on a parameter,* Tartu Riikl. Ül. Toimetised no. 448 (1978), 101–117. (Russian). MR 80h:34105

[78] Kantorovich, L. V. and Akilov, G. P., *Functional Analysis,* (second edition, revised) "Nauka", Moscow, 1977. (Russian). MR **58**#23465; English transl.: *Functional Analysis,* Pergamon Press, Oxford-Elmsford, New York, 1982. MR 83h:46002

[79] Kato, T., *Perturbation Theory for Linear Operators,* Die Grundlehren der mathematischen Wissenschaften, Band 132, Springer-Verlag, New York, 1966. MR **34**#3324

[80] Kelley, J., *General Topology,* D. Van Nostrand, Toronto-New York- London, 1955. MR **16**#1136

[81] Kolmanovskiĭ, V. B. and Nosov, V. R., *Stability and Periodic Modes of Adaptable Systems with Aftereffect,* "Nauka", Moscow, 1981. (Russian). MR 83g:34001

[82] Kozhokar′-Gonchar, M. T., *The spectrum of Cesàro operators,* Mat. Issled. **7** (1972), no. 4(26), 94–103. (Russian). MR **50**#5531

[83] Kozyakin, V. S., *On condensing and contracting operators,* Trudy Mat. Fakultet. Voronezh. Gos. Univ., no. 1 (1970), 60–70. (Russian).

[84] Krasnosel′skiĭ, M. A., *Topological Methods in the Theory of Nonlinear Integral Equations,* Gostekhizdat, Moscow, 1956. (Russian). MR **20**#3464; English transl.: *Topological Methods in the Theory of Nonlinear Integral Equations,* A Pergamon Press Book, Macmillan, New York, 1964. MR **28**#2414

[85] Krasnosel′skiĭ, M. A., *Positive Solutions of Operator Equations,* Fizmatgiz, Moscow, 1962. (Russian). MR **26**#2862; English transl.: *Positive Solutions of Operator Equations,* P. Nordhoff, Groningen, 1964. MR **31**#6107

[86] Krasnosel′skiĭ, M. A., *The Operator of Translation along the Trajectories of Differential Equations,* "Nauka", Moscow, 1966. (Russian). MR **34**#3012; English transl.: *The Operator of Translation along the Trajectories of Differential Equations,* Translations of Mathematical Monographs, Vol. 19, American Math. Society, Providence, R.I., 1968. MR **36**#6688

[87] Krasnosel′skiĭ, M. A., Kibenko, A. V., and Mamedov, Ya. D., *One-sided estimates in the conditions for the existence of solutions of differential equations in function spaces,* Azerbaĭdzhan. Gos. Univ. Uchen. Zap. Ser. Fiz.-Mat. i Khim. Nauk no. 3 (1961), 13–19. (Russian). MR **36**#4108

[88] Krasnosel′skiĭ, M. A. and Kreĭn, S. G., *On the principle of averaging in nonlinear mechanics,* Uspekhi Mat. Nauk (N.S.) **10** (1955), no 3(65), 147–152. (Russian). MR **17**#152

[89] Krasnosel′skiĭ, M. A. and Kreĭn, S. G., *On the theory of ordinary differential equations in Banach spaces,* Voronezh. Gos. Univ. Trudy Sem. Funktsional. Anal. no. 2 (1956), 3–23. (Russian). MR **19**#140

[90] Krasnosel'skiĭ, M. A. and Perov, A. I., *Existence of solutions for certain non-linear operator equations,* Dokl. Akad. Nauk SSSR **126** (1959), 15–18. (Russian). MR **21**#5153

[91] Krasnosel'skiĭ, M. A., Vaĭnikko, G. M., Zabreĭko, P. P., Rutitskiĭ, Ya. B., and Stetsenko, V. Ya., *Approximate Solution of Operator Equations,* "Nauka", Moscow, 1969. (Russian). MR **41**#4271; English transl.: *Approximate Solution of Operator Equations,* Wolters-Noordhoff, Groningen, 1972. MR **52**#6515

[92] Krasnosel'skiĭ, M. A. and Zabreĭko, P. P., *Geometric Methods of Nonlinear Analysis,* "Nauka", Moscow, 1975. (Russian). MR **58**#17976; English transl.: *Geometric Methods of Nonlinear Analysis,* Grundlehren der mathematischen Wissenschaften Vol. 263, Springer-Verlag, Berlin, New York, 1984. MR 85b:47057

[93] Krasnosel'skiĭ, M. A., Zabreĭko, P. P., Pustyl'nik, E. I., and Sobolevskiĭ, P. E., *Integral Operators in Spaces of Summable Functions,* "Nauka", Moscow, 1966 (Russian). MR **34**#6568; English transl.: *Integral Operators in Spaces of Summable Functions,* Noordhoff, Leiden, 1976. MR **52**#6505

[94] Kreĭn, S. G., Petunin, Yu. I., and Semënov, E. M., *Interpolation of Linear Operators,* "Nauka", Moscow, 1978. (Russian). MR 81f:46086; English transl.: *Interpolation of Linear Operators,* Translations of Mathematical Monographs, 54, American Math. Society, Providence, R.I., 1982. MR 84j:46103

[95] Kuratowski, K., *Sur les espaces complets,* Fund. Math. **15** (1930), 301–309.

[96] Kuratowski, K., *Topology, Vol. 1,* (new edition, revised and augmented), Academic Press, New York-London and Pańistwowe Wydawnictwo Naukowe, Warsaw, 1966. MR **36**#840

[97] Kurbatov, V. G., *The spectral radii and Fredholm radii of certain linear operators on the space of functions continuous and bounded on the real line,* in: *Collection of Papers by Postgraduates, Vol. 2,* Voronezh. Gos. Univ., Voronezh, 1972, pp. 47-52. (Russian). MR **57**#9376

[98] Kurbatov, V. G., *The spectrum of an operator with commensurable deviations of the argument and with constant coefficients,* Differentsial'nye Uravneniya **13** (1977), no. 10, 1770–1775. (Russian). MR **58**#22908

[99] Kurbatov, V. G., *The Hausdorff measure of noncompactness in L^∞,* in: *Approximate Methods for Investigating Differential Equations and their Applications,* Kuĭbyshev. Gos. Univ., Kuĭby- shev, 1982, pp. 95–97. (Russian). MR 85g:47090

[100] Lebow, A. and Schechter, M., *Semigroups of operators and measures of noncompactness,* J. Functional Analysis **7** (1971), 1–26. MR **42**#8301

[101] Leggett, R. W., *Remarks on set-contractions and condensing maps,* Math. Z. **132** (1973), 361–366. MR **49**#11278

[102] Leggett, R. W., *A note on "locally α-contracting" linear operators,* Boll. Un. Mat. Ital. (4) **12** (1975), no. 1-2, 124–126. MR **52**#15096

[103] Liptser, R. Sh. and Shiryaev, A. N., *Statistics of Random Processes,* "Nauka", Moscow, 1974. (Russian). MR **55**#4365: English transl. *Statistics of Random Processes. I. General Theory* and *II. Applications,* Applications of Mathematics, Vols. 5 and 6, Springer-Verlag, New York-Heidelberg, 1977 and 1978. MR **57**#14125 and **58**#7827

[104] Lifshits, E. A. and Sadovskiĭ, B. N., *A fixed point theorem for generalized-condensing operators,* Dokl. Akad. Nauk SSSR **183** (1968), no. 2, 278–279. (Russian).

[105] Lyal′kina, G. B., *The generalization of Schaefer's principle to condensing operators,* Perm. Gos. Univ. Uchen. Zap. no. 309 (1974), 13–19. (Russian). MR **52**#1428

[106] Lyal′kina, G. B., *Derivatives and derivatives with respect to a cone for a (k, χ)-bounded operator,* Perm. Gos. Univ. Uchen. Zap. no. 291 (1975), 14–16. (Russian). MR **56**#3688

[107] Lyusternik, L. A. and Sobolev, V. I., *Elements of Functional Analysis* (second revised edition), "Nauka", Moscow, 1965. (Russian). MR **35**#698; English transl.: *Elements of Functional Analysis,* International Monographs on Advanced Mathematics & Physics, Hindustan Publishing Corp., Delhi, 1971. MR **49**#7721

[108] Martelli, M. and Vignoli, A., *A generalized Leray-Schauder condition,* Atti. Accad. Naz. Lincei Rend. Cl. Sci. Fis. Mat. Natur. (8)**57** (1974), no. 5, 374–379. MR **54**#5918

[109] Misyurkeev, I. V., *The existence of fixed points for condensing operators that are strongly asymptotically linear with respect to a cone,* Perm. Gos. Univ. Uchen. Zap. no. 309 (1974), 20–25. (Russian). MR **52**#1429

[110] Mönch, H. and von Harten, G.-F., *The product formula for the topological degree of strict γ-contractions,* Manuscripta Math. **23** (1977/78), No. 2, 113–123. MR **57**#14028

[111] Myshkis, A. D., *Linear Differential Equations with Retarded Argument,* (second edition), "Nauka", Moscow, 1972. (Russian). MR **50**#5135

[112] Nagumo, M., *Degree of mapping in convex linear topological spaces,* Amer. J. Math. **73** (1951), 497–511 (1951). MR **13**#150

[113] Nussbaum, R. D., *The fixed point index and asymptotic fixed point theorems for k-set-contractions,* Bull. Amer. Math. Soc. **75** (1969), 490–495. MR **39**#7589

[114] Nussbaum, R. D., *The fixed point index and fixed point theorems for k-set-contractions,* PhD Dissertation, Univ. of Chicago, 1969.

[115] Nussbaum, R. D., *The radius of essential spectrum,* Duke Math. J. **37** (1970), 473–478. MR **41**#9028

[116] Nussbaum, R. D., *The fixed point index for local condensing maps,* Ann. Mat. Pura Appl. (4) **89** (1971), 217–258. MR **47**#903

[117] Nussbaum, R. D., *A generalization of the Ascoli theorem and an application to functional differential equations,* J. Math. Anal. Appl. **35** (1971), 600–610. MR **44**#7085

[118] Nussbaum, R. D., *Estimates for the number of solutions of operator equations,* Applicable Anal. **1** (1971), no. 2, 183–200. MR **45**#5839

[119] Nussbaum, R. D., *Degree theory for local condensing maps,* J. Math. Anal. Appl. **37** (1972), 741–766. MR **46**#6107

[120] Nussbaum, R. D., *Generalizing the fixed point index,* Math. Ann. **228** (1977), no. 3, 259–278. MR **55**#13461

[121] Nussbaum, R. D., *Eigenvectors of nonlinear positive operators and the linear Kreĭn-Rutman theorem,* in: Fixed Point Theory (Sherbrooke, Que., 1980), Lecture Notes in Math., 886, 1981, pp. 309-330, Springer-Verlag, Berlin-New York. MR 83b:47068

[122] Obukhovskiĭ, V. V., *Some fixed point principles for multivalued condensing operators,* Voronezh. Gos. Univ. Trudy Mat. Fak. Vyp. 4 (1970), 70–79. (Russian). MR **54**#3522

[123] Obukhovskiĭ, V. V., *Equations with multivalued operators and some of their applications,* Candidate of Science Dissertation, Voronezh, 1975. (Russian).

[124] Obukhovskiĭ, V. V. and Gorokhov, E. V., *On the determination of the rotation of a class of compactly contractible many-valued vector fields,* Voronezh. Gos. Univ. Trudy Mat. Fak. Vyp. 12 (1974), 45–54. (Russian). MR **58**#13146

[125] Obukhovskiĭ, V. V. and Skaletskiĭ, A. G., *Some theorems on the extension and quasiextension of continuous mappings,* Sibirsk. Mat. Zh. **23** (1982), no. 4, 137–141. (Russian). MR 84a:54027

[126] Petryshyn, W. V., *Structure of fixed point set of k-set-contractions,* Arch. Rational Mech. Anal. **40** (1970/71), 312–328. MR **42**#8358

[127] Petryshyn, W. V., *Note on the structure of fixed point sets of 1-set-contractions,* Proc. Amer. Math. Soc. **31** (1972), 189–194. MR **44**#3161

[128] Petryshyn, W. V., *Generalization of Schaefer's theorem to set-nonexpansive operators,* Dopovīdī Akad. Nauk. Ukrain. SSR Ser. A. (1973), no. 10, 889–891. (Ukrainian). MR **48**#4853

[129] Petryshyn, W. V. and Fitzpatrick, P. M., *Degree theory for noncompact multivalued vector fields,* Bull. Amer. Math. Soc. **79** (1973), 609–613. MR **47**#5665

[130] Petryshyn, W. V. and Fitzpatrick, P. M., *A degree theory, fixed point theorems, and mapping theorems for multivalued noncompact mappings,* Trans. Amer. Math. Soc. **194** (1974), 1–25.

[131] Pianigiani, G., *Existence of solutions of ordinary differential equations in Banach spaces,* Bull. Acad. Polon. Sci. Sér. Sci. Math. Astronom. Phys. **23** (1975), no. 8, 853–857. MR **52**#14519

[132] Potapov, A. S., *Remark on the rotation of multivalued vector fields,* in: *Collection of Works of Postgraduates,* Vyp. 2, Izd.-vo Voronezh. Gus. Univ., Voronezh, 1974, pp. 41–44. (Russian).

[133] Potapov, A. S., *On the theory of rotation of ultimately compact vector fields,* Comment. Math. Univ. Carolinae **15** (1974), 693–716. (Russian). MR **50**#10927

[134] Potapov, A. S., *K-operators and measures of noncompactness,* Voronezh. Gos. Univ. Trudy Mat. Fak. Vyp. 16 (1975), 34–40. (Russian). MR **58**#2480

[135] Potapov, A. S., *On some classes of nonlinear operators,* Candidate in Science Dissertation, Voronezh, 1975. (Russian).

[136] Potapov, A. S., Potapova, T. Ya., and Filin, V. A., *Remarks on the fixed points and eigenvectors of positive condensing operators,* Comment. Math. Univ. Carolinae **18** (1977), no. 2, 219–230. (Russian). MR **58**#23808

[137] Potapov, A. S. and Sadovskiĭ, B. N., *The distinctness of the classes K_n,* Voronezh Gos. Univ. Trudy. Mat. Fak. Vyp. 16 (1975), 40–44. (Russian). MR **58**#2481

[138] Potapov, A. S. and Sadovskiĭ, B. N., *On a fixed-point theorem for condensing operators,* in: *Operator Methods in Nonlinear Analysis,* Voronezh. Gos. Univ., Voronezh, 1982, pp. 85–89. (Russian). MR 84m:47071

[139] Rådström, H., *An embedding theorem for spaces of convex sets,* Proc. Amer. Math. Soc. **3** (1952), 165–169. MR 13#659

[140] Reich, S., *A fixed point theorem,* Atti Accad. Naz. Lincei Rend. Cl. Sci. Fis. Mat. Natur. (8) **51** (1971), 26–28. MR **46**#7984

[141] Reich, S., *Fixed points of condensing functions,* J. Math. Anal. Appl. **41** (1973), 460–467. MR **48**#971

[142] Reich, S., *A remark on set-valued mappings that satisfy the Leray-Schauder condition,* Atti Accad. Naz. Lincei Rend. Cl. Sci. Fis. Mat. Natur. (8) **61** (1976), no. 3-4, 193–194. MR **57**#17402

[143] Reinermann, J., *Forsetzung stetiger Abbildungen in Banach-Räumen und Anwendungen in der Fixpunkttheorie,* in: *Festband anlässlich der 65. Geburtstages von Erns Peschl,* Gesellsch. Math. Datenverarbeitung Bonn, Ber. No. 57, Bonn, 1972, pp. 135–145. MR **56**#6484

[144] Rodkina, A. E., *On the Cauchy problem for a neutral type equation,* in: *Collection of Articles on Applications of Functional Analysis,* Voronezh. Tekhnolog. Inst., Voronezh, 1975, pp. 155–160. (Russian). MR **57**#10167

[145] Rodkina, A. E., *The continuability, uniqueness, and continuous dependence on a parameter of the solutions of equations of neutral type,* Differentsial'nye Uravneniya **11** (1975), no. 2, 268–279. (Russian). MR **52**#8609

[146] Rodkina, A. E., *An implicit function theorem and the solution of equations of neutral type,* Differentsial'nye Uravneniya **19**, no. 9, 1632–1636. (Russian). MR 85b:34076

[147] Rodkina, A. E., *Solvability of equations of neutral type in different function spaces,* Ukrain. Mat. Zh. **35** (1983), no. 1, 64–69. (Russian). MR 84d:34072

[148] Rodkina, A. E., *On existence and uniqueness of solution of stochastic differential equations with heredity,* Stochastics **12** (1984), no. 3-4, 187-200. MR 85m:60112

[149] Rodkina, A. E., *Solvability of stochastic differential equations with deviating argument,* Ukrain. Mat. Zh. **37** (1985), No. 1, 98-103. (Russian). MR 86k:60105

[150] Rodkina, A. E. and Sadovskiĭ, B. N., *On the Krasnosel'skiĭ-Perov principle of connectedness,* Voronezh. Gos. Univ. Trudy Mat. Fak. Vyp. 4 (1970), 89–103. (Russian). MR **53**#8978

[151] Rodkina, A. E. and Sadovskiĭ, B. N., *Differentiation of the operator of translation along the trajectories of an equation of neutral type,* Voronezh. Gos. Univ. Trudy Mat. Fak. Vyp. 12 (1974), 31–37. (Russian). MR **58**#28991

[152] Rodkina, A. E. and Sadovskiĭ, B. N., *On the theory of generalized solutions of neutral type equations,* in: *Collection of Articles on Functional Analysis,* Voronezh. Tekhnolog. Inst., Voronezh, 1975, pp. 161–166. (Russian). MR **57**#6723

[153] Rudin, W., *Functional Analysis,* McGraw-Hill Series in Higher Mathematics, McGraw-Hill Book Co., New York-Düsseldorf-Johannesburg, 1973. MR **51**#1315

[154] Sadovskiĭ, B. N., *On a fixed point principle,* Funktsional. Anal. i Prilozhen. **1** (1967), no. 2, 74–76. (Russian). MR **35**#2184

[155] Sadovskiĭ, B. N., *Measures of noncompactness and condensing operators,* Problemy Mat. Anal. Slozh. Sistem No. 2 (1968), 89–119. (Russian). MR **46**#740

[156] Sadovskiĭ, B. N., *On the local solvability of ordinary differential equations in Banach space,* Problemy Mat. Anal. Slozh. Sistem No. 3 (1968), 232–343. (Russian).

[157] Sadovskiĭ, B. N., *Differential equations with uniformly continuous right hand side,* Voronezh Gos. Univ. Trudy Nauchn.-Issled. Inst. Mat. VGU Vyp. 1 (1970), 128–136. (Russian). MR **55**#797

[158] Sadovskiĭ, B. N., *Some remarks on condensing operators and measures of noncompactness,* Voronezh. Gos. Univ. Trudy Mat. Fak. Vyp. 1 (1970), 112–124. (Russian).

[159] Sadovskiĭ, B. N., *Application of topological methods to the theory of periodic solutions of nonlinear differential-operator equations of neutral type,* Dokl. Akad. Nauk SSSR **200** (1971), 1037–1040. (Russian). MR **46**#9819

[160] Sadovskiĭ, B. N., *Limit-compact and condensing operators,* Uspekhi Mat. Nauk **27** (1972), no. 1(163), 81–146. (Russian). MR **55**#1161

[161] Sadovskiĭ, B. N., *Ultimately compact and condensing operators,* Doctoral Dissertation, Voronezh, 1972. (Russian).

[162] Sadovskiĭ, B. N., *Three new results in the theory of measures of noncompactness of condensing operators,* in: *School on Operator Theory in Function Spaces,* Izd.-vo Minsk. Gos. Univ., Minsk, 1978, pp. 128–130. (Russian).

[163] Sapronov, Yu. I., *On the homotopy classification of condensing mappings,* Voronezh Gos. Univ. Trudy Mat. Fak. Vyp. 6 (1972), 78–80. (Russian). MR **56**#13263

[164] Schaefer, H., *Neue Existenzsätze in der Theorie nichtlinearer Integralgleichungen,* Ber. Verh. Sächs. Akad. Wiss. Leipzig Math.-Nat. Kl. **101** (1955), no. 7, 40pp. MR **20**#1184

[165] Sedaev, A. A., *The structure of certain linear operators,* Mat. Issled. **5** (1970), Vyp. 1(15), 166–175. (Russian). MR **43**#2540

[166] Steinlein, H., *Über die verallgemeinerten Fixpunktindizes von Iterierten verdichtender Abbildungen,* Manuscripta Math. **8** (1973), 251–266. MR **49**#3616

[167] Stuart, C. A., *The fixed point index of a differentiable (β)k-set-contraction,* J. London Math. Soc. (2) **5** (1972), 691–696. MR **47**#907

[168] Stuart, C. A., *Self-adjoint square roots of positive self-adjoint bounded linear operators,* Proc. Edinburgh Math. Soc. (2) **18** (1972/73), 77–79. MR **47**#5639

[169] Stuart, C. A. and Toland, J. F., *The fixed point index of a linear k-set contraction,* J. London. Math. Soc.(2) **6** (1973), 317–320. MR **47**#2453

[170] Szufla, S., *Some remarks on ordinary differential equations in Banach spaces,* Bull. Acad. Polon. Sci. Sér. Sci. Math. Astronom. Phys. **16** (1968), 795–800. MR **39**#595

[171] Szufla, S., *Measures of non-compactness and ordinary differential equations in Banach spaces,* Bull. Acad. Polon. Sci. Sér. Sci. Math. Astronom. Phys. **19** (1971), 831–835. MR **46**#2185

[172] Talman, L. A., *Fixed points for condensing multifunctions in metric spaces with convex structure,* Kōdai Math. Semin. Rep. **29** (1977), no.1-2, 62–70. MR **57**#3923

[173] Thomas, J. W., *The multiplicity of an operator in a special case of the topological degree for k-set-contractions,* Duke Math. J. **40** (1973), 233–240. MR **47**#4081

[174] Vaĭnnikko, G. M., *Compact Approximation of Operators and Approximate Solution of Equations,* Tartu. Gos. Univ., Tartu, 1970. (Russian). MR **42**#8353

[175] Vaĭnnikko, G. M., *Analysis of Discretization Methods. Special course,* Tartu. Gos. Univ., Tartu, 1976. (Russian). MR **58**#13699

[176] Vaĭnnikko, G. M., *Funktionalanalysis der Diskretisierungsmethoden,* B. G. Teubner Verlag, Leipzig, 1976. MR **57**#7997

[177] Vaĭnnikko, G. M. and Sadovskiĭ, B. N., *The rotation of condensing vector fields,* Problemy Mat. Anal. Slozh. Sistem No. 2 (1968), 84–88. (Russian). MR **45**#2546

[178] Vladimirskiĭ, Yu. N., *Remarks on compact approximation in Banach spaces,* Sibirsk. Mat. Zh. **15** (1974), 200–204. (Russian). MR **48**#12132

[179] Webb, J. R. L., *Fixed point theorems for non-linear semicontractive operators in Banach spaces,* J. London Math. Soc. (2) **1** (1969), 683–688. MR **40**#3392

[180] Webb, J. R. L., *On a characterization of k-set-contractions,* Atti Accad. Naz. Lincei. Rend. Cl. Sci. Fis. Mat. Natur. (81) **50** (1971), 686–689. MR **46**#6112

[181] Zabreĭko, P. P., Krasnosel′skiĭ, M. A., and Strygin, V. V., *Invariance of rotation principles,* Izv. Vyssh. Uchebn. Zaved. Matematika (1972), no. 5(120), 51–57. (Russian). MR **47**#908

[182] Zabreĭko, P. P. and Ledovskaya, I. B., *Existence theorems for equations in a Banach space and the averaging principle,* Problemy Mat. Anal. Slozh. Sistem No. 3 (1968) 122–136. (Russian).

[183] Zabreĭko, P. P. and Smirnov, A. I., *On the solvability of the Cauchy problem for ordinary differential equations in Banach spaces,* Minsk, 1979, 20pp. Deposited VINITI No. 1975-79. (Russian).

SUBJECT INDEX

Titles previously published in the series
OPERATOR THEORY: ADVANCES AND APPLICATIONS
BIRKHÄUSER VERLAG

1. **H. Bart, I. Gohberg, M.A. Kaashoek:** Minimal Factorization of Matrix and Operator Functions, 1979, (3-7643-1139-8)
2. **C. Apostol, R.G. Douglas, B.Sz.-Nagy, D. Voiculescu, Gr. Arsene** (Eds.): Topics in Modern Operator Theory, 1981, (3-7643-1244-0)
3. **K. Clancey, I. Gohberg:** Factorization of Matrix Functions and Singular Integral Operators, 1981, (3-7643-1297-1)
4. **I. Gohberg** (Ed.): Toeplitz Centennial, 1982, (3-7643-1333-1)
5. **H.G. Kaper, C.G. Lekkerkerker, J. Hejtmanek:** Spectral Methods in Linear Transport Theory, 1982, (3-7643-1372-2)
6. **C. Apostol, R.G. Douglas, B. Sz-Nagy, D. Voiculescu, Gr. Arsene** (Eds.): Invariant Subspaces and Other Topics, 1982, (3-7643-1360-9)
7. **M.G. Krein:** Topics in Differential and Integral Equations and Operator Theory, 1983, (3-7643-1517-2)
8. **I. Gohberg, P. Lancaster, L. Rodman:** Matrices and Indefinite Scalar Products, 1983, (3-7643-1527-X)
9. **H. Baumgärtel, M. Wollenberg:** Mathematical Scattering Theory, 1983, (3-7643-1519-9)
10. **D. Xia:** Spectral Theory of Hyponormal Operators, 1983, (3-7643-1541-5)
11. **C. Apostol, C.M. Pearcy, B. Sz.-Nagy, D. Voiculescu, Gr. Arsene** (Eds.): Dilation Theory, Toeplitz Operators and Other Topics, 1983, (3-7643-1516-4)
12. **H. Dym, I. Gohberg** (Eds.): Topics in Operator Theory Systems and Networks, 1984, (3-7643-1550-4)
13. **G. Heinig, K. Rost:** Algebraic Methods for Toeplitz-like Matrices and Operators, 1984, (3-7643-1643-8)
14. **H. Helson, B. Sz.-Nagy, F.-H. Vasilescu, D.Voiculescu, Gr. Arsene** (Eds.): Spectral Theory of Linear Operators and Related Topics, 1984, (3-7643-1642-X)
15. **H. Baumgärtel:** Analytic Perturbation Theory for Matrices and Operators, 1984 (3-7643-1664-0)
16. **H. König:** Eigenvalue Distribution of Compact Operators, 1986, (3-7643-1755-8)

17. **R.G. Douglas, C.M. Pearcy, B. Sz.-Nagy, F.-H. Vasilescu, D. Voiculescu, Gr. Arsene** (Eds.): Advances in Invariant Subspaces and Other Results of Operator Theory, 1986, (3-7643-1763-9)

18. **I. Gohberg** (Ed.): I. Schur Methods in Operator Theory and Signal Processing, 1986, (3-7643-1776-0)

19. **H. Bart, I. Gohberg, M.A. Kaashoek** (Eds.): Operator Theory and Systems, 1986, (3-7643-1783-3)

20. **D. Amir:** Isometric characterization of Inner Product Spaces, 1986, (3-7643-1774-4)

21. **I. Gohberg, M.A. Kaashoek** (Eds.): Constructive Methods of Wiener-Hopf Factorization, 1986, (3-7643-1826-0)

22. **V.A. Marchenko:** Sturm-Liouville Operators and Applications, 1986, (3-7643-1794-9)

23. **W. Greenberg, C. van der Mee, V. Protopopescu:** Boundary Value Problems in Abstract Kinetic Theory, 1987, (3-7643-1765-5)

24. **H. Helson, B. Sz.-Nagy, F.-H. Vasilescu, D. Voiculescu, Gr. Arsene** (Eds.): Operators in Indefinite Metric Spaces, Scattering Theory and Other Topics, 1987, (3-7643-1843-0)

25. **G.S. Litvinchuk, I.M. Spitkovskii:** Factorization of Measurable Matrix Functions, 1987, (3-7643-1843-X)

26. **N.Y. Krupnik:** Banach Algebras with Symbol and Singular Integral Operators, 1987, (3-7643-1836-8)

27. **A. Bultheel:** Laurent Series and their Pade Approximation, 1987, (3-7643-1940-2)

28. **H. Helson, C.M. Pearcy, F.-H. Vasilescu, D. Voiculescu, Gr. Arsene** (Eds.): Special Classes of Linear Operators and Other Topics, 1988, (3-7643-1970-4)

29. **I. Gohberg** (Ed.): Topics in Operator Theory and Interpolation, 1988, (3-7634-1960-7)

30. **Yu.I. Lyubich:** Introduction to the Theory of Banach Representations of Groups, 1988, (3-7643-2207-1)

31. **E.M. Polishchuk:** Continual Means and Boundary Value Problems in Function Spaces, 1988, (3-7643-2217-9)

32. **I. Gohberg** (Ed.): Topics in Operator Theory. Constantin Apostol Memorial Issue, 1988, (3-7643-2232-2)

33. **I. Gohberg** (Ed.): Topics in Interplation Theory of Rational Matrix-Valued Functions, 1988, (3-7643-2233-0)

34. **I. Gohberg** (Ed.): Orthogonal Matrix-Valued Polynomials and Applications, 1988, (3-7643-2242-X)

35. **I. Gohberg, J.W. Helton, L. Rodman** (Eds.): Contributions to Operator Theory and its Applications, 1988, (3-7643-2221-7)

36. **G.R. Belitskii, Yu.I. Lyubich:** Matrix Norms and their Applications, 1988, (3-7643-2220-9)

37. **K. Schmüdgen:** Unbounded Operator Algebras and Representation Theory, 1990, (3-7643-2321-3)

38. **L. Rodman:** An Introduction to Operator Polynomials, 1989, (3-7643-2324-8)

39. **M. Martin, M. Putinar:** Lectures on Hyponormal Operators, 1989, (3-7643-2329-9)

40. **H. Dym, S. Goldberg, P. Lancaster, M.A. Kaashoek** (Eds.): The Gohberg Anniversary Collection, Volume I, 1989, (3-7643-2307-8)

41. **H. Dym, S. Goldberg, P. Lancaster, M.A. Kaashoek** (Eds.): The Gohberg Anniversary Collection, Volume II, 1989, (3-7643-2308-6)

42. **N.K. Nikolskii** (Ed.): Toeplitz Operators and Spectral Function Theory, 1989, (3-7643-2344-2)

43. **H. Helson, B. Sz.-Nagy, F.-H. Vasilescu, Gr. Arsene** (Eds.): Linear Operators in Function Spaces, 1990, (3-7643-2343-4)

44. **C. Foias, A. Frazho:** The Commutant Lifting Approach to Interpolation Problems, 1990, (3-7643-2461-9)

45. **J.A. Ball, I. Gohberg, L. Rodman:** Interpolation of Rational Matrix Functions, 1990, (3-7643-2476-7)

46. **P. Exner, H. Neidhardt** (Eds.): Order, Disorder and Chaos in Quantum Systems, 1990, (3-7643-2492-9)

47. **I. Gohberg** (Ed.): Extension and Interpolation of Linear Operators and Matrix Functions, 1990, (3-7643-2530-5)

48. **L. de Branges, I. Gohberg, J. Rovnyak** (Eds.): Topics in Operator Theory. Ernst D. Hellinger Memorial Volume, 1990, (3-7643-2532-1)

49. **I. Gohberg, S. Goldberg, M.A. Kaashoek:** Classes of Linear Operators, Volume I, 1990, (3-7643-2531-3)

50. **H. Bart, I. Gohberg, M.A. Kaashoek** (Eds.): Topics in Matrix and Operator Theory, 1991, (3-7643-2570-4)

51. **W. Greenberg, J. Polewczak** (Eds.): Modern Mathematical Methods in Transport Theory, 1991, (3-7643-2571-2)

52. **S. Prössdorf, B. Silbermann:** Numerical Analysis for Integral and Related Operator Equations, 1991 (3-7643-2620-4)

53. **I. Gohberg, N. Krupnik:** One-Dimensional Linear Singular Integral Equations, Volume I, Introduction, 1991, (3-7643-2584-4)